Einstein Studies

Editors: Don Howard   John Stachel

Published under the sponsorship
of the Center for Einstein Studies,
Boston University

Yuri Balashov   Vladimir Vizgin
Editors

Einstein Studies in Russia

Springer Science+Business Media, LLC

Yuri Balashov
Philosophy Department
The University of Georgia
Athens, GA 30602-1627
U.S.A.

Vladimir Vizgin
Institute for the History of Science and
   Technology
Russian Academy of Sciences
Moscow, 103012, Russia

Library of Congress Cataloging-in-Publication Data

Einstein studies in Russia / Yuri Balashov and Vladimir Vizgin, editors.
    p. cm. – (Einstein studies ; v. 10)
   Includes bibliographical references and index.
   ISBN 978-1-4612-6626-6
    1. Einstein, Albert, 1879-1955. 2. Physics–History–20th century. I. Balashov, Yuri,
1960- II. Vizgin, Vladimir Pavlovich. III. Series.

QC16.E5 E518 2002
530'.0904–dc21

2002018301
CIP

AMS Classification Codes: 00B10, 00B55, 01-06, 01A60, 01A70, 01A74, 62-03, 70S10, 8103,
81P05, 83-03, 83-06, 83A05, 83C40, 83C99, 83F05

Printed on acid-free paper
©2002 Springer Science+Business Media New York
Originally published by Birkhäuser Boston in 2002   *Birkhäuser*

SPIN 10846995

ISBN 978-1-4612-6626-6   ISBN 978-1-4612-0131-1 (eBook)
DOI 10.1007/978-1-4612-0131-1

Typeset by the editors.

9 8 7 6 5 4 3 2 1

Contents

Preface

Modern scholarship in the area of Einstein Studies, broadly conceived, has a long history in Russia, going back to the mid-1960s. But the English-speaking reader knows hardly anything about it. Our goal is to remedy this by offering a selection of the best contributions by Russian scholars—historians and philosophers of science—to the Einstein Studies industry.

In 1965–67, a comprehensive four-volume edition of Einstein's works appeared in Russian (Albert Einstein, *Sobranie Nauchnykh Trudov* [*Collected Works*], 4 vols. Igor Tamm, Yakov Smorodinskiy, and Boris Kuznetsov, eds. Moscow: Nauka, 1965–67). In 1966, Nauka, the chief publisher for the USSR Academy of Sciences, launched a series called *Einshteinovskiy Sbornik* [*Einstein Studies*].

Both projects were initiated by Igor Tamm, a 1958 Nobel Prize winner (for the theory of the Cherenkov radiation), who participated in the Soviet thermonuclear project and the development of the concept of a controlled-fusion reactor (*Tocamac*). He was also a teacher of Andrei Sakharov in physics and life. In 1966–90, 14 volumes of the *Sbornik* appeared in print. After Tamm's death in 1971, Vitaly Ginzburg and Usher Frankfurt, and then Igor Kobzarev and Gennady Gorelik took over the project. *Einshteinovskiy Sbornik* focused on the history and contemporary state of fundamental twentieth century physics, as well as the creative life of the man who became the symbol of the "new physics." The series devoted about half of its materials to the Russian translations of original scientific works and historical studies.

The present collection, whose idea was suggested by one of the editors of Birkhäuser's independently conceived Einstein Studies, Don Howard, aims at the opposite. Most of the papers included here draw on materials first published in Russian in the *Einshteinovskiy Sbornik*. The book explores the historical and foundational issues in general relativity and relativistic cosmology, Einstein's contributions to early quantum mechanics, and the rise of Dirac's quantum electrodynamics. It also includes a detailed description of the physics colloquium Einstein established and coordinated in Zurich in 1912–1914 and comments on his brief interest in airfoil design in 1916.

Given the nature of this project, we strived to be faithful to the Russian originals. One unavoidable consequence of this is that some more recent scholarship is not cited in these papers.

It took a long time to bring this project to fruition. We are grateful to our contributors for their patience. Unfortunately, not all of them have lived to see this book in print. We dedicate the volume to the memory of Igor Tamm (1895–1971), Igor Kobzarev (1932–1991), and Viktor Frenkel (1930–1997).

Yuri Balashov
Vladimir Vizgin

February 2002

Acknowledgments

We are indebted to the general editors of the series, Don Howard and John Stachel, for their constant help and encouragement. Special thanks are due to Don Howard who has gone far beyond the call of duty bringing to shape the translations of several papers included here.

Gennady Gorelik has done the lion's share of work on the initial selection of articles from the *Einshteinovskiy Sbornik*. Alexei Kojevnikov has given us helpful advice on many occasions. Ann Kostant, our editor at Birkhäuser, has patiently guided us through the entire process. We extend our sincere thanks to them.

Permissions

The contributions to this collection are based, partly or in full, on the following originally published materials:

- Viktor Frenkel, "New Materials Concerning the Discussion Between Einstein and Friedmann on Relativistic Cosmology." In *Einshteinovskiy Sbornik, 1973*. V. L. Ginzburg and U. I. Frankfurt, eds. Moscow: Nauka, 1974, pp. 5–18. Copyright © 1974 by Viktor Frenkel. Used with permission of Olga Cherneva.

- Gennady Gorelik, "The Dimensionality of Spacetime and Poincaré's Quasigroup in General Relativity." In *Einshteinovskiy Sbornik, 1984–1985*. I. Yu. Kobzarev and G. E. Gorelik, eds. Moscow: Nauka, 1988, pp. 271–300. Copyright © 1988 by Gennady Gorelik. Used with permission of the author.

- Vladimir Vizgin, "The Role Played by Mach's Ideas in the Genesis of the General Theory of Relativity." In *Einshteinovskiy Sbornik, 1986–1990*. I. Yu. Kobzarev and G. E. Gorelik, eds. Moscow: Nauka, 1990, pp. 49–97. Copyright © 1990 by Vladimir Vizgin. Used with permission of the author.

- Gennady Gorelik, "The History of Relativistic Cosmology and Large Numbers' Coincidences." In *Einshteinovskiy Sbornik, 1982–1983*. I. Yu. Kobzarev and G. E. Gorelik, eds. Moscow: Nauka, 1986, pp. 302–322. Copyright © 1986 by Gennady Gorelik. Used with permission of the author.

- Yuri V. Balashov, "Uniformitarianism in Cosmology: Background and Philosophical Implications of the Steady-State Theory." *Studies in*

Contributors

Viktor Frenkel (1930–1997), worked at the Ioffe Institute of Physics and Technology of the Russian Academy of Sciences, St. Petersburg, Russia.

Gennady Gorelik, Center for Philosophy and History of Science, Boston University, Boston, MA 02215.

Vladimir Vizgin, Institute for History of Science and Technology, Russian Academy of Sciences, 1/5 Staropansky per., Moscow 103012, Russia.

Yuri Balashov, Department of Philosophy, 107 Peabody Hall, University of Georgia, Athens, GA 30602-1627.

Alexei Kojevnikov, Department of History, LeConte Hall, University of Georgia, Athens, GA 30602-1627; and Institute for History of Science and Technology, Russian Academy of Sciences, 1/5 Staropansky per., Moscow 103012, Russia.

Boris Yavelov, worked at the Russian Research Centre-Kurchatov Institute, Moscow, Russia.

Einstein Studies in Russia

Einstein and Friedmann

Viktor Frenkel

1. Introduction

In 1922, a major German physical journal published an article entitled "On the Curvature of Space," by Aleksandr Friedmann, of the Soviet Petrograd University. Two years later, it was followed by "On the Possibility of a World with a Constant Negative Space Curvature" (Friedmann 1922, 1924). These articles started a new chapter in relativistic cosmology by providing the first unambiguously evolutionary models of the universe.

In Soviet Russia, Friedmann's contribution to cosmology was "officially" acknowledged only forty years later. His groundbreaking cosmological papers were reprinted, along with his esteemed works in hydrodynamics and dynamical meteorology, in the Russian series *Classics of Science* (Friedmann 1966). The volume also contained Friedmann's popular essay *The World as Space and Time*, which was first published by Academia, one of the leading academic publishers in the Soviet Union in the 1920s (Friedmann 1923).

The focus of the present article is Friedmann's 1922 letter to Einstein accompanied by additional evidence throwing light on their debate, and the great roles played by Yuri Krutkov and Paul Ehrenfest, both of whom Einstein knew very well (see Frenkel 1970). The debate began soon after the appearance of Friedmann's first article showing the possibility of a non-stationary solution of the cosmological problem (thus laying the foundation for the theory of an expanding universe). Einstein replied to Friedmann with a note in which, as aptly observed by Fock, "he said, somewhat condescendingly, that Friedmann's results seemed suspicious to him, and that he had found a mistake in them which, when corrected, reduced Friedmann's solution to a stationary one" (Friedmann 1966, p. 401).

Great people's delusions are always instructive, especially when dealing with fundamental problems. The honesty of great men can also be

Yuri Balashov and Vladimir Vizgin, eds., *Einstein Studies in Russia.*
Einstein Studies, Vol. 10. Boston, Basel, Birkhäuser, pp. 1–15.

exemplary: the debate came to an end after the publication of Einstein's second note, in which he stressed the importance of Friedmann's work.

2. Early Reactions to the General Theory of Relativity in the Soviet Union[1]

Contrary to a popular opinion, the early 1920s marked the return of good times for Soviet physicists. Not only did they have access to major foreign journals, but they regularly published their own articles in *Zeitschrift für Physik*, one of the most prominent scientific journals at the time. Indeed, there was rarely an issue of this journal without papers by Soviet physicists. And it was this journal that published both of Friedmann's articles.

Einstein's seminal publications on general relativity, however, appeared during World War I and the Civil War in Russia. For more than six years, Russian physicists had no opportunity to keep abreast of their foreign colleagues' work.

But in the early 1920s, Einstein's works on general relativity were reported and discussed at the Petrograd University and Polytechnical Institute. A frequent speaker was Vsevolod Frederiks. In 1914–18 he had been held civilian captive in Germany. Thanks to David Hilbert's efforts, however, Frederiks had an opportunity to continue working with the famous German scientist, who was the first to provide a solid mathematical foundation for the emerging theory of general relativity. Based on his lectures read at the Petrograd University, Frederiks wrote the first Russian survey of general relativity (Frederiks 1921).[2]

Frederiks's work greatly stimulated Friedmann's interest in general relativity. In 1924, they co-authored a large work on that theory (Frederiks and Friedmann 1924). The introduction to it set forth a general plan including the fundamentals of the geometry of multidimensional spaces, electrodynamics, and the foundations of special and general relativity. Because of Friedmann's premature death in 1925, these plans did not materialize.

3. Friedman's Letter to Einstein[3]

Petrograd
Central Observatory,
Vasil'yevskiy Ostrov, Line 23, 2
Professor A. Friedmann

6 December 1922

Dear Esteemed Professor,

From a letter of a friend of mine, who is now staying abroad, I had the honor to learn that you had sent a short note to the 11th volume of *Zeitschrift für Physik*, in which you had pointed out that, if one adopted the assumptions (D_3) and (C) of my article "On the Curvature of Space,"[4] then the world equations you derived would entail that the radius of curvature of the universe should be a constant value that does not depend on time. You have obtained this result based on the fact that the vanishing of the divergence of the tensor T_{ik} is a necessary consequence of the world equations.

From this vanishing of the divergence of the tensor T_{ik} you have inferred the relation

$$(*) \qquad \frac{\partial \rho}{\partial x_4} = 0.$$

Such a relation means, naturally, that the radius of curvature R is constant and, hence, that the calculations in my work are in error.

However, I could not derive the relation (*) from the condition that the divergence of the tensor T_{ik} vanishes; the result I obtained is *consistent*[5] with the possibility of a non-stationary world. Given that such a possibility may be of some interest, permit me to present here my calculations for the divergence of the tensor T_{ik} so that they could be checked and assessed critically. Let Q_k be the kth component of the contragradient tensor representing the divergence of T_{ik}. According to the formula for the divergence,

$$Q_k = \frac{1}{\sqrt{g}} \frac{\partial \sqrt{g} g^{a\sigma} T_{ak}}{\partial x_\sigma} - \begin{Bmatrix} k\ \sigma \\ s \end{Bmatrix} g^{a\sigma} T_{as} .$$

4 Viktor Frenkel

We shall be interested in Q_4, since Q_1, Q_2, and Q_3 all vanish—precisely because equation (D_3) and assumption (C) in my article yield, for a non-stationary world:

$$\left\{\begin{array}{cc} 4 & 1 \\ & 4 \end{array}\right\} = 0, \quad \left\{\begin{array}{cc} 4 & 2 \\ & 4 \end{array}\right\} = 0, \quad \left\{\begin{array}{cc} 4 & 3 \\ & 4 \end{array}\right\} = 0, \quad \left\{\begin{array}{cc} 4 & 4 \\ & 4 \end{array}\right\} = 0 .$$

The expression for Q_4, then, is as follows:

$$Q_4 = \frac{1}{\sqrt{g}} \frac{\partial \sqrt{g} g^{\alpha\sigma} T_{\alpha 4}}{\partial x_\sigma} - \left\{\begin{array}{cc} 4 & \sigma \\ & s \end{array}\right\} g^{\alpha\sigma} T_{\alpha s}$$

$$= \frac{1}{\sqrt{g}} \frac{\partial \sqrt{g} g^{4\sigma} T_{44}}{\partial x_\sigma} - \left\{\begin{array}{cc} 4 & \sigma \\ & 4 \end{array}\right\} g^{4\sigma} T_{44} .$$

Now according to condition (C) in my article, all T_{ik} (except T_{44}) vanish. But equations (D_3) enforce $g^{4\sigma} = 0$ for all σ, except $\sigma = 4$. Thus the previous equation may be rewritten as

$$Q_4 = \frac{1}{\sqrt{g}} \frac{\partial \sqrt{g} g^{44} T_{44}}{\partial x_4} .$$

Since $g^{44} = 1/g^{44} = 1$ (in our case they are equal to the intervals established by (D_3)) and T_{44} equals $c^2 \rho g_{44}$ (hence $T_{44} = c^2\rho$), Q_4 is reduced to

$$Q_4 = \frac{1}{\sqrt{g}} \frac{\partial \sqrt{g} c^2 \rho}{\partial x_4} .$$

If, as required by your world equations, Q_4 is assumed to be zero, one obtains, not [underlined in the letter] the equation given in your article, but the following equation:

$$(**) \qquad \frac{\partial \sqrt{g}\rho}{\partial x_4} = 0.$$

Thus $\sqrt{g}\rho$ should not depend on x_4. But

$$g = -\frac{1}{c^6} R^6(x_4) \sin^4 x_1 \sin^2 x_2,$$

$$\sqrt{g} = \frac{1}{c^3} \sqrt{-1}\, R^3(x_4) \sin^2 x_1 \sin x_2,$$

and ρ, according to equation (8) in my article, is given by:

$$\rho = \frac{3A}{\kappa R^3(x_4)}.$$

From this it follows that

$$\sqrt{g}\rho = \frac{3A}{c^3 \kappa} \sqrt{-1}\, \sin^2 x_1 \sin x_2,$$

and does not, indeed, depend on x_4, as was required to prove.[6]

Dear Esteemed Professor, will you be so kind as to notify me if my calculations in this letter are correct? I have recently studied the case of a world with a constant and changing (in time) negative curvature. Sure enough, to obtain the solution for the real[7] world (the only one that may be physically and geometrically interesting), one needs to use a different expression for the interval, which I (following Bianchi, Lezioni di geometria differenziale, Band 1) put in the following form:

$$d\tau^2 = -\frac{R^2(x_4)}{x_3^2} (dx_1^2 + dx_2^2 + dx_3^2) + M dx_4^2.$$

The calculations have shown that, in this case, there can exist both a world with a constant (now negative) curvature and a world with a changing (in time) curvature.

The possibility[8] of obtaining a world with a constant negative curvature from your equations is of exceptional interest to me. This is why I am asking you not to delay your reply to this letter of mine, though I am aware that you are very busy.

If you find my calculations correct, I kindly request that you inform the editors of *Zeitschrift für Physik* about it. Perhaps in this case you will publish a correction to your earlier statement or provide an opportunity for a portion of this letter to be published.

Respectfully Yours,
A. Friedmann

4. A Brief Chronological Commentary to Friedmann's Letter

"A friend of mine who is now staying abroad" was Yuri Krutkov, Friedmann's associate at the Petrograd Polytechnical Institute. In 1922–23 he was on a business trip abroad, working in Berlin, Göttingen, and Leiden. In Leiden he was hosted by the University's department of theoretical physics, which (after Lorentz) was headed by Paul Ehrenfest.

One of the first Russian theoretical physicists, Krutkov, along with Friedmann and a few other young physicists and mathematicians, participated in a seminar that gathered around Ehrenfest in pre-revolutionary Petersburg in 1907–1912. Ehrenfest, whose wife was Russian, was fluent in this language. He, in fact, established the first school of theoretical physics in Russia and continued taking care of it after he had moved to Leiden. Russian scientists gave him a Russian equivalent of his name, "Pavel Sigizmundovich" (adding a patronymic).

In May 1923, Einstein, an honorary professor at Leiden University, visited Leiden to attend the farewell lecture of Lorentz who was going to retire (on 18 July 1923 he became 70). Einstein, as usual, stayed at the Ehrenfests'. There he met Krutkov.

One gets an idea of Krutkov's meetings and conversations with Einstein from Krutkov's surviving notes and from his letters to his sister in Petrograd. (Calculations from Friedmann's article are scattered all around

Krutkov's notebooks of May 1923). He wrote to his sister on April 29: "Einstein should come any day. And I am very interested in him" (quoted in Frenkel 1988, p. 659).[9] May 4: "I can't write anymore, however, because I have to hurry to hear a paper by Einstein. He is a very warm [gemütlich] person" (quoted in Frenkel 1970, p. 820).

Here are some other excerpts from Krutkov's notebook: "On Monday, 7 May 1923, Einstein and I read Friedmann's article in *Zeitschrift für Physik*." May 13: "I do not know why the arrival of my passport from Berlin was delayed because otherwise I would have left on the 15th with Einstein.[10] . . . Einstein is very nice." (Quoted in Frenkel 1988, p. 659.) May 18: "At five o'clock Einstein was discussing his most recent work with Ehrenfest, Droste and some Belgian. . . . I won over Einstein in an argument about Friedmann. The honor of Petrograd is saved" (quoted in Frenkel 1970, p. 820).

Einstein's second note was received by the editor of *Zeitschrift für Physik* on 21 May 1923, that is, soon after Einstein's conversations with Krutkov, which also continued in Berlin.

In August–September 1923 Friedmann visited Berlin. On August 19 he wrote to his wife: "My trip is not working out: Einstein, for example, has gone on vacation and I shall not be able to see him" (quoted in Frenkel 1988, p. 659). On September 13 Friedmann wrote to his wife about his trip to Potsdam to meet astronomers E. von Pahlen and E. Freundlich. He says, in particular: "All were impressed with my dispute with Einstein and my subsequent victory, and I find it satisfying because my papers will be accepted more readily for publication" (quoted in Frenkel 1988, p. 659).

John Stachel's article (Stachel 1987) refers to materials which are intended for publication in Einstein's *Collected Works*. Among the materials, there is a manuscript of Einstein's second note (EA 1-026) with the crossed-out final phrase saying that Friedmann's solution (for a non-stationary Universe), albeit correct mathematically, can hardly make physical sense.[11]

In Leiden, probably after a conversation with Krutkov, Einstein admitted the mathematical infallibility of Friedmann's calculations, and yet he added the same phrase; but in Berlin, having met Krutkov again, he deleted that reservation.

John A. Wheeler, who knew Einstein well, recalls (Wheeler 1968) that Einstein considered his mistaken assessment of Friedmann's calculations to be his greatest blunder. And he would probably have been extremely sorry if that phrase had been left in the note. It seems that it was precisely Krutkov who had saved him from that mistake.

We conclude with a chronological summary of important events described above:

- Friedmann's article (Friedmann 1922) was received by the editor of *Zeitschrift für Physik* on 29 June 1922.
- Einstein's first note (Einstein 1922) was received by the editor on 18 September 1922.
- Friedmann's letter to Einstein was sent on 6 December 1922.
- Krutkov's meetings with Einstein in Leiden took place on 7–18 May 1923.
- Einstein's second note (Einstein 1923) was received by the editor on 21 May 1923.

5. Appendix. Friedmann's Letter to Ehrenfest

Friedmann's extant correspondence is scanty. It consists mostly of his letters to Academicians Boris Golitsyn and Vladimir Steklov (see Friedmann 1966).

The following letter of Friedmann to Ehrenfest does not devote much space to the theory of relativity, but it gives a good idea of the circumstances in which Friedmann's seminal works on relativistic cosmology were written.[12]

August 6, 1920
Petrograd

Dear Pavel Sigizmundovich!

I take the opportunity to write a letter to you in the hope that it may reach you by some miracle. I will try to write legibly, though my handwriting has further worsened. I'd like to tell you two sorts of facts: personal matters and those concerning various projects; as far as the latter category is concerned, I can only speak for myself as, while in Perm, I had for long been isolated from the scientific community and paid more attention to administrative matters. . . .

You have heard about me from V. A. Trkal; after his departure I continued my professorship in Perm for another year, was assistant rector and in charge of administrative and economic matters. Thus I wasted nearly a year of my life. You've probably heard from Trkal that all my books and all manuscripts [underlined by Friedmann] have been lost. I had to recover my main project, which I am currently pursuing further (of which more below), from memory alone. I have just moved to Petersburg (my wife is still in Perm, and we have no children) and I am working in the Atomic Commission (like Tamarkin) on the structure of the atom, in the Physical observatory on mathematical and dynamic meteorology, and I am both lecturing and in charge of the lab classes at the Railway Institute and at the University. . . .

Let me now turn to my work. For the above-mentioned reasons, I can only tell you about my own work. Trkal may have told you why I have lost a few years: I have been engaged in the "practical" activity, which means, in my current position, that "I have been wasting my time." Nevertheless, here is what I've been doing.

1. Dynamic meteorology and, primarily, questions related to: (a) the influence of radiant heat transfer on the vertical flows and temperature gradients; (b) the influence of convection and thermal conductivity on the same parameters, and (c) the relationship among the vertical temperature gradient, the vertical flows, and the energy input (in various forms). It proved possible to clarify the cause of inversion formation and of a sharp increase in inversion temperature, to show that the introduction of the adiabatic gradient was wrong, and to replace it with a somewhat larger critical gradient characterizing the atmospheric stability. All this required an excursus into the theory of the fourth-degree curves and the application of the Cardano formula (sic!). The work has been published [Friedmann 1920], but I don't have the offprints yet. I'll send you one when I have an opportunity, as I'd like to know very much your opinion on such things. You have dealt with them, for I have seen your notes on the margins of V. Bjerknes' article on vortex formation.

At present, the Observatory's mathematical section, which I am heading, is calculating the amount of energy received by 1 m^3 of the atmosphere. It is clear to me that it is completely wrong to assume that the atmospheric processes are adiabatic.

I am now clarifying the question of whether it is possible to apply the same values for internal friction and thermal conductivity that are obtained in the laboratory to the atmosphere, or, more precisely, to the atmospheric air. Hesselberg showed long ago that the internal friction factor exceeds, by the factor of 500,000, its laboratory value. It seems to me that the same is true of the thermal conductivity factor. The point is that we study only the macro-dynamics of the atmosphere and entirely disregard micro-motions in it.

2. From dynamic meteorology I turned to the hydrodynamics of compressible fluids in which there is an input of energy. Bjerknes studied changes in the tension of the vortex line in such a fluid; I went farther and investigated the splitting of the vortex line into new separate filaments in motion with the input of energy. Here both Helmholtz's theorems are no longer valid. The splitting of the vortex line and the change of the vortex tension have proved to be determined by three quantities, which one can express, without much difficulty, in terms of pressure and density, making use of the conditional equations, which I have derived in my note published in *Comptes Rendus* in 1916 (or 1915).[13] The study of the change in vortex lines in connection with the given conditional equations has enabled me (with the help of the vector analysis, without which calculations would be utterly unmanageable) to work out the whole theory of compressible fluid with an input of energy. I am not going to expound all that, of course, and I doubt you may find this interesting since these investigations are far from the brilliant areas you are engaged in at the moment; yet your notes in the margins of Bjerknes' article indicate that you took interest in this, and this is why I am writing at length about it. For the time being, I have come up with many applications of this theory to dynamic meteorology, yet this extended hydrodynamics might find application in other areas, too. Now I want to put everything in order, meaning to use this material for my dissertation (regrettably, I haven't completed it yet, because of the difficult circumstances).[14]

3. I have been preoccupied with a number of issues in engineering:

(a) I have applied the method of asymptotic expressions to the problem of the airplane inclinometer oscillations; applied in earnest, with numerical calculations;

(b) I have worked on the theory of forced airplane vibrations;

(c) I have supervised large ballistic tables for aviation calculated by integrating differential equations using Runge's method;

(d) As the director of the factory, I have worked on a number of issues having to do with the theory of measuring instruments used in aviation;

(e) I have studied convective heat transfer.

4. I have also been working on the axiomatic structure of the restricted principle of relativity. Proceeding from two assumptions: (1) uniform motion remains uniform for a uniformly moving world and (2) the velocity of light is constant (the same in moving and stationary worlds), I have obtained, for a world with one spatial dimension, equations that are more general than the Lorentz transformations, with one more parameter. In two- and higher-dimensional spaces, the addition of velocities (the properties of the group of transformations) made it possible to recover Lorentz's formulas. This difference between the one-dimensional and other cases seems funny to me. I wonder what kind of arithmetic people would have invented if the speeds available to them had been close to that of light. I am eager to investigate the general principle of relativity, but I am short of time.[15]

Please don't be angry at me, Pavel Sigizmundovich, for my description of my work in such detail. I have changed a great deal since we parted and I have been strictly following the proverb that the cobbler should stick to his last; so it is not bragging or mere selfishness that led me to bring to your knowledge what I have been doing recently. It was nothing more than a desire to describe my life in terms of concrete accomplishments; outside scientific work, there is no life for me and, hence, no accomplishments.

I was pleased to learn that Trkal is staying with you, that he is alive and, what is more, that he is working; he is a good, kind-hearted, and able man. Please, give him best regards from me and my associates.

Today Tamarkin and I attended the session of the Atomic Commission: we made a joint report on dimensionality conditions imposing limits on quantized adiabatic invariants, and more than once we referred to the work of Tatiana Alekseevna.[16]

All the best from me and Tamarkin to the members of your family; our kindest regards to your aunt, if she hasn't forgotten us, and to Tatiana Pavlovna (she must be grown-up by now, which means that we are becoming old!) if she remembers a fat man and a thin man.[17] Are we destined to meet ever again? God knows. You can never be sure.[18]

The letter is disorderly, as I was writing in a great hurry. V. K. Lebedinsky is now a professor in Nizhni Novgorod.

Drop me a few lines, please. You cannot imagine what a festival it means for us at the Physical Institute to receive your letter.

I did not write to you about our works in the Atomic Commission; we (especially I) are learning here mainly how to put the unclear and incomplete ideas, which are numerous in the field of quanta, into strict mathematical formulas. This role of "mathematical gendarmes," so to speak, is not very pleasant, yet it happens to be useful. We are even investigating celestial mechanics (of which there is a good deal in the atom) and stability. Let me repeat that we have just started this work and are still learning.

I am not writing about Tamarkin's relatives; they will write themselves, in the near future, if not now.

I have a favor to ask you: could you send me, at least through the Physical Institute (Baumgart or someone else), *separata* [offprints] of works on dynamic meteorology that have come out in recent years (since 1914); it would be especially valuable to have the publications of the Bjernes Geophysical Institute in Leipzig. I would also like to have *Meteorologische Zeitschrift* for at least 1917, 1918, 1919, and 1920. I am sorry to bother you with all this, but without books you feel depressed and unquiet. It is our customary condition though. I live together with Tamarkin (my wife is still in Perm, lecturing in history at the Pedagogical Institute and the boarding school). We live a friendly life, all of us have got old, with grey hair and bald heads; time is running out and so are the joys of life.

Yours,
Friedmann

Acknowledgments. The author is indebted to Dr. Otto Nathan (The Albert Einstein Estate) and A. E. Engberts (Museum Boerhaave in Leiden) for their kind assistance in providing copies of Friedmann's letters to Einstein and Ehrenfest. The editors wish to thank the Albert Einstein Archives (the Jewish National and University Library, the Hebrew University of Jerusalem, Israel), The Einstein Papers Project, and the Archives for the History of Quantum Physics (Special Collections Department, University of Pittsburgh Library) for their permissions to publish their materials in English.

NOTES

[1] For a detailed account of the reception of relativity theory in Russia, see Vizgin and Gorelik 1987.

[2] For more details on Frederiks's work, see Vizgin and Frenkel's paper in this volume.

[3] EA 11-114. The editors are grateful to Dr. Ze'ev Rosenkranz, Curator of the Albert Einstein Archives, for a copy of this letter and permission to reprint it here in English and to Dr. Gayane Tavrizyan for her help in checking the translation.

[4] "Assumption (C)" of Friedmann's 1922 paper is the assumption (which has since become standard in relativistic cosmology) that the relative velocities of cosmic matter particles are small in comparison with the velocity of light. This entails the following expressions for the components of the stress-energy tensor: $T_{ik} = 0$ for i and $k \neq 4$; $T_{44} = c^2 \rho g_{44}$. "Assumption (D_3)" is the generic form used by Friedmann for the metric of a non-stationary, homogeneous, and isotropic universe:

$$ds^2 = -\frac{R^2(x_4)}{c^2}\left(dx_1^2 + \sin^2 x_1 dx_2^2 + \sin^2 x_1 \sin^2 x_2 dx_3^2 \right) + dx_4^2 .$$

Here R is a function of x_4 only, i.e., of the "cosmological time."—*Editors*

[5] Friedmann's emphasis: "Mir freilich ist es nicht gelungen, die Relation (*) uas dem Verschwinden der Divergenz des Tensors T_{ik} zu erhalten; das von mir erhaltene Resultat w i d e r s p r i g h t   n i c h t   dem Falle der nicht-stationaren Welt."

[6] Expanded and underlined in the text of the letter: ". . . u n d   i s t   w i r k l i c h   v o n   x_4   u n a b h ä n g i g ,   w a s   a u c h   g e f o r d e r t   w i r d ."

[7] "Materielle" is crossed out by hand in the typed text and replaced by "reelle."—*Editors*.

[8] "Das Verhandensein [obviously a typo; should be "Vorhandensein"] der Möglichkeit" —*Editors*.

[9] Quotations in this paragraph are from the Archives of the USSR Academy of Sciences, Leningrad Div., File 759, Sub. 4 (papers of Yu. A. Krutkov in the file of A. N. Krylov), File 946, Sub. 1 (file of Yu. A. Krutkov). Krutkov used to write the greater part of his letters in German.

[10] Einstein actually left Leiden somewhat later, and Krutkov did so with one day's delay.

[11] "It follows that the field equations, besides the static solution, permit dynamic (that is, varying with the time coordinate) spherically symmetric solutions for the spatial structure *to which a physical significance can hardly be ascribed*" (Stachel 1987, p. 65, italic in the original).

[12] The letter was written in Russian. The editors thank Dr. W. Gerald Heverly, Curator of the Special Collections Department of the University of Pittsburgh Libraries, for a copy of this letter.

[13] Friedmann's article "Turbulence in a Fluid of Variable Temperature" was published in *Comptes Rendus* in 1916 (Friedmann 1916).

[14] Friedmann defended his dissertation, *A Study in the Hydromechanics of Compressible Fluids*, in Petrograd in 1922. It was later published as a book (Friedmann 1934).

[15] The "restricted" and "general principles of relativity" refer, no doubt, to the special and general theories of relativity.

[16] Tatiana Alekseevna Afanasieva-Ehrenfest, the wife of P. Ehrenfest.

[17] Tatiana Pavlovna van Aardenne-Ehrenfest, the daughter of P. Ehrenfest. The "fat and thin men" are Ya. Tamarkin and A. Friedmann respectively, as they were called in 1910 by Ehrenfest's small daughter.

[18] It should be noted that the letter was sent from the starving Petrograd. Later, in 1924, Friedmann and Ehrenfest met both in the USSR and abroad.

REFERENCES

Einstein, Albert (1922). "Bemerkung zu der Arbeit von A. Friedmann 'Über die Krümmung des Raumes'." *Zeitschrift für Physik* 11: 326.

— (1923). "Notiz zu der Arbeit von A. Friedmann 'Über die Krümmung des Raumes'." *Zeitschrift für Physik* 16: 228.

Frederiks, Vsevolod Konstantinovich (1921). "Obshchiy printsip otnositel'nosti Einshteina." ["Einstein's General Principle of Relativity"] *Uspekhi fizicheskikh nauk* 2: 162–188.

Frederiks, Vsevolod Konstantinovich, and Friedmann, Aleksandr Aleksandrovich (1924). *Osnovy teorii otnositel'nosti* [*The Basics of Relativity*], part 1. Leningrad.

Frenkel, Yakov Il'yich (1923). *Teoriya otnositel'nosti.* [*The Theory of Relativity*] Petrograd.

Frenkel, Viktor Yakovlevich. (1970). "Yuriĭ Aleksandrovich Krutkov." *Soviet Physics. Uspekhi* 13: 816–825.

— (1988). "Aleksandr Aleksandrovich Fridman (Friedmann): A Biographical Essay." *Soviet Physics. Uspekhi* 21: 645–685.

Friedmann, Aleksandr Aleksandrovich (1916)."Sur les tourbillons dans un liquide à température variable." *Comptes rendus hebdomadaires des séances de l'Académie des sciences* 163: 219–222.

— (1920). "O raspredelenii temperatury s vysotoy pri nalichnosti luchistogo teploobmena Zemli i Solntsa." ["On the Relation of Temperature to Altitude in the Presence of Radiation Heat Exchange Between the Earth and Sun"] *Izvestiya Glavnoy Geofizicheskoy Observatorii* 2: 42–44.

— (1922). "Über die Krümmung des Raumes." *Zeitschrift für Physik* 10: 377–386.

—— (1923). *Mir kak prostranstvo i vremya*. [*The World as Space and Time*] Petrograd: Academia.

—— (1924). "Über die Möglichkeit einer Welt mit konstanter negativer Krümmung des Raumes." *Zeitschrift für Physik* 21: 326–332.

—— (1934). *Opyt gidromekhaniki szhimaemoy zhydkosti*. [*A Study in the Hydromechanics of Compressible Fluids*] Moscow: Gostekhizdat.

—— (1965). *Mir kak prostranstvo i vremya*. [*The World as Space and Time*] Moscow: Nauka.

—— (1966). *Izbrannye trudy*. [*Selected Works*] Moscow: Nauka.

Stachel, John (1987). "'A Man of my Type'—Editing the Einstein Papers." *British Journal for the History of Science* 20: 57–66.

Vizgin, Vladimir Pavlovich, and Gorelik, Gennady Efimovich (1987). "The Reception of the Theory of Relativity in Russia and USSR." In *The Comparative Reception of Relativity*. T. Glick, ed. Boston: Reidel, pp. 265–326.

Wheeler, John A. (1968). *Einsteins Vision; Wie steht es heute mit Einsteins Vision alles als Geometrie aufzufassen?* Berlin and New York: Springer-Verlag.

The Problem of Conservation Laws and the Poincaré Quasigroup in General Relativity

Gennady Gorelik

1. Introduction

The role played by conservation laws in physics is well known. One can hardly imagine a physics text that makes no use, one way or the other, of the concepts of energy, momentum, and angular momentum engendered by the respective conservation laws. (I will also use the term *moment* for the total 4-moment of momentum, i.e., the angular momentum and the velocity of the center of inertia.)

The physico-mathematical essence of conservation laws, as consequences of a theory's symmetries, was clarified in the second decade of the twentieth century by Felix Klein, David Hilbert, and Emmy Noether on the basis of the special theory of relativity (SR). This approach was intended to be used in the general theory of relativity (GR). Ironically, it so happened that the status of conservation laws was soon questioned in GR itself (see Vizgin 1972, 1981).

In creating his general theory of relativity, Einstein used the law of conservation of energy-momentum as one of his main tools. In the final theory, however, conservation laws became a problem, rather than a consequence of the theory. The problem arose from the idea, fundamental to GR, of geometrizing physical interactions, the idea that led to the notion of space-time with variable curvature. After short but heated debates, Einstein found a solution to this problem in a very important, albeit special case of an "island configuration," an isolated system where the geometry is noticeably curved only in a finite region, whereas at infinity, it is asymptotically flat (Einstein 1918). Einstein's solution is based on the so-called pseudotensor of the gravitational field's energy-momentum. Most

Yuri Balashov and Vladimir Vizgin, eds., *Einstein Studies in Russia.*
Einstein Studies, Vol. 10. Boston, Basel, Birkhäuser, pp. 17–43.

specialists in gravitation still believe that the pseudotensor approach is the best possible one (see, e.g., Faddeev 1982).

Such a limited solution has, however, been unable to satisfy everyone. Some theorists believe that the situation with conservation laws in GR is so unsatisfactory that they reject Einstein's theory of gravitation as a whole (Logunov 1998). The criticism of such views has so far avoided the problem of conservation laws in the general formulation, that is, in the case of arbitrary geometries admitted by GR. Yet the existence of the problem is clearly revealed when one attempts to establish a correspondence between SR and GR using conserved quantities as a "bridge." Those specialists who do not question GR have also evinced dissatisfaction with this situation. Penrose (1982), for one, put the energy-momentum-moment problem at the very top of his list of unsolved problems in classical GR.

Below we suggest a novel analysis of the problem of conservation laws in GR.

We take the field theory in SR as our point of departure. The situation with conservation laws in this case is well known and quite clear. But the special relativistic case is "degenerate" from the point of view of GR. Curved space-time "splits" various properties that are equivalent in SR, making some of them independent and rendering some others meaningless. Therefore, not just any formulation of conservation laws that is "natural" and customary in SR can be taken as basic if one's target is GR.

The Noether theorem (the so-called *first* theorem), which completed the theory of conservation laws, affords the deepest explanation of them in SR, and this theorem will be used as a basis for the analysis of the problem in GR. (The so-called *second* Noether theorem is irrelevant to conservation laws in SR and will not be discussed in this paper.)

The key question of this paper is whether there is a connection, in the sense of the correspondence principle, between the inferior (pseudo-) conservation laws in GR and the ten absolutely clear conservation laws in SR. More specifically, is there a formal construction, definable on a generic space-time of GR, that corresponds to the ten-parameter Poincaré group in SR?

After a brief account of the history of the pseudotensor approach we undertake a Noether-type approach to the general problem of conservation laws in GR, based on the concept of the Poincaré quasigroup (Gorelik 1988).

2. The Noether Theorem and Conservation Laws in Special Relativity

The soundest foundation for discussing conservation laws in a sufficiently developed theory is provided by the Noether theorem (see Bogolyubov and Shirkov 1980, Vizgin 1972). This theorem establishes a link between the theory's symmetries (the invariance of its action) and its conservation laws:

Each one-parameter symmetry (a one-parameter set of transformations of the theory's variables keeping the action invariant) corresponds to a separate conservation law.

Let us briefly recall the Noether-type formulation leading to conservation laws for the field $u_a(x^i)$, where a is an index or a set of indices, x^i are coordinates in space-time, and $i = 1, \ldots, n$ (for generality, let us allow an arbitrary number n of dimensions of space-time). The field dynamics is described by the Lagrangian $\mathcal{L}[u]$ and the corresponding action, the integral of the Lagrangian over the space-time region Ω:

$$A = \int_\Omega \mathcal{L}[u] d^n x. \tag{1}$$

The principle of stationary action, $\delta A = 0$, then gives the field equations

$$\delta\mathcal{L}/\delta u_a = 0. \tag{2}$$

Let the theory have an s-parameter symmetry Σ^s; that is, transformations of coordinates $x^i \rightarrow x'^i$ and of the field variables $u_a \rightarrow u'_a$, corresponding to this symmetry, do not change the action:

$$A(x',u') = A(x,u). \tag{3}$$

For infinitesimal symmetry parameters Δ^A ($A = 1, \ldots, s$), the action is invariant with regard to the transformations $x^i \rightarrow x^i + \delta x^i$, $u_a \rightarrow u_a + \delta u_a$, where

$$\delta x^i = X_A^i \Delta^A, \tag{3a}$$

$$\delta u_a = \Psi_{aA} \Delta^A, \tag{3b}$$

and X_A^i and Ψ_{aA} are definite quantities reflecting the structure of the symmetry Σ^s and the properties of the field variables u_a. Then, according to the Noether theorem, $\mathcal{L}$, u_a, X, and Ψ give rise to the quantities

$$\Theta_A^i = \Theta_A^i(\mathcal{L}, u, X, \Psi), \tag{4}$$

which yield, via the field laws (2), s equations of the form

$$\partial_i \Theta_A^i = 0, \quad A = 1, \ldots, s; \quad \partial_i \equiv \partial/\partial x^i. \tag{5}$$

Equations (5) are the conservation laws *in differential form*. The name is appropriate because integrating (5) over some space-time region Ω limited by the hypersurface $\partial\Omega$ leads, by Gauss's theorem, to the relations

$$\oint_{\partial\Omega} \Theta_A^i \, d\sigma_i = 0, \tag{6}$$

which are sometimes called *integrated balance equations*. And these, in turn, entail conservation laws in the conventional sense. To see this, consider Minkowski (or Newtonian) space-time, and, as the hypersurface $\partial\Omega$, choose a four-dimensional cylinder with the lower and upper bases corresponding to the three-dimensional volume V at the times t_1 and t_2. Then Equations (6) take the form of the balance relations

$$C_A(V, t_2) - C_A(V, t_1) = \int_{t_1}^{t_2} Q_A dt, \tag{7}$$

where

$$C_A(V, t) \equiv \int_V \Theta_A^0 \, dV, \tag{7a}$$

$$Q_A(\partial V, t) \equiv \oint_{\partial V} \Theta_A^\alpha \, ds_\alpha, \quad \alpha = 1, 2, 3, \tag{7b}$$

that is, the change of the quantity Θ_A in the volume V is equal to the flow of that quantity through the surface ∂V bounding the volume.

If we move the surface ∂V to spatial infinity and assume that the fields $u_a(x)$ (as well as Θ_A) vanish on the surface (the assumption of a closed system), we obtain a conservation law in its pure form:

$$C_A(t_2) = C_A(t_1),\tag{8}$$

with the integration in (7a) performed over the entire three-dimensional space. It is Equations (7) and (8) that justify the name, "conservation laws," for Equations (5), from which they follow.

The energy-momentum-moment conservation laws in classical mechanics and in SR can be obtained in the manner just described, and they turn out to correspond to the space-time symmetries—the invariance of the action with respect to translations and rotations of the reference system (that is, to the space-time homogeneity and isotropy).

The Noether theorem indicates that these laws are grounded precisely in the properties of space-time itself, rather than in the specific properties of any physical systems (though, of course, the particular expressions of the conserved quantities Θ_A are also determined by the Lagrangian, $\mathcal{L}$, that is, by the properties of the system in question). Hence the general character of conservation laws in classical mechanics and SR. This is also the reason for regarding the energy-momentum-moment conservation laws as *space-time conservation laws*. We will talk, henceforth, only about such conservation laws, and therefore the adjective will be omitted. (The Noether theorem also gives rise to other conservation laws (e.g., of electric charge), corresponding to transformations that do not involve the space-time coordinates and are characterized by $X = 0$, $\Psi \neq 0$.)

The central symmetry of SR is the invariance with respect to transformations from one inertial reference system to another. This symmetry is described by the ten-parameter Poincaré group, $\mathcal{P}^{10}$, and, therefore, generates ten "conserved" quantities $\Theta_A{}^i$:

$$\partial_i \Theta_A^i = 0, \quad A = 1, \ldots, 10.\tag{9}$$

The quotation marks here are meant to stress a simple, albeit important point that conservation laws depend on the assumption about the behavior of the field at infinity.

The number of conservation laws in SR is equal to the dimensionality of the Poincaré group and is determined by the number of space-time dimensions, n:

$$s = n(n + 1)/2 = 10 \quad \text{for} \quad n = 4. \tag{10}$$

At first glance, the quantities $\Theta_A{}^i$ (Equation (9)) look somewhat unusual, since the two indices have quite different characters: index i ranges over four space-time coordinates, whereas index A ranges over ten independent symmetries. Usually, however, one deals with the energy-momentum tensor T^{ik}, symmetric in the indices ik, and—not that usually—the tensor of moment $M_{l}{}^i{}_m$, which is anti-symmetric in the indices lm. Nevertheless, the set of all such quantities T and M constitutes precisely the $\Theta_a{}^i$:

$$\Theta_A^i = \{T^{ik}, M_{lm}^i\}.$$

The possibility of such a division is grounded in the specific symmetry properties of Minkowski space-time, but for the Noether scheme itself, these are just additional conditions.

3. Conservation Laws at the Time of the Creation of General Relativity

The history of conservation laws in GR began even before the creation of this theory (Vizgin 1972, 1981). While thinking about the construction of a relativistic theory of gravitation in 1907, Einstein identified, in the principle of equivalence, the seed from which the theory grew over the next eight years. It was not until 1913, however, that an adequate mathematical expression of this principle emerged: a Riemannian geometrization of gravitational interaction. In Einstein's strenuous and tormenting efforts from 1913 to 1915 to build a theory of gravitation on this basis, a major role was played by the energy-momentum conservation law borrowed from electrodynamics and SR, in the form

$$\partial_i \Theta^{ik} = 0. \tag{11}$$

Speaking about the postulates that should figure in the foundations of a relativistic theory of gravitation, Einstein mentioned, first and foremost, the "satisfaction of the laws of conservation of momentum and energy" (Einstein 1913, p. 1250). It was just then that Einstein discovered the crux of the problem of conservation laws in GR: the apparent incompatibility of general covariance with the energy-momentum conservation law (or, more exactly, with the equation that Einstein referred to as a conservation law). At first, because of his inadequate knowledge of Riemannian geometry, Einstein decided to sacrifice general covariance in favor of the conservation law. But after two years of painful search for noncovariant field equations, Einstein decided to place his trust in the power of Riemannian geometry and to jettison the universal validity of conservation laws.

4. Einstein's Attitude toward Conservation Laws

Einstein's attitude toward conservation laws was shaped by his primarily physical (rather than mathematical) approach to formulating GR. And the law of energy conservation was undoubtedly something to be respected by him, due to its historical merits in physics and its general validity. More-over, Einstein had, in fact, no other tool for constructing a theory that could match the power of the law of energy conservation.

Guided by the idea that the relativistic gravitation theory should be a generalization of the special principle of relativity (asserting the admissibility of any, not only inertial, reference systems), Einstein arrived, fairly early in the process, at the conclusion that space-time coordinates in GR could not be assigned an operational or objective metrical significance. Hence the motto of general covariance.

At the start of the path towards GR, energy conservation was for Einstein one of the main arguments against the general covariance of the field equations. But even after completing GR, when Einstein returned to generally covariant field equations, he was not inclined to abandon the conservation law. He sought to import the relationship (11) from electrodynamics and SR into gravitation theory, generalizing this relationship or deducing it from the field equations. This effort led to the introduction of the so-called energy-momentum pseudotensor of the gravitational field t^{ik}. A relation somewhat similar to (11) was introduced for the sum of this pseudotensor with the energy-momentum tensor of matter

$$\partial_i[(-g)(T^{ik} + t^{ik})] = 0. \tag{12}$$

This equation was derived from the fundamental property of covariant field equations

$$D_i T^{ik} = 0 \tag{13}$$

or (which is equivalent, given Einstein's equations $R^{ik} - (\frac{1}{2})g^{ik}R = \kappa T^{ik}$) from the general property of Riemannian geometry:

$$D_i(R^{ik} - \frac{1}{2}g^{ik}R) = 0. \tag{14}$$

The covariant derivative $D_i = \partial_i + \Gamma_i$, being a characteristic feature of a generally covariant theory, differs from the ordinary one by the component Γ_i, inevitable in a curved geometry. Yet because of this addition, Gauss's theorem cannot be used to turn Equation (13), which seems to be a natural covariant generalization of the differential conservation law of SR (11), into a balance equation similar to (6) and a conservation law similar to (8).

Einstein worked out his compromise pseudotensor solution (12) under the influence of two essentially methodological principles: confidence in the law of energy conservation and confidence in general covariance. The starting point was the covariantly formulated equation (13). By a formal transformation, it was turned into Equation (12) expressing the fact of vanishing of the ordinary divergence, yet the energy-momentum of the gravitational field was to be described by the quantity t^{ik} made up of the gravitational potentials (the metric) g^{ik}. But this quantity depends on the choice of the coordinate system in a noncovariant and nontensorial manner: hence the name, the t^{ik} *pseudotensor*.

The noncovariant character of the t^{ik} was immediately demonstrated in the examples adduced by Schrödinger and Bauer, where, in certain coordinates, the t^{ik} become zero for a definitely non-zero field but do not vanish in empty Minkowski space-time. In combination with the motto of GR (all coordinate systems are permissible), these examples appeared to be fatal to the pseudotensor approach (see Cattani and De Maria 1993).

The way out of this difficult situation suggested by Einstein in 1918 was as follows: The ambiguity in the energy-momentum of the gravitational field should be resolved for the integrated values $\int t^{ik}dV_i$ by limiting the domain of applicability of the pseudotensor approach. Attention was restricted to island situations, where all matter is concentrated in a certain finite spatial volume, and, outside this volume, Cartesian coordinates (Galilean, in Einstein's words) were employed. Since Einstein's approach to conservation laws in GR (worked out by him in 1918) continues to be accepted, let us have a closer look at this work.

Einstein begins his 1918 paper with this dramatic statement: "Although the general theory of relativity has found acceptance among the majority of theoretical physicists and mathematicians, almost all of my colleagues object to my formulation of the law of conservation of energy-momentum" (Einstein 1918, p. 448). But this circumstance did not undermine Einstein's confidence in the viability and importance of the conservation law. In defending his point of view, Einstein begins thus:

> Just like the law of conservation of momentum, which is formed out of three similar conservation equations, the law [of conservation of energy] was, in its original formulation, an integral law. The special theory of relativity blended all four conservation laws into a unified differential law, which asserts the vanishing of the divergence of the "energy tensor." This differential law is equivalent to the integral laws abstracted from experience; it is here, alone, that its significance lies. (Einstein 1918, p. 448)

And somewhat later he says: "*Experience* clearly compels us to seek a differential law that is equivalent to the integral laws of conservation of momentum and energy" (Einstein 1918, p. 449).

Einstein points out that his pseudotensor formulation runs into his colleagues' "objections because . . . they expect all physically significant quantities to be capable of being conceived as scalars or components of tensors" (Einstein 1918, p. 449). In refuting this objection, Einstein wanted to show that, with the aid of the pseudotensor equation (12),

> the concepts of energy and momentum are established just as strictly as we are accustomed to demand in classical mechanics. The energy and momentum of a closed system are completely determined, independently of the choice of the coordinate system, if only the state of motion of the system (considered as a whole) is given relative to the coordinate system. (Einstein 1918, pp. 449–450)

In support of the limitation to the island situations (closed systems), Einstein says: "In order for us to be able to speak of the energy and momentum of a system, the density of energy and momentum must vanish outside a certain region" (Einstein 1918, p. 450). And he goes on to show that if, outside this region, only Cartesian coordinates are used, then the integral values of energy and momentum (throughout the occupied region) do not depend on the choice of the coordinate system inside the region: "Thus, contrary to what is now our customary way of thinking, we come to ascribe more reality to an integral than to its differentials" (Einstein 1918, p. 452).

Yet restriction to an island situation could not satisfy Einstein himself (who, by 1918, had already established relativistic cosmology), and more than half of his paper is devoted to integral conservation laws for a closed universe.

There is no indication in Einstein's works that he was aware of the connection between symmetry and conservation principles. (He discussed the importation into GR, not of all ten conservation laws of SR, but only of those pertaining to energy and momentum). This is not surprising. It was only in 1911 that Gustav Herglotz established, for the first time, a link between ten conservation laws and ten symmetries of the Poincaré group in the context of the mechanics of continuous media in SR (Vizgin 1972), which was rather remote from Einstein's domain of interest. And Noether's work, in which the symmetry-conservation link was elaborated in a general form, appeared only in 1918, when Einstein had already worked out his pseudotensor solution.

Had Einstein realized the Noether nature of any conservation law when he was thinking about the problem of conservation laws in GR, he would have had to look for some ten-parameter symmetry in general Riemannian space. According to contemporary experts (Trautman 1962; Schmutzer 1970, 1979), such a strategy is doomed to failure. This pessimistic view presupposes, however, that in GR the symmetries suitable for the Noether theorem can only be the *symmetries of the movement* of space itself (those associated with the Killing vectors). And such symmetries indeed single out only very special geometries and are nonexistent in the generic Riemannian space.

Another way of understanding symmetries in GR, the approach applicable to the generic Riemannian case—the ten-parameter *symmetry of description*—will be discussed in §9.

5. Conservation Laws in General Relativity after 1918

After Einstein's work of 1918, discussion of the problem of conservation laws in GR faded for nearly forty years. It looked as though the absence of a general solution to the problem and the absence of a Noether inter-pretation did not worry theorists. Moreover, when conservation laws were going through hard times in the domain of relativistic quantum mechanics in the 1920s and 1930s, "theoretical" aid was proffered from GR (Gorelik and Frenkel 1994).

Interest in the problem of conservation laws in GR flared up again in the late 1950s, giving rise to an abundant literature and diverse physico-mathematical treatments, with the Noether theorem now taking the central place in the ensuing discussions (Trautman 1962, Schmutzer 1970). A solu-tion that could satisfy specialists did not come forth, however. Over the past seventy years, the pseudotensor approach has not undergone any fundamental changes. The ambiguity in the expression for Einstein's pseudotensor was identified and other versions were suggested; the law of conservation of angular momentum was introduced on the same pseudotensor basis (Landau and Lifshitz 1973, Trautman 1962, Schmutzer 1970); conditions at "Galilean" infinity were specified for the island system (Faddeev 1982), and some general verbal arguments for the non-localizability of gravitational energy were set forth.

Yet many questions remain unanswered. There is no explanation for why, in a generally covariant theory, the asymptotically flat space outside the island system must necessarily be described in Cartesian coordinates (and not, say, in spherical coordinates). No explanation has been given yet why conservation laws can be applied only to two extreme situations in GR: the island system in empty (flat) space and the universe as a whole. After all, the differential conservation law in SR (5) produces an integral conservation law (8) only in a degenerate, idealized situation, and, in the generic case, it gives a balance equation (7): The change of the quantity Θ_A in the volume V equals the flow of that quantity through the boundary of V.

And finally, there are two other unanswered questions related to the concepts that are central to this essay. First, how can one establish a connection, in terms of the correspondence principle, between the inferior pseudotensor conservation laws of GR and the ten absolutely clear conservation laws of SR? And, second, what does the Noether theorem, the most general and fundamental basis for conservation laws, have to do with the pseudotensor laws?

6. The Pseudotensor Approach from the Noether Point of View

Instead of repeating the arguments pro and con, let us have a look at the pseudotensor proposal from the Noether point of view. One would think that Equation (11), $\partial_i \Theta^{ik} = 0$, is beyond doubt as a point of departure. Both the quantities Θ^{ik} and Equation (11) itself can be obtained in SR by the Noether procedure. But matters stand quite differently with regard to Equation (13), $D_i T^{ik} = 0$. The tensor T^{ik} is obtained in GR, not from a Noether formulation, but from the variation of the Lagrangian of matter for the metric tensor:

$$T^{ik} = \delta \mathcal{L} / \delta g_{ik}.$$

This variation is, in fact, part of the derivation of the equations of motion for the system consisting of the gravitational field and matter, that is, Einstein's equations (Landau and Lifshitz 1973). Tensor T^{ik} plays the role of the source in these equations. The reason for calling T^{ik} the energy-momentum tensor for matter consists in the fact that, in a flat geometry, the expression for T^{ik} turns into a (spatially symmetrized) energy-momentum tensor Θ^{ik} (obtained by the Noether procedure applied to the group of translations of Minkowski space), whereas Equation (13) for GR turns into a differential form of the conservation law of SR.

But even though Equation (13) seems to be a general relativistic generalization of (11), it cannot be called a conservation law, since it has no Noether interpretation. Besides, from the Noether point of view, the two indices in the special relativistic tensor Θ^{ik} are not quite equivalent. Their seeming equivalence is a consequence of the "accidental" fact that the Poincaré group contains a subgroup of translations, which can be parameterized by a 4-vector. Strictly speaking, the quantity Θ^{ik} should be designated Θ^{iA}, where $A = 1, \ldots, 4$ (see Equations (9), with only a subset of them forming (11)). Rotations, which the Noether approach associates with the 4-moment of momentum, cannot be parameterized by a 4-vector, but they can be parameterized by the antisymmetric tensor ω^{kl} (the Lorentz rotations in the xt-planes, i.e., relative motions of reference systems, are also included among the rotations). This is why the moment is described, in SR, by the quantities Θ^{iA}, (where $A = (kl)$, $1 \le k < l \le 4$), or by the third-rank tensor M_{kl}^{i}, which is anti-symmetrical in the indices kl. All of the quantities conserved in SR can be expressed in the form Θ^{iA}, where $A = 1, \ldots, 10$. Translations and rotations are independent symmetries of the space-time description in SR, therefore energy-momentum and the moment of momentum are independent (and equal in their status) quantities.

This makes it hard to understand the desire to solve the energy-momentum problem in GR separately from and independently of the problem of angular momentum. In fact, there is a unified problem of *energy-momentum-moment* in GR, and therefore, it is necessary, in this theory, to look for the quantities Θ^{iA}, where $A = 1, \ldots, 10$. However, it is impossible to identify any asymmetry of the indices (in order to turn one index into $A = 1, \ldots, 10$) in the tensor T^{ik}, the right part of the Einstein equations. Due to their nature, the indices in T^{ik} are absolutely symmetrical.

Division of the quantities Θ^{iA} into two sets (the energy-momentum tensor and the moment tensor), covariant with respect to the Poincaré group, reflects precisely the composition of that group. The impossibility of such a division in GR in the generic situation should not be surprising.

The transition from Newtonian mechanics to SR led to the unification of the concepts of energy and momentum into that of energy-momentum; the concepts of angular momentum and the velocity of the center of inertia have also been united. The transition from SR to GR leads to the unification of all of these quantities (see §§9–11).

The desire to interpret the equation $D_i T^{ik} = 0$ as a conservation law in GR must surprise one no less than, say, the desire to attach some special meaning in general Riemannian space to the lines $x^i = \kappa^i \tau$, $\kappa^i = const.$ (and to call these lines "straight") in an arbitrary system of coordinates, only because it holds true for Cartesian coordinates in Minkowski space.

7. The Problem of Conservation Laws in General Relativity and the Correspondence Principle

The concepts of energy, momentum and angular momentum, and the appropriate conservation laws are indispensable to the rest of "nongravitational" physics, providing the most general, basic means of description. At the same time, there exist quite different opinions about the importance of conservation laws in GR.

The most common view simply denies that there is any problem here, the argument being that, because of the special character of the gravitational field, energy-type concepts and the appropriate conservation laws lose their meaning in most situations admitted by GR. At the same time, the Einstein equations are held to give a complete description of any situation.

In the opinion of other specialists, the problem of conservation laws in GR remains unresolved and presents a challenge for the theory (Penrose 1982, Trautman 1962, Schmutzer 1970).

One of the simple arguments used by the advocates of the first position consists in the following: The equivalence principle underlying GR allegedly dooms to failure any attempt to introduce a concept of the gravitational field energy. Any gravitational field is meant to disappear for the observer who steps into Einstein's elevator, a freely-falling reference system; in that case, the field energy must disappear, too.

Does this argument point to the impossibility of introducing the notion of gravitational energy in GR? No more so than in classical mechanics, where a transition to the reference system moving with a given body, thus "eliminating" the motion of this body, does not prevent the introduction of the concept of kinetic energy. Such an argument, however, indicates quite clearly that, in GR, the question of conservation laws is intimately linked to the concept of reference systems; see §8.

The role played by the equivalence principle in GR has been extensively debated (see, e.g., Synge 1960, Fock 1979, Ginzburg 1979). The differing attitudes to the principle are prompted mainly by the different frameworks in which it is considered. Within the framework of Newtonian mechanics, the equivalence principle has a quite definite meaning, and, from the viewpoint that "reveals" GR's links with classical physics and experiment, the equivalence principle is really a fundamental property of gravitation. If it is viewed from within GR, however, it appears meaningless, at least if the concept of an accelerated reference system has not yet been introduced. Within the theoretical framework of GR the equivalence principle "dissolves" into the very notion of geometrizing gravity.

This situation can be compared with the discussion of the principle of spatial isotropy between a supporter of Aristotle's doctrine of space and an advocate of the Newtonian conception of space, in which the principle of isotropy is similarly "dissolved."

Those who say that energy-type concepts lose all meaning in GR sometimes point to the tendency for long-standing notions to die off in the course of scientific progress. Thus, the transition from classical mechanics to quantum mechanics, for example, led to the extinction of the notion of (observable) particle trajectories. In quantum mechanics, however, it is shown *how and when* (in terms of the correspondence principle) the concept of a trajectory may become justified.

In the case of conservation laws in GR the situation is different. It is not known how to effect a transition (in the sense of correspondence) to conservation laws in the case of small deviations of geometry from the flat case; isolated distributions of matter do not exhaust all possibilities, for the curvature may be globally small but without Euclidean properties at infinity as, for example, in the case of constant large-radius curvature.

What are the general concepts and structures of GR that transform into conservation laws in SR? An answer to this question is necessary irrespective of one's attitude to the problem of conservation laws in GR.

Since the situation with conservation laws in SR is quite clear, and since SR is (logically and historically) a natural starting point and a limiting case for GR, it is useful to resort to the correspondence principle in analyzing the issue of conservation laws. One ought to clarify in this context *what* can or should be called a conservation law in GR; that is to say, what quantities, properties, and facts known in SR can be generalized in GR.

The connection between a theory's symmetries and its conservation laws established by the Noether theorem greatly facilitates understanding the essence of the problem. In classical mechanics and in SR, there are ten space-time conservation laws: of energy (1), of momentum (3), of angular momentum (3), and of the velocity of the center of inertia (3). In accordance with the Noether theorem, the number of conservation laws is directly determined by the fact that classical mechanics is based on the ten-parameter Galilean group, while SR rests on the ten-parameter Poincaré group.

In classical mechanics, the laws of nature or, more to the point, the action is invariant with regard to the following transformations of the Cartesian system of spatial coordinates and time $(t, x, y, z) \rightarrow (t', x', y', z')$: displacement of the origin of time (one parameter, Δt), rectilinear displacement of spatial coordinates (three parameters, $\Delta \mathbf{x}$), rotations of the system of coordinates (three parameters; e.g., three Euler angles), and transformation to another inertial system (three parameters, the components of relative velocity).

In SR, a similar role is played by the Poincaré group, which includes four space-time translations and six rotations (three purely spatial and three Lorentz "rotations"). In each case, the breakdown of the number of conserved quantities, $10 = 1+3+3+3$ and $10 = 4+6$, reflects the structure of the corresponding group, and their total number, 10, is determined by the number of space-time dimensions (see §§9, 10).

In terms of the correspondence principle and the Noether theorem, the question may now be posed as follows: What kind of structure in GR corresponds to the ten-parameter Poincaré group describing the geometry of Minkowski space? Sure enough, this is not the group of all continuous coordinate transformations: not only because it is an infinite-parameter group (though this, by itself, is sufficient to dismiss such a proposal), but also because such a group can be easily introduced even in Minkowski space-time.

The locally flat character of the geometry of space-time in GR is usually not connected with the understanding of the Poincaré group as the limit of a certain structure defined in general Riemannian space (see, for example, Misner, Thorne, and Wheeler 1973). At the same time, the description of the geometry of Minkowski space-time based on the properties of the Poincaré group proved very fruitful in physics. It is believed that in a generic Riemannian space (lacking movement symmetries), there is no natural way of defining a "natural" finite-parameter family of transformations of coordinates (Trautman 1962) and that it makes sense to speak only of the infinite-parameter group of all smooth transformations of coordinates. Yet, as the subsequent analysis will show, the space-time symmetries can be linked not only to motions in space, but also to the observer's displacements, that is, to the displacements of an appropriately defined reference system.

In the case of a homogenous space, and in particular in SR, we know of two equivalent interpretations of spatial transformations: *active* and *passive*. According to the active interpretation, the system of coordinates does not change, while the geometrical or physical system under consideration, for example, the region in which the field variables do not vanish, is transformed. The active viewpoint may alternatively be characterized as follows: The observer (reference system) stands still, whereas the entire space moves "dragging" all objects along with it. According to the passive viewpoint, space and the objects stand motionless while the observer (reference system) moves. As a result of each of these procedures, the coordinates in the spatial region under consideration undergo change, but the difference between the active and passive approaches cannot even be expressed in "internal" terms, that is, not supposing some embedding space.

In a generic Riemannian (arbitrarily curved) space, the active approach is impossible: In the general situation, space cannot be shifted along itself, and a complex enough system (for example, of $n+2$ points in n-dimensional space) cannot be appropriately displaced without internal changes (in the above example, without changing the distances between the points). However, the passive approach—the "movement" of an observer, that is, of a properly defined reference system—leads, as we shall see, for n-dimensional Riemann space, to an $n(n+1)/2$-parameter set of coordinate transformations, the set possessing a quasigroup structure generalizing the structure of the Poincaré group (§9).

But before considering this suggestion in detail, let us discuss the connection between conservation laws and the concept of reference system.

8. Conservation Laws and the Concept of Reference System

The problem of conservation laws in GR is inseparable from the issue of space-time reference systems. The fairly common neglect of the concept of reference system in GR could be justified if physicists were only interested in scalar quantities, which do not depend on the reference system; but energy, at any rate, does not belong among such quantities. There is a clear connection between conservation laws and reference systems in SR, where the conserved quantities are defined by the invariance of the action with regard to Poincaré transformations connecting various inertial reference systems. From a mathematical point of view, one selects a set of standard coordinate systems in Minkowski space (those that allow a physical interpretation in terms of reference systems), and the coordinates are generated by a quite rigorous procedure.

The mixing of the concepts of reference system and coordinate system in general relativity—which is justly criticized—has nevertheless a reasonable foundation. After all, a reference system in physics is, first and foremost, a concrete way of establishing coordinates for space-time points. Such a physical incarnation of a coordinate system is called below a *coordinate reference system.*

It is true that such clarifications may seem pointless in GR, since there is an opinion, traceable to Einstein, that, in the Riemannian geometry of GR, it is not possible to impart a metrical (measurable, physical) significance to coordinates. Yet, for all the heuristic importance of this idea for Einstein, it is wrong. Suffice it to recall that the very first coordinate system introduced into Riemannian geometry by Riemann himself (1854) had a quite definite metrical significance and was applicable to arbitrary Riemannian spaces. Other similar ways of introducing coordinates are also known. One of these—perhaps logically the simplest—will be used in the next section to characterize the symmetries of general Riemannian space.

In SR, inertial reference systems are typically described in Cartesian coordinate systems in Minkowski space. Space-time in SR can surely be described by means of any coordinates (spherical, cylindrical, etc.), but as far as conservation laws are concerned, only Cartesian coordinates are usually considered.

Is such a restriction necessary? A restriction to the class of *standard* coordinate systems is certainly necessary; this much is required to bring out the ten-parameter ($10 = n(n+1)/2$, with $n = 4$) transformations describing the space-time symmetry in SR (homogeneity and isotropy). But restriction just to Cartesian coordinates is not necessary. For example, one may employ spherical coordinates or any other class of *standard*, that is, equivalently defined, coordinates. Each of these classes of coordinates is fit for describing the space-time symmetry in SR.

In SR, the inertial coordinate reference systems are called "privileged," and not without reason. The notion of a privileged class of coordinate systems in GR will be linked here only to the possibility of a general, standard, and constructively described (physically realizable) way of introducing coordinates. In this sense, there may be many classes of "privileged" coordinates.

The need for a preliminary restriction and standardization in the way of describing coordinates is quite natural. To identify the invariant properties of some object, various observers have to study it with the possibly strictest limitations in their techniques of investigation. All observers must follow standard methods so that the results of their investigations disclose the object's properties, rather than the properties of the observers.

Therefore, in GR it is essential first to determine the class of standard coordinate reference systems. After all, conservation laws emerge, according to the Noether theorem, for any action, including the general relativistic action, only if a finite-parameter set of coordinate transformations is identified in a physically sensible way. This is why, in pursuing the Noether approach to conservation laws, one has to identify a natural, finite-parameter family of coordinates in general Riemannian space.

9. The Poincaré Quasigroup

Let us start with a simple question: Why are there exactly *ten* conservation laws in SR? The answer is: Because space-time has *four* dimensions.

Indeed, a Noether-type connection between conservation laws and space-time symmetries provides a relationship between the number of symmetries in the n-dimensional Minkowski space-time and n itself: there are n independent translations along n axes and $C_n^{\ 2} = n(n-1)/2$ independent rotations; altogether this yields $s = n(n+1)/2$ independent symmetry transformations. If $n = 4$, then $s = 10$.

The origin of the ten-parameter Poincaré group describing space-time symmetries in SR is similar. This group, like any other, may be represented in a purely algebraic fashion, but physicists do not always take care to distinguish between the Poincaré group and one of its representations: the linear representation in a 3+1-dimensional Minkowski M^{3+1} space. This representation is formed by linear transformations from one Cartesian system of coordinates in M^{3+1} to another, which conserve the metrical structure—the expression for the metrical interval

$$I = \Delta s^2 = -c^2(\Delta t)^2 + (\Delta x)^2 + (\Delta y)^2 + (\Delta z)^2.$$

It should be emphasized that the linear representation is just one of many possible representations of the Poincaré group, and that it cannot be extended to Riemannian space, for this representation relies upon the linear structure of M^{3+1}, which is not available in curved space.

So the conservation laws in SR are generated by the symmetries of the metrical space-time structure, the number of the symmetries being determined by the number of space-time dimensions. The space-time in question is a linear space. The concept of linear space is also sufficient for constructing a finite-parameter (in the n-dimensional case, an $n(n+1)/2$-parameter) set of coordinate systems, or, in physical terms, a set of inertial reference systems in SR. It is through the Noether theorem that this ten-parameter set of coordinate reference systems leads to ten conservation laws.

Let us now turn to the Riemannian geometry of GR, where we also want to identify a natural $n(n+1)/2$-parameter set of coordinate reference systems in an arbitrary n-dimensional Riemannian space.

The obvious and seemingly crucial objection to such an undertaking hinges on the fact that the presence of a natural, finite-parameter set of coordinate transformations must reflect some sort of space-time symmetry, a property that does not change from point to point. And such a property, it would seem, must be nonexistent in a space with variable curvature. And yet, even in an arbitrarily curved space-time of GR, there is a property that does not change from point to point: the *space-time dimensionality*. The task now is to turn this trivial observation into a productive strategy.

First of all, we cannot use the concept of linear dimensionality, which ensures the connection between conservation laws and the dimensionality in SR. We also cannot use the topological concept of dimensionality, since it ignores the metrical space-time structure, which is basic to GR.

Yet a metrical concept of dimensionality is available. Consider a generalization of the following geometrical fact:

In a Euclidean (pseudo-Euclidean) n-dimensional space, the position of an arbitrary point can be determined solely in terms of its distances (intervals) to n fixed points. (Gorelik 1978, 1979)

In a Riemannian space, a suitable measure of "distance" can be provided by the interval (or the world function; see Synge 1960) $I(p,p')$ between two arbitrary space-time points p and p'. The description of space-time in terms of I is equivalent to the conventional description employing the metric tensor g_{ik} or the intervals ds^2 between infinitely close points:

$$I(p,p') = (\int_p^{p'} ds)^2, \quad ds^2 = g_{ik} dx^i dx^k. \tag{15}$$

Here integration from p and p' is done along a *geodesic*.

Let us now fix a certain point b in space. This defines the function $x(p) \equiv I(p,b)$ on this space. Fixing n such points $\{b^i\}$, $i = 1, \ldots n$, in an n-

dimensional space provides a full-blown *basis* and, hence, generates a definite coordinate system: each point p is assigned n numbers $x^i(p) = I(p,b^i)$. These coordinates, introduced with the aid of the spatial metric, can be called *metric coordinates*. In the four-dimensional Minkowski space, any four points in a generic position (that is, not lying on the same plane) provide a basis.

Consider the set of all such metric coordinate systems generated by various bases. Since each coordinate system is completely determined by the position of its basis points, the transformation from one such system to another can be accomplished by indicating the positions of new basis points $\{b''^i\}$ in the old basis $\{b^k\}$, that is, by supplying n^2 numbers $I(b''^i,b^k)$, $i,k = 1, \ldots , n$, which can be represented in the form of a matrix.

Not all of these n^2 parameters are equally essential. To exclude "coordinate" effects, one should use standard, normalized bases, with the intervals between basis points fixed, for example, in such a way that they coincide with the intervals between the following points in Minkowski space: (0000), $(0a00)$, $(00a0)$, $(000a)$. Each of these bases is characterized by the basis diameter a, which can be made infinitesimal thus effecting a local correspondence of GR with SR.

The use of normalized bases reduces the number of parameters by $n(n-1)/2$. So the set of bases (or of metric coordinate systems) is characterized by $n^2 - n(n-1)/2 = n(n+1)/2$ parameters. Thus in four-dimensional space, there are ten such parameters, just as in a Poincaré group.

Each coordinate reference system thus constructed is obtained in the same way and can be introduced in an arbitrarily curved space.

The $n(n+1)/2$-parameter set of bases (or of coordinate reference systems) introduced via the above procedure generates a *quasigroup* structure. Indeed, fix a particular basis b_0 (the lower index will now range over bases). Then any other basis, as mentioned, corresponds to a certain $n \times n$ matrix. The sequence of transformations from basis b_0 to basis b_1 and then from b_1 to b_2 is equivalent to a transformation from b_0 directly to b_2. This defines the group *product* of two matrices (not in the sense of matrix algebra, of course). This product is determined by the geometry of a given space-time $\mathbb{R}^n$, specifically, by a function of $2n$ variables, the function defining the interval between two points p and p' in terms of the intervals between these points and the points of both bases:

$$I(p,p') = \Gamma(\ldots (p,b^i)\ldots; \ldots I(p',b^k)\ldots). \qquad (16)$$

The function Γ provides a global characterization of the geometry of $\mathbb{R}^n$.

If $B^{ik} = I(b_1{}^i, b_2{}^k)$ is the matrix effecting the transformation from basis b_1 to b_2, the *inverse* matrix (in the group sense) should naturally be the transposed matrix B^T responsible for the transformation from b_2 to b_1. The product and inverse operations define an $n(n+1)/2$-parameter quasigroup in space $\mathbb{R}^n$: the *Poincaré quasigroup* $\mathcal{P}\mathbb{R}^n$ of this space (Gorelik 1981, 1988).

The mathematical notion of group has long been customary in physics. A quasigroup, on the other hand, is a relatively new object. What makes it different is the *nonassociative* character of the law of multiplication; that is, for three quasigroup elements A, B, C, in general, $(AB)C \neq A(BC)$ (Bruck 1971). As algebraic objects, quasigroups have been known in mathematics for about sixty years. Yet the significant role played by them in differential geometry (Sabinin 1981) has only recently been recognized. Nonassociativity is the price to pay for the shift from highly symmetrical spaces (e.g., those of constant curvature) to spaces with variable curvature.

In Minkowski space, the quasigroup $\mathcal{P}\mathbb{R}^n$ becomes isomorphic to the ordinary Poincaré group: $\mathcal{P}M^n = \mathcal{P}^{(n)}$. Indeed, every transformation from the Poincaré group in the standard linear representation is in one-to-one correspondence with a certain transformation of spatial unit vectors (0100), (0010), and (0001). The basis formed by the initial point of the unit vectors and by their ends, corresponds, one-to-one, to a set of spatial unit vectors. And every transformation of unit vectors corresponds, one-to-one, to a transformation of the basis.

The set of transformations of one basis into another in Minkowski space—those expressed in terms of mutual distances (intervals) of the points constituting the bases—forms a *nonlinear representation* of the Poincaré group. This representation is more complex than the conventional linear representation, but it *allows* a generalization in the case of curved space.

10. What Kind of Symmetry is Described by the Poincaré Quasigroup?

It is widely believed that the notion of symmetry can be expressed mathematically by an associated group (Weyl 1952, Vizgin 1972), yet the concept of symmetry is consistent with a more general mathematical structure. There are no apparent reasons, for example, for precluding a set of transformations from being associated with a symmetry, if the set is closed under composition, contains a unit (identity) transformation, and, for each transformation, contains the inverse one. But this is precisely what a quasigroup is. The demand for associativity (converting a quasigroup into a group) seems no more imperative for the description of symmetry than the demand for commutativity (valid only for the simplest symmetries).

Although the requirement that the set of coordinate transformations must form a group is frequently mentioned in connection with the Noether theorem, it plays no role there. The structure of a set of transformations remains, of course, essential for their physical interpretation and for the specification of the corresponding conserved quantities.

What kind of symmetry of the arbitrarily curved space $\mathbb{R}^n$ is described by the quasigroup $\mathcal{P}\mathbb{R}^n$? A generic curved space-time $\mathbb{R}^n$ retains a property that does not change from point to point: the number of its dimensions. It is an aspect of the space-time homogeneity that is not affected by the transition from flat to curved space-time. The quasigroup $\mathcal{P}\mathbb{R}^n$ represents a symmetry of a *description* of space $\mathbb{R}^n$ by an observer (in the four-dimensional case—a coordinate reference system produced by four space-time points with fixed intervals between them). This symmetry can be looked upon as a relation among space-time descriptions of one and the same physical system by different, albeit standard, observers (coordinate reference systems).

Does the use of special, "privileged" coordinates contradict the principle of general covariance underlying GR? Not at all. What really matters in GR is the possibility to have a construction realizable in an *arbitrary geometry* rather than the possibility of using *arbitrary coordinates*. After all, the latter possibility does not hinder conservation laws in Minkowski space.

When the invariance of a certain object is at issue, one should be clear about instrumental procedures—their standardization and the "calibration" of the instruments themselves—employed in describing that object. In our case, the object in question is curved space-time.

In speaking of an arbitrary reference system in SR, one usually means (explicitly or not) "any inertial Cartesian reference system." This is why in GR, too, it is necessary to seek a constraint on reference systems, the constraint "equal in power" to the restriction to inertial and Cartesian systems in SR. In a privileged class of coordinate systems in GR, there must be "as many elements" as there are in the class of inertial reference systems in SR. In the same way, reference systems in GR must be capable of being constructed "everywhere." If a reference system is related to some space-time structure, then it must be possible to relate the reference system with a similar structure located at any other position in space-time.

To describe a curved (as well as flat) space, one is free to use any coordinates. Yet the task of describing a physical situation with the aid of conserved quantities implies a restriction on coordinate systems. The demand for such a restriction is not unique to GR. In SR, the very expressions for the conserved quantities contain information about the type of coordinate reference systems employed. This information is packed into the quantities X and Ψ describing coordinate transformation; see Equations (3a, b).

Likewise, in GR, the conserved quantities, which are engendered by the ten-parameter Poincaré quasigroup and the general relativistic action, contain information about the type of coordinates employed, and also about the geometry of a given space-time $\mathbb{R}^n$. This information is packed into the function Γ; see Equation (16). Of course, the conserved quantities of SR also encapsulate information about the geometry of Minkowski space-time, but only in a trivial sort of way, because this geometry is fixed and global.

11. The Ten Noether Laws of Conservation of Energy-Momentum-Moment in GR

Thus the desired $n(n+1)/2$-parameter set of transformations of coordinate systems (or, remembering the need for a physical interpretation, of coordinate *reference* systems) is built into an arbitrarily curved Riemannian space-time and has the structure of a quasigroup.

The general relativistic action

$$A = \int (Rg^{\frac{1}{2}} + \mathcal{L}_m) d^4x, \tag{17}$$

which is invariant with respect to arbitrary transformations of coordinates, is invariant, in particular, with regard to transformations from $\mathcal{P}\mathbb{R}^n$. This is why, thanks to the Noether theorem for the action (17), the $n(n+1)/2$-parameter set of coordinate transformations $\mathcal{P}\mathbb{R}^n$ generates $n(n+1)/2$ conservation laws of the type $\partial_i \Theta_A{}^i = 0$, $A = 1, \ldots, n(n+1)/2$. It seems natural to refer to the quantity $\Theta_A{}^i$ as the *density of energy-momentum-moment*.

In Minkowski geometry, $n(n+1)/2$ quantities Θ_A can be easily divided into n and $n(n-1)/2$ values—energy-momentum and moment—since the Poincaré group contains translation and rotation subgroups. In the case of a general Riemannian space, whose symmetries are described by a *quasigroup*, the division of the conserved quantities Θ_A into similar sets is a special task that cannot be accomplished uniquely.

The ten conserved components of energy-momentum-moment, $\Theta_A{}^i$, $A = 1, \ldots, 4(4+1)/2$, generated by the Poincaré quasigroup $\mathcal{P}\mathbb{R}^4$, satisfy the Noether-type conservation laws

$$\partial_i \Theta_A{}^i = 0,$$

but in the generic Riemannian space $\mathbb{R}^4$, these differential laws can take on an integral form only as balance equations

$$\oint \Theta_A^i d^3\sigma_i = 0.$$

In general, however, it is not possible to turn these balance equations into the conventional conservation laws, as in SR (see §2, Equations (6–8)), since in GR, the supposition that "the field at infinity must vanish" radically restricts the space-time geometry. Recall that this problem was actually discussed in 1918 by Einstein, who supported the restriction to island situations (see §4).

The density of energy-momentum-moment $\Theta_A{}^i$ is a *nonlocal* quantity, since it is determined by the entire geometry (via the function Γ); this is why $\Theta_A{}^i$ cannot, in general, have simple transformation properties (similar to tensor properties in SR). And the ten energy-momentum-moment quantities do not, in general, decompose into smaller sets like energy-momentum and angular momentum.

However, the quantities $\Theta_A{}^i$ have resulted from the same general Noether approach as in the rest of physics.

12. Conclusion

The problem of conservation laws in GR has a dramatic history. The gravitational energy-momentum pseudotensor emerged as early as 1913, two years before the real birth of GR. And yet, many decades after the emergence of GR, experts do not regard the pseudotensor solution to be completely satisfactory.

The drawbacks of the pseudotensor approach are especially obvious from the viewpoint of the correspondence principle and in light of the Noether theorem.

The theory's symmetry, required by the Noether theorem, can be identified, in the case of GR, with the dimensional homogeneity of space-time: the number of dimensions of Riemannian space is the same at each point. The description of the dimensionality of Riemannian space $\mathbb{R}^n$ in metric terms makes it possible to introduce the Poincaré quasigroup $\mathcal{P}\mathbb{R}^n$, with the number of parameters, $n(n+1)/2$, connected with the number of dimensions in the same way as in the case of the Poincaré group in SR. And in the limiting case of Minkowski space, this quasigroup becomes isomorphic with the Poincaré group.

By the standard Noether procedure, this symmetry of the generic Riemannian space-time $\mathbb{R}^n$ turns up ten laws of conservation of energy-momentum-moment, or, more precisely, ten balance equations.

The conservation laws, which are based on the Poincaré quasigroup, have an advantage over the pseudotensor laws, not only because "pseudo" smacks of "sham" and "quasi" suggests "having resemblance to." Philology aside, the quasigroup approach also has methodological advantages.

Abandonment of associativity makes the quasigroup structure less definite and more flexible than the group structure. Groups are distinguished in a discrete way. Their differences cannot be made infinitely small. For example, the group of circular motions does not collapse with the group of rectilinear motions with the increase of the circle radius. But quasigroups can change continuously, turning into each other. The Poincaré group is sharply separated from the de Sitter group and other groups realizable in the homogeneous spaces of GR. But the Poincaré quasigroup connects and unites all these cases. This suggests yet another way of looking at the transition from SR to GR.

And finally, the Noether-type approach to conservation laws in GR provides a novel basis for the study of concrete situations arising in this theory, where the conserved quantities may be useful.

Acknowledgments. I dedicate this paper to the memory of Igor Yur'yevich Kobzarev (1932–1991) who extraordinarily combined professional work in theoretical physics with expertise in the history of science and profound interest in the humanities.

REFERENCES

Bogolyubov, Nikolay Nikolaevich, and Shirkov, Dmitri Vasil'yevich (1980). *Introduction to the Theory of Quantized Fields*, 3rd ed. New York: Wiley.

Bruck, Richard Hubert (1971). *A Survey of Binary Systems.* Berlin: Springer.

Cattani, Carlo, and De Maria, Michelangelo (1993). "Conservation Laws and Gravitational Waves in General Relativity 1915–1918." In *The Attraction of Gravitation: New Studies in the History of General Relativity*. John Earman, Michel Janssen, and John D. Norton, eds. Boston: Birkhäuser, pp. 63–87.

Einstein, Albert (1913). "Zum gegenwärtigen Stande des Gravitationsproblems." *Physikalische Zeitschrift* 14: 1249–1266.

—— (1918). "Der Energiesatz in der allgemeinen Relativitätstheorie." *Königlich Preussische Akademie der Wissenschaften* (Berlin). *Sitzungsberichte*: 448–459.

Faddeev, Lyudvig Dmitrievich (1982). "The Energy Problem in Einstein's Theory of Gravitation." *Soviet Physics. Uspekhi* 25: 130–142.

Fock, Vladimir Aleksandrovich (1979). "Fizicheskie printsipy teorii tyagoteniya Einshteina." ["Physical Principles of Einstein's Theory of Gravity"] In *Einshtein i filosofskiye problemy fiziki XX veka [Einstein and Philosophical Problems of Twentieth-Century Physics]*. K. Delokarov, ed. Moscow: Nauka.

Ginzburg, Vitaly Lazarevich (1979). *O teorii otnositel'nosti. [On the Theory of Relativity]* Moscow: Nauka.

Gorelik, Gennady Efimovich (1978). "O razmernosti prostranstva, osnovannoy na metrike." ["On the Dimensionality of Space Based on Metric"] *Vestnik Moskovskogo Universiteta* 5: 58–62.

—— (1979). "Razmernost' prostranstva v fizike i topologii." ["Dimensionality of Space in Physics and Topology"] *Istoriya i Metodologiya Estestvennykh Nauk* 21: 47–57.

—— (1981). "Zakony sokhraneniya v OTO i printsip sootvetstviya." ["Conservation Laws in GR and the Correspondence Principle"] *Istoriya i Metodologiya Estestvennykh Nauk* 26: 110–118.

—— (1983). *Razmernost' prostranstva: Istoriko-metodologicheskiy analiz. [Dimensionality of Space: A Historical and Methodological Analysis]* Moscow: Moscow State University Press.

—— (1988). "Razmernost' prostranstva-vremeni i kvazigruppa Puankare v OTO." ["Space-Time Dimensionality and the Poincaré Quasigroup in GR"] In *Einshteinovskiy sbornik, 1984–1985*. I. Yu. Kobzarev and G. E. Gorelik, eds. Moscow: Nauka, pp. 271–300.

Gorelik, Gennady Efimovich, and Frenkel, Viktor Yakovlevich (1994). *Matvei Petrovich Bronstein and Soviet Theoretical Physics in the Thirties*. Boston: Birkhäuser.

Landau, Lev Davidovich, and Lifshits, Evgeny Mikhailovich (1973). *Teoriya polya*. Moscow: Nauka. Trans.: *The Classical Theory of Fields*. Oxford and New York: Pergamon Press, 1975.

Logunov, Anatoly Alekseevich (1998). *Relativistic Theory of Gravity*. Commack, NY: Nova Science Publishers.

Misner, Charles, Thorne, Kip, and Wheeler, John Archibald (1973). *Gravitation*. San Francisco: W. H. Freeman.

Pauli, Wolfgang (1958). *Theory of Relativity.* G. Field, trans. New York: Pergamon Press.

Penrose, Roger (1982). "Some Unsolved Problems in Classical General Relativity." In *Seminar on Differential Geometry.* Shing-Tung Yau, ed. Princeton, New Jersey: Princeton University Press, pp. 631–668.

Riemann, Bernhard (1854). "Über die Hypothesen welche der Geometrie zu Grunde liegen." *Köngliche Gesellschaft der Wissenschaften und der Georg-Augusts-Universität* (Göttingen). *Mathematische Classe. Abhandlungen* 13 (1867): 133–152. Lecture, Göttingen, 10 June 1854.

Sabinin, Lev Vasil'evich (1981). *Analiticheskiye kvazigruppy i geometriya.* [*Analytic Quasigroups and Geometry*] Moscow: P. Lumumba University Press.

Schmutzer, Ernst (1970). *Symmetrie und Erhaltungssätze in der Physik.* Berlin: Akademie Verlag.

—— (1979). *Relativitätstheorie Aktuell.* Leipzig: Teubner.

Synge, John Lighton (1960). *Relativity: The General Theory.* Amsterdam: North-Holland; New York: Interscience.

Trautman, Andrzej (1962). "Conservation Laws in General Relativity." In: *Gravitation: An Introduction to Current Research.* L. Witten, ed. New York: Wiley, pp. 169–198.

Vizgin, Vladimir Pavlovich (1972). *Razvitiye vzaimosvyazi printsipov invariantnosti s zakonami sokhraneniya v klassicheskoy fizike.* [*Evolution of Interrelation between Invariance Principles and Conservation Laws in Classical Physics*] Moscow: Nauka.

—— (1975). "Printsip simmetrii." ["The Principle of Symmetry"] In: *Metodologicheskiye printsipy fiziki: istoriya i sovremennost'.* Moscow: Nauka, pp. 268–342.

—— (1981). *Relyativistskaya teoriya gravitatsii: istochniki i formirovaniye, 1900–1915.* [*Relativistic Theory of Gravitation: Sources and Formation, 1900–1915*] Moscow: Nauka.

Weyl, Hermann (1952). *Symmetry.* Princeton, New Jersey: Princeton University Press.

The Role Played by Mach's Ideas in the Genesis of the General Theory of Relativity

Vladimir Vizgin

> Though unable now to say how clear Mach's views of the relativity of inertia are, I can say yet quite definitely that I have been greatly influenced by his ideas.
>
> Albert Einstein

1. Introduction

General theory of relativity (GR) was almost entirely created by Einstein. He takes credit for the major breakthroughs in its development: the principle of equivalence (1907–1911), the early theories of static fields consistent with this principle (1911–1912), the tensor-geometrical concept of gravitation (1912–1915), and the generally covariant equations of the gravitational field (1915). Among Einstein's contemporaries who influenced him and thus contributed to the creation of GR (Max Abraham, Gunnar Nordström, Marcel Grossmann, Erwin Freundlich, David Hilbert, and others), Ernst Mach holds a very special position.

From 1909 onwards, Einstein wrote many times, in his papers, biographical notes, and letters, about the impact of Machian ideas on him. The complex of these ideas was, for Einstein, one of the important starting points in his work on the theory of relativity. Einstein is notorious for being parsimonious in his references and not very accurate in mentioning his predecessors. But he consistently referred to Mach's works in virtually all his papers on the theory of gravity written in 1912–1916 (the time when he laid down the basics of GR) and in a number of later works.

Yuri Balashov and Vladimir Vizgin, eds., *Einstein Studies in Russia.*
Einstein Studies, Vol. 10. Boston, Basel, Birkhäuser, pp. 45–89.

One could easily compile a list of rather enthusiastic statements Einstein made on Mach and on the importance of his concepts for the creation and development of relativity. Most such statements date from a later time, when Einstein was already aware of Mach's sharp criticism of that theory found in the preface to *Die Prinzipien der physikalischen Optik* (1913).[1] At that time, Einstein was departing, both philosophically and methodologically, from Mach's radical phenomenalism.[2] We will confine ourselves to two statements made in 1916 and 1949.

After quoting the passages from Mach's *Mechanics* containing criticism of the concepts of absolute space and time and elaboration of Mach's own idea of inertia, Einstein wrote in 1916:[3]

> These quotations show that Mach clearly recognized the weak points of classical mechanics, and thus came close to demanding a general theory of relativity—and this almost half a century ago! It is not improbable that Mach would have hit on relativity theory when in his time—when he was in fresh and youthful spirit—physicists would have been stirred by the question of the meaning of the constancy of the speed of light. . . . The contemplations on Newton's experiment with the pail demonstrate how close his mind was to the demands of relativity in a wider sense (relativity of accelerations). (Einstein 1916a, p. 103, trans. by A. Engel)

In his *Autobiographical Notes* Einstein spoke about how he had been influenced, not only by Mach's ideas on relativity and inertia ("Mach conjectures that in a truly reasonable theory inertia would have to depend upon the interaction of the masses, precisely as was true for Newton's other forces, a conception that for a long time I considered in principle the correct one" (1979, p. 27)), but also by the general critical pathos and methodological concepts of the Austrian thinker: "It was Ernst Mach who, in his *History of Mechanics*, upset this dogmatic faith [in mechanics as the foundation of physics]; this book exercised a profound influence upon me in this regard while I was a student. I see Mach's greatness in his incorruptible scepticism and independence" (1979, p. 19).[4]

Various aspects of the relationship among Einstein, Mach, and the relativity theory have been extensively discussed in the literature. One should mention a series of works by Friedrich Herneck (1966, 1974, 1983, 1984a, 1984b) and Helmut Hönl (1966a, 1966b), who were the first to publish Einstein's four letters to Mach. Hönl also devoted considerable attention to Mach's principle, the quintessence of Mach's ideas on inertia, and especially to its destiny in subsequent advances of GR and relativistic cosmology. This aspect of the problem is considered in a great number of works with a large variety of opinions, sometimes conflicting as to the essence and status of this principle in GR. Besides the above-mentioned articles by Hönl, there are works by Hubert Goenner (1970), M. Reinhardt (1973), Robert Dicke (1964), John Wheeler (1964), Dennis Sciama (1969), Jayant Narlikar (1979), and Hans-Jürgens Treder, as well as by Russian researchers B. G. Kuznetsov (1967), Ya. B. Zel'dovich and I. D. Novikov (1971, 1975), and N. P. Konoplyova (1978). The importance of Machian ideas for the formation of GR has been widely debated in works on the history of the theory (Hiebert 1970, Pyenson 1974, Illy 1979, Stachel 1979, Vizgin 1981, Sanchez 1981, Pais 1982) and in the literature on Mach (Heller 1964, Blackmore 1972, Holton 1973, Herneck 1983, 1984b, Wolters 1987, Barbour and Pfister 1995). The philosophical and methodological aspects of the problem are considered in Holton 1973, Stein 1977, Zahar 1977, 1981, Delokarov 1979, Feyerabend 1980, 1984, Haller and Stadler 1988. Some of Holton's claims regarding Mach's role in the development of relativity were criticized by Gereon Wolters (1987). His book merits special attention. Apart from providing vast material on the relationship between Einstein and Mach, Wolters argues, as mentioned above, for the implausibility of Mach's acute hostility to relativity. There are also weighty arguments in favor of a new dating of Einstein's famous "Christmas" letter to Mach.[5] Most researchers, including the present author (Vizgin 1981), considered this letter to be written in late December 1912 (or early January 1913). According to Wolters, however, it should be dated December 1913 (or early January 1914).

Though the influence exerted by Mach's ideas on the origin of GR was appreciated by many researchers (including the author of this paper), the 150th anniversary of Mach's birthday provides another opportunity to return to this question, the more so because there are some, albeit not numerous, new materials and interpretations. Besides, we intend to show that the impact of Machian ideas and methodology was fruitful for Einstein at all stages of the development of GR (1907–1915). The first such stage was the analysis of logical and theoretical difficulties of classical mechanics and the theory of gravitation. Here we compare Mach's and Einstein's understanding of these difficulties. Then we will try to reconstruct Mach's influence on Einstein in the latter's work on the equivalence principle (1907). The next stage has to do with the static field theory (1912). Incidentally, within the framework of this theory, Einstein demonstrated, for the first time, how the ideas usually associated with Mach's principle can actually be implemented. At that time, Einstein's first references to Mach's works appeared. The introduction of the tensor-geometrical concept of gravitation (the Einstein–Grossmann theory of 1913–1914) was a key achievement in the development of GR. And here, as we will show, Mach's ideas, primarily those related to Mach's principle, were essential for Einstein.

These ideas also provided an important touchstone for Einstein's subsequent formulations of the correct equations of the gravitational field.

The format of this paper does not enable us to address a number of important questions, such as:

1. The function of Mach's ideas (Mach's principle) in the foundations of relativistic cosmology (Einstein 1917);

2. The evolution of Einstein's attitude toward Mach's philosophical and methodological doctrines, as well as to Mach's principle;

3. The historical development of the ideas associated with Mach's principle in GR, relativistic cosmology, and in twentieth-century physics in general.

Moreover, we will leave almost entirely aside the intriguing question of Mach's attitude to relativity theory. (See, in this regard, the above-mentioned work by Wolters (1987).) In conclusion, we will briefly consider how some of the issues just listed were regarded at the time when the fundamentals of GR had already been fully elaborated.

2. Difficulties of Classical Mechanics

2.1 GENERAL OBSERVATIONS

One of the main factors in the genesis of relativity theory and, in particular, of its initial phase, was a growing awareness of the inadequacy of classical mechanics and, more broadly, of certain aspects of classical physics in general. As far as GR is concerned, Newton's law of gravitation, along with classical mechanics as a whole, became, at the turn of the 19th century, a subject of critical analyses. Empirical difficulties (such as an anomalous precession of Mercury's perihelion in the order of 41–43" per century), however, were not crucial. Logical and conceptual problems seemed more substantial. Revision of the classical theory of gravitation was prompted directly by emergence of the special theory of relativity (SR) and the need to bring Newton's law of gravitation in agreement with it. Whereas Henri Poincaré and Hermann Minkowski, among other theorists, found more or less suitable Lorentz-covariant generalizations of this law (1906–1909), Einstein surmised, upon taking up this problem, that the approach based on

SR was not sufficient. At that point, further progress demanded a more profound critical analysis of classical mechanics and the classical theory of gravitation, as well as of SR. Obviously, that was the second wave of criticism of the foundations of classical physics. It was largely linked with the first wave aimed at a Lorentz-covariant reformulation of classical theories. Einstein used to say that the works of Mach, primarily his *Mechanics*, provided an instructive example of such criticism. The logical and theoretical difficulties of Newton's theory of gravitation, existing in a common theoretical framework with classical mechanics, can be summarized in the form of the following seven "isms":[6] geometrical absolutism,[7] geometrical apriorism (postulation of the Euclidean geometry of space or of the Galilean–Newtonian symmetry of space-time), empiricism (in dealing with the equality of inertial and gravitational masses), as well as the other four difficulties having to do primarily with the gravitational theory: phenomenologism, isolationism, finitism, and instantism. The first two have a fairly common character: phenomenologism reflects the lack of a corpuscular or etheric interpretation of Newton's law of gravitation (a drawback especially noticeable in the background of successes scored by the atomic and kinetic theories of matter). Isolationism is, in essence, the denial of any links between gravitation and other interactions and physical phenomena. Finitism has to do with the gravitational paradox of a cosmological character, that is, with problems involved in applying Newton's law to infinite space. And instantism refers to the action-at-a-distance nature of gravitation, whose reputation was strongly undermined by the Faraday–Maxwell program in electrodynamics.

Historical and scientific analyses clearly show that geometrical absolutism and the related absolutism of inertia, as well as empiricism (the puzzle of the equality $m_{in} = m_g$), instantism,[8] and, to a somewhat lesser extent, geometrical apriorism were most important for Einstein in his work on GR. Phenomenologism and isolationism were of no importance for him. To be more precise, isolationism overlapped with instantism (affinity between gravitation and electromagnetism) and empiricism (relationship between gravitational and inertial properties of matter), whereas phenomenologism in the generally accepted sense was not considered as a drawback at all. Finitism did not attract Einstein's attention until 1917, when he applied GR to the cosmological problem.

Although Mach was much less interested in gravitation and his criticism was aimed mainly at the fundamentals of mechanics in general, his perception of the chief defects of classical physics was close to that of Einstein. He found such defects in Newton's treatment of absolute space and time, and inertia (geometrical absolutism and absolutism of inertia). Mach's keen interest in inertia also prompted him to speculate on those aspects of empiricism that concerned a mysterious kinship (and even identity) between inertial and gravitational masses. Mach's criticism of these aspects of classical physics was especially insightful and, so to speak, "coherent," as it combined the above "isms." His critique was also constructive: Mach's concepts of general relativity, of the lack of operational and physical ground for the notions of absolute space and time, of the relativity of acceleration and of inertia (more precisely, of the dependence of bodies' inertial properties on remote masses of the universe), and some other ideas, were regarded by Einstein as valuable starting points or guidelines on the way to a relativistic theory of gravitation.

One can find, in Mach's speculations, quite a few non-trivial statements testifying to his understanding of classical instantism and geometrical apriorism. Yet these observations, standing somewhat aloof and scattered around, were hardly essential for Einstein.

To support these observations, we shall quote some of Mach's statements. The main sources are his *Mechanics* (the fourth prerelativistic edition of 1901), *History and Root of the Principle of the Conservation of Energy* (1872), and *Knowledge and Error* (1905). Einstein, as a rule, referred to *Mechanics*. Many of Mach's fundamental ideas were formulated and developed for the first time in the mid-1860s and summarized in the Prague paper "The Principle of Conservation of Work" (1871) which appeared next year as a separate small book supplemented with important notes (quoting from and commenting upon some of Mach's articles of the 1860s). The year 1909 saw the second German edition of *The Principle* which Mach sent to Einstein; this started their correspondence. The quotations below from *Knowledge and Error* are part of the articles that were published mostly in the early 1900s and eventually composed the book.

2.2 GEOMETRICAL ABSOLUTISM AND SOME ASPECTS OF EMPIRICISM

In his notes to *The Principle* Mach makes references to his lectures entitled "On Certain Basic Questions of Physics," which he gave in 1868. He criticizes the classical formulation of the principle of inertia and develops his idea of replacement of absolute motion by the motion of relatively remote masses of the universe: "I remarked many years ago that there is in this law a great indefiniteness; for which body it is, with respect to which the direction and velocity of the body in motion is determined, is not stated" (Mach 1911, p. 75). On noting that in 1869 Carl Neumann advanced a similar objection against mechanics (Neumann 1870), Mach wrote:

> Obviously it does not matter whether we think of the earth as turning round on its axis, or at rest while the celestial bodies revolve round it. Geometrically these are exactly the same case of a relative rotation of the earth and of the celestial bodies with respect to one another. . . . But if we think of the earth at rest and the other celestial bodies revolving round it, there is no flattening of the earth, no Foucault's experiment, and so on—at least according to our usual conception of the law of inertia. Now, one can solve the difficulty in two ways: Either all motion is absolute, or our law of inertia is wrongly expressed. Neumann preferred the first supposition, I, the second. The law of inertia must be so conceived that exactly the same thing results from the second supposition as from the first. By this it will be evident that, in its expression, regard must be paid to the masses of the universe. (Mach 1911, pp. 76–77)

And he continues:

> We learn that *all* bodies, each with its share, are of importance in the law of inertia. . . . Now, what share has every mass in the determination of direction and velocity in the law of inertia? No definite answer can be given to this by our experience. We only know that the share of the nearest masses vanishes in comparison with that of the farthest. (Mach 1911, p. 79)

Somewhat later in the same notes Mach quotes his article of 1866 entitled "On the Development of Notions of Space." The quotation indicates that the fictitious status of space and time, or at least a radically relativistic (or relational) concept of spatio-temporal relations, was the leading idea for him:

> The spatial relations of material particles can, indeed, only be recognized by the forces which they exert on one another. . . . The physical space which I have in mind—and which, at the same time, contains time in itself[9]—is thus nothing other than *dependence of phenomena on one another*. A complete physics, which would know this fundamental dependence, would have no more need of special considerations of space and time, for these latter considerations would already be included in the former knowledge. (Mach 1911, pp. 89–90)

Mach developed these ideas further in *Mechanics*. After providing broad quotations from Newton's *Principia* concerning the notions of absolute time and space he notes that here

Newton has again acted contrary to his expressed intention only to investigate *actual facts*. No one is competent to predicate things about absolute space and absolute motion; they are pure things of thought, pure mental constructs, that cannot be produced in experience. All our principles of mechanics are, as we have shown in detail, experimental knowledge concerning the relative positions and motions of bodies. (Mach, 1907, p. 229)

Mach then turns to his famous analysis of Newton's bucket experiment and shows how abandonment of geometrical absolutism and expansion of relativity from uniform to arbitrarily accelerated motions (Einstein referred to this as the "relativity of acceleration") lead to a new understanding of inertia which Einstein later called the "relativity of inertia" (Mach 1907, pp. 231–232).[10] Ascribing the inertial movement of a body to a reference system (rather than to absolute space) related to remote stars—more precisely, to the center of mass of the universe

$$\frac{d^2}{dt^2}\left[\frac{\Sigma m_i r_i}{\Sigma m_i}\right] = 0 \, ,$$

where r_i stands for the distance of the moving body from the mass m_i—led Mach to define the body's mass as a "dynamic property characterizing the interaction among the bodies in the universe" (Konoplyova 1978, p. 223), in other words, to what came to be known subsequently as Mach's principle.

So, the abandonment of the geometrical and inertial absolutisms brought Mach to the idea of a profound kinship between inertia and gravitation, even though he did not express this idea sufficiently clearly. He obviously connects inertia, or the inertial mass of a body, with the influence exercised on it by all (but primarily, by remote) masses of the universe. Consequently, there appears, albeit vaguely, the prospect of explaining the empiricism of classical theory as resulting from the coincidence (proportionality) of inertial and gravitational masses.[11]

2.3 GEOMETRICAL APRIORISM

We have already cited Mach's earlier statements (made in the mid-1860s). It would be more natural to interpret them as an attack, not so much on the geometrical absolutism of classical mechanics, as on geometrical apriorism (or on both of these "isms").

In *Knowledge and Error* Mach speaks about the feasibility of using non-Euclidean geometry in physics, even though he finds Euclidean geometry to be quite sufficient and, because of its simplicity, preferable. He writes:

> We are able, therefore, to represent the facts of spatial observation with all possible precision by both Euclidean geometry and the geometries of Lobachevsky and Riemann, provided in the two latter cases we take the parameter k large enough. [Mach refers here to the geometries of spaces with a constant negative or positive curvature $1/k$.] Physicists have as yet found no reason for departing from the assumption $k = \infty$ of Euclidean geometry. (Mach 1976, p. 322)

Here Mach is more cautious than some 30–40 years earlier when he wrote that "we need not necessarily represent to ourselves molecular-processes spatially, at least not in a space of three dimensions" (1911, p. 86). So, Mach admitted in principle the possibility of using spaces of higher dimensions (>3) and of non-Euclidean spaces. "The possibility of analogous experiences [i.e., those testifying to non-Euclidean metrics] in three-dimensional space the physicist cannot as a matter of principle reject," he wrote (1976, p. 323), although he believed that the realization of such a possibility was highly improbable.

2.4 INSTANTISM

Generally speaking, in Mach's works we find no clear hints to the effect that the influence of remote masses of the universe on any massive body, the mass of which is determined exactly by this impact, propagates with a certain speed, that is, has a local, field-like character. Modern analyses of the nature of this influence (see, e.g., Zel'dovich and Novikov 1971, Konoplyova 1978) lead one to conclude that it must be instantaneous action at a distance. Incidentally, here lay the principal difficulty for Einstein in the development of a relativistic field theory of gravitation: he sought to overcome the geometrical inertial absolutism, on the one hand, and the instantism of gravitation, on the other, and to combine a Machian concept of general relativity and the relativity of inertia with the field program and SR. And when he thought he had succeeded (i.e., when GR had been built) it soon transpired that the realization of Mach's program was largely an illusion.

For all that, despite the entire range of questions surrounding Mach's principle, his works contain fairly high assessments of the local action program in electrodynamics, regrets for the fact that gravitation retains an action-at-a-distance character, and consideration of some attempts to overcome gravitational instantism. Thus in *Mechanics* (beginning with the fourth edition) he wrote:

> Faraday's unbiased and ingenious conceptions and Maxwell's mathematical formulation of them again turned the tide in favor of the forces of contact [i.e., short-range forces]. Diverse difficulties had raised doubts in the minds of astronomers as to the exactitude of Newton's law, and slight quantitative variations of it were looked for. After it had been demonstrated, however, that electricity travels with finite velocity, the question of a like state of affairs in connexion with the analogous action of gravitation again naturally arose. (Mach 1907, pp. 534–535)

Mach went on to discuss some attempts to introduce the speed of propagation of gravitation into Newton's theory of gravitation, in particular, the idea of negative masses, which must be repelled from ordinary masses (thus likening gravitational theory to electrodynamics) and some other ideas, referring to reviews by Drude and Wien.

Presumably, Hertz's work (not only his work on the theory of electromagnetic field but also his *Principles of Mechanics* (1894)) influenced Mach's approach to the concept of field. Beginning with the third edition of *Mechanics*, Mach wrote with admiration about Hertz's project, which Mach associated with a field program and with his intention to exclude the notion of force from mechanics:

> It is not difficult to analyse the psychological circumstances which led Hertz to his system. After inquirers had succeeded in representing electric and magnetic forces that act at a distance as the results of motions in a medium, the desire must again have awakened to accomplish the same result with respect to the forces of gravitation, and if possible for all forces whatsoever. The idea was therefore very natural to discover whether the concept of force generally could not be eliminated. (Mach 1907, pp. 553–554)[12]

2.5 EINSTEIN'S STATEMENTS ON MACH (1913–1916)

We have already mentioned that Einstein's first direct references to Mach appeared in 1912. After that time, in nearly every one of Einstein's articles on the development of GR, one can find statements on Mach's critique of mechanics, its analysis, agreement with it, and comments on its stimulating importance for constructing a relativistic theory of gravitation. Let us give these statements in chronological order.

1913

> The theory sketched here overcomes an epistemological defect that attaches not only to the original theory of relativity [the case in point is the first version of GR, the Einstein–Grossmann theory], but also to Galilean mechanics, and that was especially stressed by E. Mach. It is obvious that one cannot ascribe an absolute meaning to the concept of acceleration of a material point, no more so than one can ascribe it to the concept of velocity. Acceleration can only be defined as relative acceleration of a point with respect to other bodies. This circumstance makes it seem senseless to simply ascribe to a body a resistance to an acceleration (inertial mass of the body in the sense of classical mechanics); instead, it will have to be demanded that the occurrence of an inertial resistance be linked to the relative acceleration of the body under consideration with respect to other bodies. It must be demanded that the inertial resistance of a body could be increased by having unaccelerated inertial masses arranged in its vicinity.[13] (Einstein 1914a, p. 290)

Here Einstein briefly reproduces the logical chain of Mach's reasoning that has to do with his critique of the geometrical absolutism of classical mechanics. Inertial absolutism is another form of geometrical absolutism; abandonment of the latter leads to the relativity of acceleration and, along with it, to the relativity of inertia.

The term "the relativity of inertia" was coined by Einstein who used it in his Vienna paper of 21 September 1913 (Einstein 1913, pp. 1260–1262). He notes that in a relativistic theory of gravitation, even in the static approximation, the inertia of a body increases "due to cumulation of masses in its vicinity" and therefore "we have no choice but to view the inertia of a point as being *caused* by the existence of the other masses" (1913, p. 1260). And he goes on to stress that this point of view was first set forth, with all acuteness and clarity, by Mach in his history of mechanics. "I shall call the conception sketched here," he continues, "'the hypothesis of the relativity of inertia'." It was, in essence, one of the first formulations of Mach's principle. The next paragraph of this paper has exactly the spirit of Machian analysis:

> To avoid misunderstandings, let me repeat here that, just as Mach, I do not think that the relativity of inertia is a logical necessity. But a theory in which the relativity of inertia is preserved is more satisfactory than the current theory, because, on the one hand, the state of motion of the inertial system introduced into the latter does not depend on the states of observable objects, and thus is not caused by anything accessible to observation, while, on the other hand, it is to determine the behavior of the material point. (Einstein 1913, p. 1261)

Shortly before his Vienna paper Einstein wrote in his third letter to Mach (of 25 June 1913) that if, in 1914, the expected eclipse confirmed the deviation of sun rays, predicted on the basis of the principle of equivalence, "then—in spite of Planck's unjustified criticism—your brilliant investigations on the foundations of mechanics will have received a splendid confirmation. For it follows of necessity that *inertia* has its origin in some kind of *interaction* of the bodies, exactly in accordance with your argument about Newton's bucket experiment." (Einstein to Mach, 25 June 1913, Einstein 1993, Doc. 448, pp. 531–532)

1914

In the article completed in February Einstein acknowledged the great complexity of his theory (that is, the Einstein–Grossmann theory), as compared to scalar theories. "But in return," he pointed out, "it eliminates an epistemological weakness that hitherto attached to mechanics and that has long been felt by perspicacious epistemologists,[14] especially by Ernst Mach" (Einstein 1914b, p. v). And in the next paragraph Einstein again endorses Mach's criticism of the geometrical and inertial absolutism of classical physics:

> The law of motion of the material point, and therewith the whole of mechanics, indeed the whole of theoretical physics, were based by Galileo and Newton on the concept of acceleration. But a simple analysis shows that acceleration is accessible to observation only as *relative acceleration* with respect to other bodies, that we are only able to define a *relative* acceleration. It is therefore doubtful that the Galilean–Newtonian law of motion, which says that bodies exert a resistance to acceleration, says something about an acceleration in itself (absolute acceleration, not relative acceleration). The new theory of gravitation avoids this inconsistency; according to this theory, inertia shows up as a resistance against the relative acceleration of bodies. (Einstein 1914b, pp. v–vi)

Turning to the analysis of certain thought experiments of the Machian type, to those that illustrate the above "isms" of classical mechanics, Einstein wrote in a scientific-popular article published in the spring of 1914 in the Italian journal *Scientia*:

> Classical mechanics, as well as the theory of relativity in the narrower sense, which has been discussed briefly in the foregoing, suffer from an undeniable fundamental defect that is accessible to epistemological arguments. The weaknesses of our world picture to be discussed here have already been

uncovered with perfect clarity by E. Mach in his penetrating investigations into the foundations of Newtonian mechanics, so that I cannot claim that what I am saying here in this respect is new. (Einstein 1914c, p. 344)

In the last (November 1914) extensive description of the Einstein–Grossmann theory (i.e., that tentative form of GR that still lacked generally covariant equations of the field) Einstein wrote, somewhat modifying his description of Mach's criticism, that Newton's proof of the absolute character of rotational motion, relying on the emergence of centrifugal forces in the rotating reference system, is not valid, "as has, in particular, been demonstrated by E. Mach." And he goes on to say:

This is because we need not necessarily derive the existence of centrifugal forces from a motion of K' [i.e., the rotating system]; instead, we can just as well derive them from the averaged rotational movement of distant ponderable masses in the environment, but relative to K', thereby treating K' as "at rest." If the Newtonian laws of mechanics and gravitation do not allow for such interpretation, it may well be founded in deficiencies of this theory. (Einstein 1914d, pp. 1031–1032, trans. by A. Engel)

1915–1916

The series of four November 1915 articles, which are laconic but abound with new results, say virtually nothing about the shortcomings of mechanics and do not mention Mach's name. We now know about a tense and dramatic competition with Hilbert, in which they were written. It was not the right time for Einstein to engage in "epistemological arguments," though in some of these articles one may feel the influence of Mach's ideas. (We shall look at this later.)

But in a detailed and methodical account of GR, completed in mid-March 1916, Einstein devotes much space to critical analysis of the fundamentals of classical mechanics, specifically, to the discussion of geometrical absolutism. As previously, he refers to Mach. He describes, more thoroughly, a thought experiment with two liquid balls, one rotating and the other being at rest (he first considered this example in spring 1914) and, in addition, makes use of the principle of causality in what is essentially its Machian form.

The thought experiment with liquid balls was discussed on many occasions. As Hönl aptly noted, the experiment was "merely a *modification* of Mach's objection to Newton's interpretation of the experiment with a rotating vessel" (1966a, p. 29). Therefore, we will cite only some fragments of Einstein's new argumentation making use of the afore-mentioned principle of causality. In discussing the reason for the flattening of the rotating liquid ball, Einstein notes that no answer to the question about that reason

> can be admitted as epistemologically satisfactory, unless the reason given is an *observable fact of experience*. The law of causality has not the significance of a statement as to the world of experience, except when *observable facts* ultimately appear as causes and effects. (Einstein 1916b, p. 771; trans. by W. Perrett and G. B. Jeffery)

This is an understanding of causality in a typically Machian style incorporating the principle of observability into the essence of the matter! Suffice it to compare Einstein's statement with those Mach made in his Prague paper of 1871 (see Mach 1911, pp. 73–74).

Here we'll have to interrupt the chronology of Einstein's statements on Mach and, at the same time, on the main ("epistemological") weaknesses of classical mechanics: its geometrical absolutism and the closely related absolutism of inertia. Reconstructing this chronology further would certainly be of interest, as in 1917, 1918, 1921, and later on, Einstein said a good deal on the issue under discussion, while enhancing his argumentation and understanding of Mach's criticism of mechanics. The foundations of GR had already been laid by that time, however.

3. The Principle of Equivalence

3.1 EINSTEIN'S VIEWS ON THE CONNECTION BETWEEN THE PRINCIPLE OF EQUIVALENCE AND MACH'S IDEAS

Whereas our analysis of the perception of the difficulties of classical mechanics by Mach and Einstein, as well as Einstein's use of Mach's criticism as a point of departure in building GR, drew exclusively on the texts of both classics of relativism, the role played by Mach's ideas in the genesis of the principle of equivalence can be revealed only indirectly, because Einstein's early works of 1908 and 1911, where this principle was developed, contain no references to Mach. At the same time, the obituary of 1916 includes Einstein's direct statements to the effect that the idea of the principle of equivalence originates with Mach.

Recapitulating, in 1916, his own recent steps, Einstein pointed out that Mach's considerations on Newton's bucket experiment reveal a clear understanding of the fact that the equality of inertial and gravitational masses invites a generalization of the principle of relativity ("relativity in a wider sense"), for "we are not in a position to decide by experiments if the falling of a body relative to a coordinate system is caused by the presence of a gravitational field or by a state of acceleration of the coordinate system" (1916a, p. 103, trans. by A. Engel).

In Einstein's quoted article of 1914 (Einstein 1914d, p. 1031) we find the following explanation of this statement. At first glance, says Einstein, the centrifugal forces in the rotating system prove the absoluteness of rotation. But, according to Mach, he continues, these forces of inertia can be interpreted as a result of the influence (attraction) of the system of remote stars, if the first system is taken to be at rest and the second to be rotating. But what are these forces of "influence" or "attraction" of the remote mass? For Einstein, there is no doubt that gravitation is at issue. There is only one step from here to identifying inertia and gravitation. Subsequently Einstein uses this idea of identity for "kinematizing" (and geometrizing) gravitation. And Mach—speaking, of course, in vaguer terms about gravity and the equality of inertial and gravitational masses— "dynamizes" the kinematics of relative motion and, along with it, the notion of inertia.

Like the principle of equivalence, Mach's considerations regarding the concept of general relativity and his analysis of Newton's bucket experiment served, to a certain extent, as an argument in favor of the expansion of the special principle of relativity. Let us recall how Einstein came to the principle of equivalence. He tried to apply the relativistic approach (drawing on SR) to gravitation and hit upon the fact of the equality of inertial and gravitational masses: "This law, which may also be formulated as the law of the equality of inertial and gravitational mass, was now brought home to me in all its significance. I was in the highest degree amazed at its existence and guessed that in it must lie the key to a deeper understanding of inertia and gravitation," Einstein recalled in 1933 (Einstein 1954, p. 287). The relativistic interpretation of this equality led him to the principle of equivalence, which required the introduction of accelerated reference systems as permissible ones. On the other hand, thought experiments of the Machian type, primarily the experiment with two rotating liquid balls, which Einstein considered in his 1916·article, suggested to him, along with the principle of equivalence, "the need for an extension of the postulate of relativity" (Einstein 1916b, p. 771).

At the same time, it should be emphasized that Mach does not speak directly about the equality of masses ($m_{in} = m_g$) and, at any rate, does not attach fundamental importance to this equality. It is natural that Einstein, who thoroughly studied *Mechanics* and other works of Mach, was "greatly surprised" that such a law, namely, the equality $m_{in} = m_g$, existed. Mach put forward the idea of inertia being caused by the masses of the universe without explicitly connecting this causal action with gravitation. Moreover, he considered the relativity of rotational motion as having a global character and related to a field of centrifugal forces that can hardly be interpreted as a field of gravitation. It may seem, therefore, that the closeness (noted by Einstein) of Mach's reasoning on the equivalence of the forces of inertia and the forces engendered by the influence of the remote masses in the universe, on the one hand, and Einstein's principle of equivalence, on the other, is a strained interpretation.

In 1933, when recalling the situation in the theory of gravitation after the creation of SR, Einstein wrote:

> From the purely kinematic point of view there was no doubt about the relativity of all motions whatever; but physically speaking, the inertial system seemed to occupy a privileged position, which made use of coordinate systems moving in other ways appear artificial.
>
> I was of course acquainted with Mach's view, according to which it appeared conceivable that what inertial resistance counteracts is not acceleration as such but acceleration with respect to the masses of the other bodies existing in the world. There was something fascinating about this idea to me, but it provided no workable basis for a new theory. (Einstein 1954, p. 286)

Einstein then recapitulated the discovery of the principle of equivalence as if it were an event independent of Mach's idea and noted that it was precisely that principle that had created an acceptable ground for a new theory.

Anyway, in addition to what has already been said (about the relationship of a host of ideas associated with Mach's principle, on the one hand, and the principle of equivalence, on the other), one could mention at least two weighty arguments in favor of the possible influence of Mach's ideas on Einstein's development of the equivalence principle:[15] (1) Mach's project of "kinematizing" classical mechanics has a profound kinship with Einstein's idea of "kinematizing" (geometrizing, in case of a four-dimensional approach) of the physical influence, primarily of gravitation (one should not forget, at the same time, that Einstein's idea was firmly grounded in the equivalence principle); (2) in introducing the principle of equivalence Einstein was guided by the relativistic program that had much in common with Mach's relativistic program.

3.2 MACH'S PROJECT OF KINEMATIZING MECHANICS

The gist of Einstein's principle of equivalence is the idea of kinematizing a homogenous gravitational field, that is, interpreting the latter, in an inertial reference system, as a kinematic effect of motion of a uniformly accelerated system. The equivalence situation in Einstein's sense is, in essence and form, very close to Mach's claim of parity between the rotating reference system and the "rotating sky of fixed stars," on the one hand, and the gravitational field, on the other, which is equivalent to the field of centrifugal forces arising in the first system. Both situations are specific thought experiments based, essentially, on the equality $m_{in} = m_g$ (Mach makes no direct reference to this equality, it remains rather implicit in his writings) and the principle of observability. And both situations provide strong grounds for the extension of the special principle of relativity. Let us quote Mach again (compare this quotation with his other statements reproduced earlier):

> For me only relative motions exist . . . and I can see, in this regard, no distinction between rotation and translation. When a body moves relatively to the fixed stars, centrifugal forces are produced; when it moves relatively to some different body, and not relatively to the fixed stars, no centrifugal forces are produced. I have no objection to calling the first rotation "absolute" rotation, if it be remembered that nothing is meant by such a designation except *relative rotation with respect to the fixed stars*. Can we fix Newton's bucket of water, rotate the fixed stars, and *then* prove the absence of centrifugal forces?
>
> The experiment is impossible, the idea is meaningless, for the two cases are not, in sense-perception, distinguishable from each other. I accordingly regard these two cases as the *same* case and Newton's distinction as an illusion. (Mach 1907, pp. 542–543)

Einstein's and Mach's situations differ in yet another respect: the former has a local character and the latter a global and cosmological one. This cosmological character of Mach's reasoning becomes especially perspicuous when he attempts to convey his understanding of the principle of inertia in mathematical terms:

> Instead of saying, the direction and velocity of a mass μ in space remain constant, we may also employ the expression, the mean acceleration of the mass μ with respect to the masses $m, m', m'' \ldots$ at the distances $r, r', r'' \ldots$ is $= 0$, or $d^2(\Sigma mr/\Sigma m)/dt^2 = 0$. The latter expression is equivalent to the former,

as soon as we take into consideration a sufficient number of sufficiently distant and sufficiently large masses. (Mach 1907, p. 234)

In other words, the motion of any mass μ is considered, not with respect to absolute space, but with respect to the center of mass of the universe.

Incidentally, it follows from Mach's relationship $d^2(\Sigma mr/\Sigma m)/dt^2 = 0$ that the mass of a certain body μ is determined exactly by remote and heavy masses:

$$\mu = \frac{|\Sigma mr - C_1 - C_2 t|}{|\rho|},$$

where ρ is the distance of mass μ from the center of mass of the universe and C_1 and C_2 are certain constants determined by the position and velocity of that center. The above is none other than the original, Mach's, form of Mach's principle.[16]

Mach concludes the discussion of his notion of inertia with a somewhat vague phrase which, however, suggests that he did assign a fundamental role to the notion of acceleration. This phrase also contains ideas pertinent to the principle of equivalence:

The considerations just presented show that it is not necessary to refer the law of inertia to a special absolute space. On the contrary, it is perceived that the masses that in the common phraseology [first and foremost, in accordance with the law of universal gravitation] exert forces on each other as well as those that exert none [i.e., the remote masses in the universe], stand with respect to acceleration in quite similar relations. We may, indeed, regard *all* masses as related to each other. (Mach 1907, pp. 235–236)

Apparently, if the equality of the inertial and gravitational masses holds, one can speak of "perfect uniformity" here. Mach does not state this much in clear terms, though he was quite aware of this. (See, for example, Mach's discussion of "the measurement of a body's mass by its weight" (1907, p. 195).)

The above leads Mach to the following conclusion: "That *accelerations* play a prominent part in the relations of the masses, must be accepted as a fact of experiment" (Mach 1907, p. 236). He explains this "prominent part" of the notion of acceleration as follows:

In all processes of nature the *differences* of certain quantities u play a determinative rôle. Differences of temperature, of potential function, and so forth, induce the natural processes, which consist in the equalisation of these differences. The familiar expressions d^2u/dx^2, d^2u/dy^2, d^2u/dz^2, which are determinative of the character of the equalisation, may be regarded as the measure of the departure of the condition of any point from the mean of the conditions of its environment—to which mean the point tends. The accelerations of masses may be analogously conceived. (Mach 1907, p. 236)

Mach pointed out the fundamental, even unique, status of acceleration in mechanics as early as 1867 in his article "On the Definition of the Notion of Mass" (see Mach 1911, pp. 80–85). In that context, he sought to bring the initial definitions and principles closer to experience: to abandon the definition of mass as a quantity of matter and replace it by a definition in terms of the observable (measurable) kinematic notion of acceleration and to turn Newton's second law into the definition of force. This definition would thereby acquire a purely kinematic meaning and become a secondary notion, while Newton's three principles would be replaced by the following ones:

> *Theorem of experience.*—Bodies placed opposite to one another communicate to each other accelerations in opposite senses in the direction of their line of junction. The law of inertia is included in this. . . .
> *Theorem of experience.*—The mass-values remain unaltered when they are determined with reference to other forces and to another body of comparison which behaves to the first one as an equal mass.
> *Theorem of experience.*—The accelerations which many masses communicate to one another are mutually independent. The theorem of the parallelogram of forces is included in this. (Mach 1911, pp. 84–85)

It should be noted that in *Mechanics*, Mach formulated the second experimental principle somewhat differently:

> The mass-ratios of bodies are independent of the character of the physical states (of the bodies) that condition the mutual accelerations produced, be those states electrical, magnetic, or what not; and they remain, moreover, the same, whether they are mediately or immediately arrived at. (Mach 1907, p. 243)

Exact definition is given, not of mass, but of the mass-ratio of a pair of bodies: "The mass-ratio of any two bodies is the negative inverse ratio of the mutually induced accelerations of those bodies" (Mach 1907, p. 243).

So the basic definitions and laws of mechanics were reformulated in terms of a singularly fundamental measurable quantity, acceleration. Mach supposed that all mechanics could be easily restructured on that basis, that the "propositions above set forth satisfy the requirements of simplicity and parsimony," and that "no doubt can exist with respect to any one of them either concerning its meaning or its source; and we always know whether it asserts an experience [Mach's "experimental principles"] or an arbitrary convention [definitions of mass and force]" (Mach 1907, p. 244).

Mach's axiomatization of mechanics was subjected to criticism on more than one occasion and, irrespective of a high assessment of the kinematic principle underlying the entire construction, this axiomatic has not found recognition. Indeed, it has a number of serious defects, even in comparison with Newton's traditional formulation (see, in this regard, Bunge 1966, Konoplyova 1978).

Nonetheless, Mach continued this work in his lectures "On Certain Basic Questions of Physics" delivered in the summer of 1868 (see Mach 1911, pp. 75–76). In these lectures, Mach developed, for the first time, his concept of the relativity of acceleration and his idea of relating a body's inertia to the influence exerted on it by remote masses of the universe.

These lectures comprised the analysis of the notions of absolute space and time, relativity, and inertia, the key role given to the notion of acceleration being obvious here too (see Mach's statement (quoted above) on the "prominent part" of acceleration).

The fundamental role of the notion of acceleration in the formulation of Einstein's equivalence principle; the juxtaposition in it of the forces of gravitation and inertia; the connection of the equivalence principle with the extension of relativity; and finally, the idea of kinematization of physical interaction—all this is very consonant with Mach's idea of kinematizing mechanics and his concepts of relativity and inertia.

After the emergence, in 1894, of Hertz's *Principles of Mechanics*, Mach included a section on it in Chapter II of his own book. Hertz's mechanics especially impressed him by its explicitly "kinematic" character:

> Though Hertz's criticism of existing systems of mechanics cannot be accepted in all their severity, his own novel views must be regarded as a great step in advance. Hertz, after eliminating the concept of force, starts from the concepts of time, space, and mass alone, with the idea in view of giving expression only to that which can actually be observed. (Mach 1907, p. 550)

Hertz's ideas clearly belonged in the framework of kinematization and geometrization of mechanics, the framework whose roots can be traced back to Descartes who "would grant no other properties to matter than extension and motion" and who "sought to reduce all mechanics and physics to a geometry of motions" (Mach 1907, p. 553).

Notably, in the introduction to the 4th edition of *Mechanics,* Mach wrote that, in W. K. Clifford, he had found "a thinker of kindred aims and points of view" (1907, p. xv). It will be remembered that Clifford was a major forerunner of the concept of geometrization of physical properties.

3.3 MACH'S RELATIVISTIC PROGRAM AS A PROTOTYPE OF EINSTEIN'S RELATIVISTIC PROGRAM

The reconsideration of the fact of the equality of inertial and gravitational masses on the basis of the relativistic program[17] for restructuring physics and the kinematic-geometrical concept considered above were, in our opinion, of crucial importance in the discovery of the equivalence principle (Vizgin 1981). In fact, one can speak of two versions of the relativistic program: (1) exclusively in the sense of SR, that is, as a task of Lorentz-covariant reformulation of physics and bringing the basic equation of physics in agreement with the requirements of SR; (2) in the sense of extending SR, primarily through enlarging the class of permissible reference systems.

Virtually all the physicists who adopted SR shared the provisions of the relativistic program in the first sense. In the first two or three years after 1905, this group included, apart from Einstein, German physicists Max Planck, Arnold Sommerfeld, Max von Laue, Max Born, Minkowski, Gustav Herglotz, Nordström, Jakob Laub and others who were later joined by theorists from other countries. This program produced the relativistic mechanics of simple systems, relativistic hydrodynamics and the theory of elasticity, relativistic generalizations of thermodynamics and of the kinetic theory of gases. The relativistic solid state theory, relativistic electrodynamics of continuous media, and some other relativistic disciplines were in the making at that time. At first it seemed there would be no special problems with a relativistic generalization of Newton's theory of gravitation. One possible way of doing it had already been outlined by Poincaré: a direct Lorentz-covariant generalization of the law of universal gravitation. The same path was actually pursued by Minkowski. But Einstein, as early as 1907, arrived at the conclusion that the Poincaré–Minkowski approach was limited. He sought to find a Lorentz-covariant field generalization for the Poisson equation but realized that the corresponding generalization of the equations of motion did not agree with the fact of the equality of masses ($m_{in} = m_g$). On noting this remarkable relationship (it was, as already mentioned, unexpected for him), Einstein surmised that herein lay the key to resolving the problem.

He viewed it as a fundamental empirical law expressing, in a very simple and symmetric form, the identity of gravitation and inertia. This fact, remarkably, had been well known since Newton, lying, so to speak, on the surface. Ernest Cassirer (1921) wrote that even Kant, in his *Metaphysical Foundations of Natural Science*, admonished Newton for his failure to explain the fact of the equality of masses within the framework of mechanics, the fact that "remains accidental and mysterious" (1921, p. 71). Many scientists, from G. Lesage to H. A. Lorentz and P. Langevin, tried to explain this fact and thereby dispense with the empiricism of the classical theory of gravitation. Their explanations drew mostly on the ether-mechanistic conception of gravitation, which did not withstand the test of time, or on the electromagnetic field program (i.e., on reducing gravitation to electromagnetism), which was also encountering insurmountable obstacles.

Einstein, however, was not prone to such an "explanatory" approach. Simplicity and a fundamental character of the empirical equality $m_{in} = m_g$ required placing it at the very basis of the relativistic gravitational theory. And Einstein believed that, to impart a theoretical status to this relationship, one had to accommodate it in the relativistic program. It then became clear that Einstein's interpretation of this program was marked by high flexibility and informality. Let us point out the main characteristic features of this interpretation (see Vizgin 1981 for more detail). It included, sure enough, the special principle of relativity or, more precisely, a demand for Lorentz-covariance. Yet new empirical circumstances, according to Einstein, could give reasons for enlarging the class of permissible reference systems, which would mean going beyond the framework of SR.

Furthermore, though the speed of light enjoys fundamental importance in SR and this theory rules out instantaneous action at a distance, a more formal rendition of the relativistic program allows searching directly for a Lorentz-covariant generalization of Newton's law (the Poincaré–Minkowski approach). The notion of a physical field, however, had a special importance for Einstein. And his program, as evidenced in particular by his subsequent search for a field theory, was not simply a relativistic one, but rather a relativistic-field program.

Another highly important point. Those who followed Einstein's footsteps considered the methodological techniques, which he used in creating SR, as "scaffolding," as something dispensable, in comparison with the main result of the theory, its Lorentz-covariance. Einstein, on the other hand, did not separate the relativistic program from the arsenal of methodological techniques and idealizations that helped create SR. This arsenal included the principle of observability requiring an operational definition of fundamental notions. In this context, the method of thought experiment proved to be very important. The relativistic specificity of thought experiments having an operational character boiled down to a skillful employment of such notions as ideal rulers, clocks, reference systems, light signals, and so on.

Furthermore, in SR, many phenomena that were earlier regarded to be dynamic (e.g., Lorentz contraction) assumed a kinematic status. This means that the idea of kinematic reductionism was in principle very germane to the relativistic program. For example, the consideration of electrostatic phenomena in a certain reference system K from the viewpoint of an inertial system K' moving relative to the former leads to the emergence of a magnetic field.

Finally, the last item, which is characteristic not so much of the relativistic program (even in Einstein's understanding) as of Einstein's methodology in general, is what might be called "postulatory-explanatory inversion." This method consists in taking a certain fact or relationship that was earlier sought to be explained as a postulate and seeing where its theoretical development may lead. This method contains a large portion of healthy phenomenologism aimed against the preponderance, since the 19th century, of fruitless ether-mechanistic explanatory constructions. Let us now list the characteristic features of Einstein's relativistic program: admission of the possibility of extending the class of inertial reference systems; the fundamental character of the field concept; relativistic methodological techniques including the operational grounding of the fundamental theoretical notions with the aid of ideal clocks, rulers, reference systems, light signals, and so on (this presupposes the use of the principle of observability in thought experiments with ideal objects); kinematization (geometrization, in the four-dimensional framework) of traditional dynamic constructs; and the "phenomenological approach" assuming the form of explanatory-postulatory inversion.

We will now turn to Mach's version of the relativistic program and show its considerable kinship with Einstein's program. Did Mach have any prototype of such a program? Our analysis so far indicates that he did. Mach saw logical and theoretical shortcomings of classical mechanics and suggested, as early as in the 1860s, a definite approach to its restructuring that contained a good deal of relativism and kinematism. In this sense, one may well speak of a Machian prototype of the relativistic program. Sure enough, Mach's relativistic program was vague, his account of the achievements in electrodynamics and electron theory was inadequate, and the field concept played no role in it. But certain principles figuring prominently in Einstein's relativistic program, such as relativism, kinematism, the aim at an operational definition of fundamental notions (the observability principle), thought experimenting, and the "phenomenological approach," were all much in the style of the Machian program.

Relativism and kinematism have already been discussed in detail. As for the relativistic methodological techniques, including the principle of observability, thought experimenting, and the "phenomenological approach," it is evident that Mach was one of their leading forefathers. Suffice it to glance at his quoted statements containing criticism of geometrical absolutism or at Mach's principle and his kinematic concept of mechanics. It should be recalled that he was one of the first to introduce the term "thought experiment" (*Gedankenexperiment*).[18]

Here is one of his early formulations of the principle of observability:

> If the hypotheses are so chosen that their subject (*Gegenstand*) can never appeal to the senses and therefore also can never be tested, as is the case with the mechanical molecular theory, the investigator has done more than science, whose aim is facts, requires of him—and this work of supererogation is an evil. (Mach 1911, p. 57)

It will be noted that later Mach also drew attention to the "theoretical ladenness" of experiment, observation, and "facts": "Observation and theory are not sharply separable, since almost any observation is already influenced by theory" (Mach 1976, p. 120). Mach's analysis of Newton's bucket experiment is a model of thought experiment that combines the principle of observability and operating with such ideal objects as "the sky of fixed stars" or a coordinate system related to them. Already in his articles of the 1860s, Mach tried to link space-time notions with "their sensibly conceivable measures" (Newton's expression). He sought to express spatial and temporal relationships as "dependence of phenomena on one another" (1911, p. 89). Pointing out—after Lobachevski, Helmholtz, and others—the dependence of our geometrical notions on the behavior of solid bodies, Mach noted prophetically that "It is even likely that light waves in vacuo will furnish future physical standards of length and time, in terms of wave length and period of oscillation respectively and that these basic standards will be more appropriate and generally comparable than any others" (Mach 1976, p. 348).

We have already noted that, unlike most critics of the classical gravitational theory, Mach and Einstein did not think that the "phenomenologism" of this theory, that is, the absence of its atomistic, kinetic or etheric-mechanistic interpretation, was its weakness. For Mach, with all his hostility towards atomistics, it was quite natural. His reconstruction of the fundamentals of mechanics had a distinctively phenomenological motivation: if there was no way of doing without the notions of mass and force, then, according to Mach, their "experimental" aspect was to be clearly separated from the "metaphysical" one.

Let us now return to the equivalence principle. On encountering the equality of masses and having realized its fundamental character, Einstein followed the phenomenological tradition; without seeking to explain it in an etheric or mechanistic way, he, on the contrary, thought it necessary to use it as a foundation. Einstein attempted to include this empirical fact in the framework of theory on the basis of the relativistic program (which, as noted, he understood informally and flexibly enough) and the kinematic conception of mechanics (or of physical interactions in general) related to that program (or even contained in it). The well-known (in particular, to Mach) kinematic interpretation of the equality of masses in a gravitational field, dating back to Galileo, as the independence of acceleration g in this field of the mass and internal structure of a test body meant, from the viewpoint of the Mach–Einstein relativistic program, a license to regard the homogenous field in the inertial system and the field of inertial forces in a uniformly accelerated reference system (moving with acceleration g) as being equivalent. Incidentally, this interpretation was also prompted by Machian considerations of the relativity of rotational motion, in which the field of centrifugal forces in one rotating coordinate system turned out to be equivalent to the "field" of forces exerted by "fixed stars" rotating relative to the first system. If the only observable manifestation of a homogenous gravitational field in mechanics is the presence of one and the same acceleration in all bodies, then, from the viewpoint of the relativistic program (and of the kinematic conception), along with the idea of identity of gravitation and inertia, it is natural to interpret this effect as a transition to a uniformly accelerated reference system. This immediately entailed the requirement to go beyond the framework of SR, that is, to enlarge the class of allowed reference systems, which well agreed with Mach's "generalized relativity" (or his "relativity of acceleration"). For the advocate of the relativistic program in a narrow ("special relativistic") sense, that would mean a collapse of the entire program. For Einstein, however, this pointed to the need to transgress SR, even if one's goal is a relativistic theory of a perfectly homogenous gravitational field. The extension of the class of permissible reference systems thus fitted into the more flexible version of the Mach–Einstein relativistic program.

Likewise, the advocate of the classical (dynamic, not kinematic or geometrical) notion of physical interaction would regard the reduction of interaction to a mere kinematic effect as absurd. For Einstein, however, who took the kinematic interpretation of Machian mechanics close to heart, such a reduction paved a way to a new theory.

To sum up, there is every reason to believe that Mach's ideas were (or, at any rate, could be) of great help to Einstein in his development of the equivalence principle, the central physical tenet of GR.

4. Tensor-Geometrical Concept of Gravitation

The core of GR is the tensor-geometrical concept of gravitation according to which gravitation is identified with space-time geometry in the form of the identification of the potential of a gravitational field with the metric tensor of Riemannian space-time g_{ik} ($ds^2 = g_{ik}dx_i dx_k$ is the metric of this space) and of equations of geodesics with equations of motion of a material point in the gravitational field. This concept found its first theoretical embodiment in the framework of the Einstein–Grossmann theory which appeared in May–June 1913. After the introduction of the equivalence principle, that theory was the next, to a large extent decisive, step towards GR.

In (Vizgin 1981 and 1985), we have provided a detailed account of a five year long period preceding this step. Now we will try to identify the "Machian aspect" of Einstein's work during that period.

4.1 EINSTEIN IN PRAGUE

In the spring of 1911 Einstein moved to Prague and joined German University as a professor of theoretical physics. Mach had been professor and then rector of this university for 30 years (till 1895). His influence had been great here, and his former students, physicist A. Lampa and mathematician Georg Pick, who determined the physics and mathematics curricula at the university, sent an invitation to Einstein. The invitation was preceded (in 1909) by a brief correspondence between Mach and Einstein, which revealed the former's interest in SR. According to Wolters (1987), Mach learned about this theory from the paper that Minkowski had given in Köln in the Autumn of 1908. He saw in it the development of the ideas he had been working on since the mid-1960s. At exactly that time, Mach had prepared for publication the second edition of *The Principle*, which came out in 1909. In the introduction to this edition Mach wrote unambiguously: "I subscribe, then, to the principle of relativity which is also firmly upheld in my *Mechanics* and *Wärmelehre*" (Mach 1911, p. 95).

Before moving to Prague, that is, between 1907 and the middle of 1911, Einstein achieved no progress in his work on gravitational field theory. The main reason was that the principle of equivalence, which had unexpectedly opened up a promising perspective in the development of the concept of gravity, had also brought difficulties that proved nearly insurmountable: the perceived need for extending SR combined with the uncertainty as to the scale and character of this extension; the use of accelerated reference systems deprived the coordinates of their metric meaning thus allowing change in geometry; finally, at that point, Einstein had no clear idea of how to apply the equivalence principle to non-homogenous fields.

In June 1911, while in Prague, Einstein finished a new work on the principle of equivalence in which he stressed the inherently local character of that principle:

> Of course, one cannot replace an *arbitrary* gravitational field by a state of motion of the system without a gravitational field, just as one cannot transform to rest all the points of an arbitrarily moving medium by means of a relativistic transformation. (Einstein 1911, p. 899, footnote)

Herein lay the problem of applying the idea of kinematization (or geometrization, in the four-dimensional approach) to arbitrary non-homogenous gravitational fields.

In February–March 1912, encouraged by Abraham's research, which had obtained a scalar differential equation of a gravitational field (a generalization of the Poisson equation) incorporating Einstein's relationship between the potential and the speed of light (this relationship was a direct consequence of the equivalence principle), Einstein developed two versions of the theory of static (non-homogenous) fields. Both theories relied on the generalization of the Poisson equation and the corresponding generalizations of the equations of motion of a material point. The first theory used the velocity of light c, and the second its square root $\sqrt{c}$ as the potential. Throughout the summer and fall, Einstein and Abraham were debating this subject. As subsequent developments in the gravitational theory revealed, both of them, having chosen a scalar and, besides, a non-covariant approach with respect to Lorentz transformations, turned out to be on the wrong track.

Dissatisfaction with the chosen approach and a desire to expand it can, however, be felt in Einstein's two articles on the theory of static fields. Thus Einstein was becoming increasingly convinced that geometry in accelerated reference systems becomes non-Euclidean and, as an example, considered the "Ehrenfest disk," that is, a rotating reference system.

Einstein further points to the right way of deriving equations of motion of a material point in the gravitational field, in the form of a variational principle for a four-dimensional space metric with a variable speed of light

$$\delta \int ds = 0.$$

Then Einstein (in debates with Abraham) arrives at the conclusion that it is difficult (or even impossible) to bring scalar and vector theories of the gravitational field in agreement with the principle of equivalence. These were all precursors of an approach that subsequently led to the tensor-geometrical concept of gravitation. A few of the major "precursors" of that kind were associated either with Mach's influence or, less directly, with his students, in particular, with Pick.

The range of questions having to do with rotating reference systems, and thereby with the "Ehrenfest disk," is typically Machian. Ehrenfest, with whom Einstein met during the Prague period (their correspondence started in April 1911), used a thought experiment with an absolutely rigid rotating disc, on which geometry becomes non-Euclidean, in the attempt to prove that the idealization of absolute rigidity was out of place in SR. For Einstein, however, this thought experiment signaled the need to invoke non-Euclidean geometry to take account of non-inertial reference systems. We have already mentioned that Mach insisted on extending the principle of relativity to rotational motion. Dynamically, the equivalence of the rotating earth and "the sky of fixed stars" was achieved in this case by explaining the inertia of a certain body by the influence of the stars on it. True, Mach said nothing about the change in geometry in rotating reference systems, though in principle he admitted a possibility of using non-Euclidean geometry in physics. The idea of the dependence of the geometry of real space on physics was dear to him, as was the belief that the character of geometry could only be settled by experiment (see Mach 1907, pp. 492–494).[19]

Of key importance in the creation of the tensor-geometrical concept of gravitation was the idea of the fundamental character of the geometrical four-dimensional formulation of the principle of inertia used by Planck and Minkowski in relativistic mechanics ($\delta \int ds = 0$). To obtain a Riemannian space-time, one only has to replace the special relativistic interval in this formula, $ds = (c^2 dt^2 - dx^2 - dy^2 - dz^2)^{\frac{1}{2}}$, where $c = c_0 = const.$, by a similar expression $c = c\,(x, y, z, t)$, in which c becomes a function of space-time. The metric approach was coming to the fore in place of the coordinate approach.

Notably, the variational reformulation of the principle of inertia figured prominently in Mach's mechanics. Thus, in discussing the fundamentals of Hertz's mechanics and pointing out its affinity with his ideas Mach wrote:

> Hertz, after eliminating the concept of force, starts from the concepts of time, space, and motion alone, with the idea in view of giving expression only to that which can actually be observed. The sole principle which he employs may be conceived as a combination of the law of inertia and Gauss's principle of least constraint. Free masses move uniformly in straight lines. If they are put in connexion in any manner they deviate, in accordance with Gauss's principle, as little as possible from this motion; their *actual* motion is more nearly that of *free* motion than any other *conceivable* motion. Hertz says the masses move as a result of their connexion in a *straightest* path. Every deviation of the motion of a mass from uniformity and rectilinearity is due, in his system, not to a force but to rigid connexion with other masses. (Mach 1907, pp. 550–551)

Recall that Hertz's principle of the "straightest path," which Mach explains here, was typically stated in the form

$$\delta \int ds = 0.$$

where ds is the interval of the Riemannian multidimensional configuration space of a mechanical system. Comparison of the special relativistic form of the principle of inertia with the Hertz principle prompted a transition to Riemannian space-time. It should be noted, in passing, that Einstein never referred to Hertz in this connection.

Einstein was thus led, at the end of his Prague period, to turn to the four-dimensional Riemannian geometry, a geometrical structure that naturally emerges from the consideration of accelerated reference systems and, correspondingly, of arbitrary gravitational fields. In Prague Einstein had close friendly relations with the mathematician Pick, Mach's former student and disciple who was also the author of a series of works on differential geometry in non-Euclidean spaces.[20] According to Philipp Frank (1949), it was precisely Pick who suggested that Einstein use Riemannian geometry in the gravitational theory. Frank also indicates that, in his private conversations with Einstein, Pick referred more than once to Mach's statements that could be interpreted as anticipating the theory of relativity. Pick's letter to Mach of 2 August 1912, says:

> Although it was obvious to me, I learnt the fact that the physicists who were developing a new theory had experienced a great influence of your ideas from one of them, namely Einstein, who told me about it himself.
> [Dass die Physiker, welche die neue Theorie ausgebildet haben, von Ihren Gedanken wesentlich beeinflusst waren, wüsste ich, wenn es mir nicht sächlich einleuchten würde, von einem derselben, Einstein, aus persönlicher Mitteilung.] (Blackmore and Hentschel 1985, p. 104)

Einstein's article "Is There a Gravitational Effect Which Is Analogous to Electrodynamic Induction?" written in April–May 1912 (Einstein 1912) provides direct evidence that Mach's ideas were, at the time, at the center of Einstein's attention. This was, incidentally, Einstein's first article with direct references to Mach. Here, too, he considered, for the first time, two "Machian" consequences of the relativistic theory of gravitation or, more precisely, of its scalar "static" version. The article also contains Einstein's first formulation of Mach's principle. Einstein showed that the presence of a massive shell K increases the inertial mass of the material point P inside that shell in accordance with the formula

$$m' = m + kMm/Rc_0^2,$$

where m and m' are the masses of the material point P with and without regard for the shell K of mass M, R is the shell's radius, k the gravitational constant, and c_0 the speed of light in vacuum. Point P is located at the center of the shell K. The result obtained, Einstein wrote, "suggests that the *entire* inertia of a mass point is an effect of the presence of all the other masses, which is based on a kind of interaction with the latter" (Einstein 1912, p. 39). This is precisely Einstein's first formulation of Mach's principle. Einstein adds in a footnote: "This is exactly the same point of view that E. Mach advanced in his astute investigations on this subject." And he refers to Mach's *Mechanics*. Besides, Einstein calculated the force on the mass m exerted by an accelerated shell

$$F = \frac{3}{2} k \frac{mM}{Rc^2} \Gamma,$$

where Γ is acceleration of the shell. Einstein showed that the expressions for these effects and the relationships corresponding to them must be modified in a consistent relativistic theory of arbitrary gravitational fields. Yet the fact that "Machian effects" were present in the theory of static field satisfying the equivalence principle provided a guarantee that this principle and the path chosen by Einstein were correct.

4.2 EINSTEIN–GROSSMANN THEORY

In all likelihood, Einstein's entire Prague period paved the way to a decisive step towards the relativistic theory of gravitation, with Mach's ideas having a great direct or indirect effect on him. In September 1912, he moves to Zurich where he makes this step, not yet the final one, with the help of mathematician Marcel Grossmann, his friend from student years. Einstein and Grossmann's groundbreaking article (1913), which was completed in April–May 1913, contained the foundations of the tensor-geometrical conception of gravitation and, at the same time, the traces of Einstein's Prague quest for this conception. One finds there, in particular, the equations of motion of a material point in the gravitational field in variational metric form (the Hertz–Planck form), the idea of using non-Euclidean Riemannian geometry of space-time, and the idea of a tensor character of gravitational potentials. Finally, Einstein showed, based on his previous theory of the static field, the dependence of the inertial mass of a material point on the gravitational potential. He noted that "this is consonant with Mach's daring idea that inertia has its origin in an interaction between the mass point under consideration and all of the other mass points." "For if we accumulate masses in the vicinity of the mass point under consideration," he continued, "we thereby decrease the gravitational potential c, thus increasing the quantity m/c that is determinative of inertia" (Einstein and Grossmann 1913, p. 6).

Again although Mach is mentioned in this article only once, in connection with the relativity of inertia, or Mach's principle, the presence of which in the Einstein–Grossmann theory appeared to be its important "epistemological" (Einstein's term) advantage, retracing the logic of the discovery of the tensor-geometrical gravitational concept in the *Entwurf* and the Prague articles preceding that discovery points to a greater and broader influence of Mach's ideas.

First, the equivalence principle is a focal point of the entire tensor-geometrical conception of gravitation, and, as we have sought to show, it grew largely on the "Machian soil."

Second, the theory of static fields already involved Mach's notion of the dependence of the inertia of a material point "here" on the masses of the surrounding bodies "there" (in Wheeler's terminology). Einstein considered this a true sign that he was on the right track and, hence, what had to be done next was to find an appropriate way of applying the approach, not only to static, but to arbitrary fields.

And third, at least four Machian concepts came together at that point to suggest a method of generalizing non-Euclidean geometry to the Riemannian and to identify the Riemannian metric tensor with the gravitational potential.

(1) Presented in the Hertz–Planck form, the equation of motion of a material point in a gravitational field immediately invited, through Hertz's mechanics (that was so close to Mach's heart), a Riemannian metric

$$ds^2 = g_{ik}dx_i dx_k.$$

(2) Mach's idea of general relativity suggested a generalization of ds^2 that was characteristic of the static theory,

$$ds^2 = -dx^2 - dy^2 - dz^2 + c^2 dt^2,$$

where $c = c(x, y, z, t)$, to the generic Riemannian metric

$$ds^2 = g_{ik}dx_i dx_k.$$

The scalar potential $c(x, y, z, t)$ was thereby promoted to the tensor g_{ik}, which, at the same time, turned out to be a metric tensor of Riemannian space-time.

(3) This agreed with Mach's notion of kinematization of dynamic quantities (in the four-dimensional approach, kinematization is equivalent to geometrization).

(4) Finally, Mach's project of implementing non-Euclidean geometry in physics assumed a definite physical form. Incidentally, the kinematization of dynamics, as has been repeatedly mentioned, is equivalent to the "dynamization" of geometry. In a way, this idea is Machian too. It would be opportune to recall, for example, Mach's statement to the effect that "spatial relations of material particles can, indeed, only be recognized by the forces which they exert on one another" (1911, p. 89).[21]

Thanks to Grossmann's aid, Einstein mastered Riemannian geometry and tensor analysis (the second part of the *Entwurf* written by Grossmann contained a brief description of the mathematical apparatus applied to the four-dimensional geometry of space-time). The theory appeared to be a powerful, mathematically elaborate scheme specially adapted for the representation of physical laws conforming to the principle of general relativity. Yet the authors failed to bring the emerging generally covariant equations of a gravitational field with the Ricci tensor R_{ik} in agreement with the physical requirements of correspondence (that is, the existence of the Newtonian limit), of conservation of energy-momentum and, in particular, with the demands of the principle of causality. This led them to abandon generally covariant field equations and to restrict themselves to linearly covariant field equations. As a result, it took Einstein another two and a half years of intense studies to arrive at the general theory of relativity in late November 1915. And Machian ideas continued to play an important role in his quest.

4.3 EINSTEIN'S THIRD LETTER TO MACH

But let us return to the spring–summer of 1913. Einstein, together with Grossmann, had just finished the *Entwurf* and immediately sent it, along with an accompanying letter, to Mach.[22] The latter says more about "Machian effects" than the article itself. Apart from the dependence of the inertial mass of a body on its distance from other bodies, two other "inductive" effects are mentioned there. The letter also points to the great influence of Mach's ideas on Einstein exactly at the time when he was developing the tensor-geometrical concept of gravitation. The full text of the letter dated 25 June 1913 is below.

> Highly esteemed Colleague,
>   You have probably received a few days ago my new paper on relativity and gravitation, which is now finally completed after unceasing toil and tormenting doubts. Next year, during the solar eclipse, we shall learn whether light rays are deflected by the sun, or in other words, whether the underlying fundamental assumption of the equivalence of the acceleration of the reference system, on the one hand, and the gravitational field, on the other, is really correct.
>   If yes, then—in spite of Planck's unjustified criticism—your brilliant investigations on the foundations of mechanics will have received a splendid confirmation. For it follows of necessity that *inertia* has its origin in some kind

of *interaction* of the bodies, exactly in accordance with your argument about Newton's bucket experiment.

You will find a first consequence in this sense on the top of page 6 of the paper [i.e., Einstein and Grossmann 1913]. Beyond that, the following results have been obtained:

1. If one accelerates an inertial spherical shell *S*, then, according to the theory, a body enclosed by it experiences an accelerating force.

2. If the shell S rotates about an axis passing through its center (relative to the fixed stars ("Restsystem")), then a Coriolis field arises inside the shell, i.e., the *place* of the Foucault pendulum is being carried along (though with a practically immeasurably small velocity).

It gives me great pleasure to be able to tell you about this, all the more so because Planck's criticism always seemed to me to be most unjustified.

With kindest regards, I remain very respectfully yours,

Einstein

Thank you sincerely for sending me your book.

(Einstein to Mach, 25 June 1913, Einstein 1993, Doc. 448, pp. 531–532)

4.4 EINSTEIN'S VIENNA REPORT

In September 1913 Einstein gave two major talks on the *Entwurf*. The first one was presented on September 9 at an annual meeting of the Swiss society of natural scientists in Frauenfeld (Einstein 1914a).[23] The second presentation, more extensive and thorough as regards the development of Machian ideas, was made on September 23 at the 85th meeting of the society of German natural scientists and physicians in Vienna (Einstein 1913).

The concluding paragraph of the Frauenfeld talk outlined Mach's reasoning from the relativity of acceleration to the relativity of inertia. In the Vienna report, this problem is treated in an extensive section (§9). It should be noted that the term "relativity of inertia" appeared for the first time in the former lecture. Having described how the inertial resistance of a body increases with accumulation of masses in its vicinity and how this effect vanishes when the surrounding masses accelerate with the body, Einstein points out that "this behavior of the inertial resistance, which we may call *relativity of inertia*, actually follows from [the new theory]" and adds: "This circumstance constitutes one of the strongest pillars of the theory sketched" (Einstein 1914a, p. 290).

At this stage of development of the gravitation theory, Mach's concept of inertia provided for Einstein a sort of touchstone for the tensor-geometrical concept of gravitation and, hence, for the theory in general.

This is why the Vienna report says nothing about the range of Machian ideas in the initial paragraphs, which outline the formulation of the problem and the theory's basic assumptions.

From the equation of motion of a material point in Newtonian approximation Einstein obtains expressions for the momentum and energy of that point in the gravitational field.[24] These expressions make it feasible to conclude that the accumulation of masses "effects an *increase* in the inertia of the mass point in question." Then follows a characteristic statement whose analogs are found in earlier works (in particular, in Einstein 1912 and Einstein and Grossmann 1913) and which constitutes the essence of Mach's principle:

> If the inertia of a body can *increase* due to cumulation of masses in its vicinity, then we have no choice but to view the inertia of a point as being *caused* by the existence of the other masses. Thus inertia appears to be caused by a sort of interaction between the mass point to be accelerated and all of the other mass points. (Einstein 1913, p. 1260)

Einstein goes on to connect this assertion (Mach's principle) with the relativity of acceleration, calling the former "the hypothesis of the relativity of inertia" and attributes the concept of inertia thus described to Mach.[25] He also stresses that the relativity of inertia is not a logically necessary requirement, and that theories meeting this requirement are preferable epistemologically (see Einstein's statement quoted earlier in this connection).

Then, based on gravitational equations in second approximation (field equations and the equations of motion of a material point),[26] Einstein derives two other Machian effects about which he wrote to Mach in June. Though Einstein did not expect these effects to be confirmed experimentally in view of the small value of the gravitational constant, he thought them to be natural outcomes of the hypothesis of the relativity of inertia and regarded their existence as a major argument in favor of the new theory of gravitation and, *a fortiori*, of the tensor-geometrical theory of gravitation. In comparing this theory with the scalar Lorentz-covariant theory of the gravitational field (Nordström's theory), Einstein notes that the latter "is rather simple and . . . satisfies the main requirements to be imposed upon a theory of gravitation but does not include the relativity of inertia among its consequences" (Einstein 1913, p. 1262). He saw in this a major shortcoming of the theories of this kind, competing with the Einstein–Grossmann theory or GR. The last word in this dispute, as Einstein mentioned in his letter to Mach and at the end of his Vienna lecture, belongs, of course, to experiment.

In subsequent works (both before and after the creation of GR) Mach's ideas figured prominently both at the "input" and "output." At the "input," Einstein drew on Machian-style considerations on general relativity, the need to extend SR in describing gravitational fields, and the principle of equivalence. At the "output," he considered the Machian effect of the relativity of inertia as an essential additional argument in favor of the theory (see, for example, Einstein 1914d, 1916b, 1922). Sure enough, both of these aspects were interconnected via the idea of the relativity of acceleration, though the effects of the relativity of inertia were "calculated" by solving the equations of the theory and constituted, in this sense, a definite result following from the theory, rather than a heuristic consideration or "scaffolding" employed in elaborating the theory.

4.5 EINSTEIN'S FOURTH LETTER TO MACH

Until recently, many researchers, including the author of the present article, have considered this letter to be the third one. Written on Christmas, it has no exact date. Wolters sounds convincing, I think, when he argues that it was written during the Christmas days of 1913/1914 rather than of 1912/1913 (Wolters 1987).

This is remarkable in several respects. First, in a very laconic and expressive way Einstein expounds in it the Einstein–Grossmann theory focusing on the meaning of the tensor-geometrical concept of gravitation. In so doing he, as in the previous letter, emphasizes the stimulating role of Mach's criticism of the foundations of mechanics. Besides, he links Mach's ideas with the above concept, noting also that these ideas, in essence, had been so far the only weighty argument in favor of the "new theory," an argument of an epistemological character.

This letter was probably written in reply to Mach's letter which evinced "a friendly interest in the new theory," while noting at the same time the mathematical difficulties impeding the study of that theory (Wolters 1987). Indeed, the *Entwurf*, which obviously was sent to Mach in the summer of 1913, drew upon a mathematical apparatus that was extremely complex and unusual for physicists. Let us quote this letter in full.

> Highly esteemed Colleague,
>
> Your friendly interest in the new theory makes me very happy. Unfortunately, the mathematical difficulties which one encounters in pursuing these ideas are enormous for me as well. I am tremendously pleased that the

development of the theory brings to fore the depth and importance of your investigations on the foundation of classical mechanics. To this day, I still cannot understand how Planck, whom I have otherwise learned to prize like no one else, could show so little understanding for your endeavors. Incidentally, he also disapproves of my new theory. I cannot blame him for that. Because that epistemological argument is the only thing that I can thus far advance in favor of my new theory. It seems to me absurd to ascribe physical properties to "space." The totality of masses produces the $g_{\mu\nu}$-field (gravitational field), which in turn governs the course of all processes, including the propagation of light rays and the behavior of measuring rods and clocks. First of all, everything that happens is referred to four *completely arbitrary* space-time variables. If the principles of momentum and energy conservation are to be satisfied, these variables must then be specialized in such a way that only (completely) *linear* substitutions shall lead from one justified reference system to another. The reference system is, so to speak, tailored to the existing world with the help of the energy principle, and loses its nebulous aprioristic existence.

In the near future I will send you several presentations of the topic in which the formal element is kept to a minimum, the essential content being emphasized as much as possible. But I am not very successful in separating the substance from the form in these abstract things.

With best wishes for the New Year, your devoted

A. Einstein

(Einstein to Mach, second half of December 1913, Einstein 1993, Doc. 495, pp. 583–584)

Referring to the theory's statements in which "the formal element is kept to a minimum," Einstein apparently had in mind his article "On the Problem of Relativity" written for *Scientia* (Einstein 1914c). Mach's concepts in it are expressed clearly and at length. Einstein stresses once again that the entire range of ideas (in particular, the thought experiment with two masses rotating with respect to each other) that have to do with the relativity of rotation and acceleration was developed by Mach "so that I cannot claim that what I am saying here in this respect is new" (Einstein 1914c, p. 344). The exposition of the theory in this article, incidentally, is close in essence and form to the brief formulations which we find in the letter (see especially pp. 347–348). Einstein also notes that the arbitrary continuous transformations have to be limited to those conforming with the law of conservation of energy-momentum and that this limitation does not bring one back to classical mechanics or SR, "because according to this generalized theory of relativity, physical properties peculiar to privileged spaces no longer exist" (Einstein 1914c, p. 348).

5. Concluding Remarks

Thus, at all stages of the construction of GR, Mach's ideas were important for Einstein in one way or the other. Einstein did not always explicitly refer to Mach whose influence in these cases reveals itself only in the reconstruction of Einstein's reasoning. Of particular importance for him was the range of Mach's ideas arising out of the latter's relativistic critique of the foundations of Newtonian mechanics and the Machian concept of inertia. Einstein consistently referred to Mach in this respect since 1912. Equally essential for the founder of relativity were Mach's methodological concepts: his adherence to the principle of observability, to the operational approach, and to thought experiment techniques.

Machian themes, specifically, Mach's principle, were also instrumental in the genesis of Einstein's relativistic cosmology. In (Einstein 1917), Mach's principle assumed a distinctively cosmological character. In 1918, Einstein first put in circulation the expression "Mach's principle" and gave its clear definition (1918, p. 242). Along with considerations of relativity and equivalence, Mach's principle, in Einstein's words, was the third basic proposition of the GR. He formulated it as follows: "The G-field [i.e., the gravitational field described by the metric tensor] is *completely* determined by masses of the bodies . . . ; this means that the G-field is conditioned and determined by the energy tensor of matter" (1918, p. 242; trans. by J. Barbour). In the footnote to this passage, he wrote: "I have chosen the name 'Mach's principle' because this principle has the significance of a generalization of Mach's requirement that inertia should be derived from an interaction of bodies." Einstein goes on to explain that, in general, Mach's principle may be not fulfilled in GR. But if field equations are modified to include the cosmological λ-term, they yield a solution in the form of a static (spherical or elliptical) closed world, in which Mach's principle is realized. It thus became clear that, drawing on the ideas close to Mach's principle, Einstein built a theory (GR) which does not, in general, involve this principle. Consequently, all the issues raised by Mach's principle have preserved their basic meaning only in the cosmological context.

NOTES

[1] Recently Gereon Wolters offered convincing arguments against the authenticity of this text published by Mach's elder son Ludwig in 1921. Wolters argues that the introduction was forged by Ludwig. See Wolters (1987, pp. 404–405).

[2] The growing philosophical and methodological schism between Einstein and Mach was described, for example, by Holton (1973).

[3] Unless otherwise indicated, quotations from Einstein are given in A. Beck's translation.

[4] In the quoted obituary of Mach, Einstein noted this general methodological aspect of Mach's ideas: "Es ist deshalb durchhaus keine müßige Spielerei, wenn wir darin geübt werden, die längst geläufigen Begriffe zu analysieren und zu zeigen, von welchen Umständen ihre Berechtigung und Brauchbarkeit abhängt, wie sie im einzelnen aus den Gegebenheiten der Erfahrung herausgewachsen sind. Dadurch wird ihre allzu große Autorität gebrochen" (Einstein 1916a, p. 102). Einstein believed Mach was an outstanding master of such analysis.

[5] The one-year difference in dating this letter is quite substantial both in regard to the genesis of GR (the letter contains an outline of the Einstein–Grossmann theory, the first version of GR which lacked generally covariant equations of the gravitational field) and in respect of the analysis of Mach's attitude to the theory of relativity during that period. The text of the letter points unambiguously to Mach's benevolent attitude to Einstein's GR.

[6] See Vizgin 1981, pp. 29–42. These "isms" were discussed, in one way or another, at the turn of the 20th century, although various critics gave them unequal weight in different periods (in the 1870s and 1880s, in 1890–1900, immediately after the emergence of SR, and later).

[7] Inertial mass was considered an intrinsic property of bodies, a sort of "absolute" property independent of anything external to the body.

[8] Closely related to instantism is another weakness that Einstein identified with the division of energy into two parts, kinetic and potential (see Einstein 1979).

[9] This agrees with the view that Mach recognized the concept of the four-dimensional space-time, about which he wrote in *Knowledge and Error* (see Mach 1976, pp. 348–349).

[10] We do not give here the full and well-known quotation. It begins as follows: "Newton's experiment with the rotating vessel of water simply informs us, that the relative rotation of the water with respect to the sides of the vessel produces *no* noticeable centrifugal forces, but that such forces *are* produced by its relative rotation with respect to the mass of the earth and the other celestial bodies" (Mach 1907, p. 232).

[11] We will look into this question in discussing Mach's influence on Einstein's principle of equivalence.

[12] As for finitism, Mach, to all appearances, overlooked the formulation of the gravitational paradox in C. Neumann's work (1870), which was well known to him, but he noticed and praised Hugo Seeliger's article (1896) "who has shown the incompatibility of a rigorous interpretation of Newton's law with the assumption of an unlimited mass of the universe" (Mach 1907, p. xiv).

[13] Einstein's paper, from which this passage is quoted, was based on a lecture delivered on 9 September 1913 at the annual meeting of the Schweizer Naturforschende Gesellschaft in Frauenfeld.

[14] Such "perspicacious epistemologists" (*scharfsinnigen Erkenntnistheoretik-ern*) and critics of Newtonian classical concepts also include C. Huygens, G. W. Leibniz, and G. Berkeley (see, for example, Vasiliev 1923, Kuznetsov 1967, Stein 1977).

[15] This influence could be indirect. Einstein used to talk about the tremendous impression Mach's *Mechanics* produced on him during his student years. He wrote about this impression in his letter to M. Besso of 6 January 1948 and then added: "Wie weit sie (Mach's "Mechanik" and "Wärmelehre") auf meine eigene Arbeit gewirkt haben, ist mir offen gestanden nicht klar. . . . Aber wie gesagt bin ich nicht imstande, das im unbewußten Denken Verankerte zu analysieren" (Speziali 1972, p. 391).

[16] We are following Konoplyova (1978, pp. 223–225) here.

[17] We are using the notion of a program in the broadly Lakatosian sense. For a detailed account of the relativistic program understood in this sense, see Vizgin 1981.

[18] See his 1897 article "Thought Experiments" later included in *Knowledge and Error* (Mach 1976, pp. 134–147).

[19] He also gives there an example of two-dimensional beings living on a sphere and wishing to understand the geometry of their world: "A thinking being is supposed to live on the surface of a sphere, with no other kind of space to institute comparisons with. His space will appear to him similarly constituted throughout. He might regard it as infinite, and could only be convinced of the contrary by experience. Starting from any two points of a great circle of the sphere and proceeding at right angles thereto on other great circles, he could hardly expect that the circles last mentioned would intersect. So, also, with respect to the space in which we live, only experience can decide whether it is finite, whether parallel lines intersect in it, or the like" (Mach 1907, p. 493, footnote).

[20] Pick studied non-Euclidean differential geometry particularly actively in 1911–1912. One could reasonably suppose that this study was stimulated by conversations with Einstein.

[21] Cf. another statement of Mach: "The physical space which I have in mind—and which, at the same time, contains time in itself—is thus nothing other than *dependence of phenomena on one another*" (Mach 1911, p. 89).

[22] The *Entwurf* was published by Teubner as a booklet, which came out before 25 June 1913, and later by *Zeitschrift für Mathematik und Physik*. The letter quoted below makes it clear that the work was sent to Mach in the booklet form (see Wolters 1987). The correspondence between Einstein and Mach was again resumed by Mach who sent to Einstein the 7th edition of *Mechanics* published in 1912.

[23] Einstein's talk in Frauenfeld was supplemented by Grossmann who spoke on the mathematical foundations of the theory.

[24] The relevant equations of the Vienna paper are as follows:

$$I_x = m\left(1 + \frac{\kappa}{8\pi}\int \frac{\rho_0\,dv}{r}\right)\frac{\dot{r}}{c},$$

$$E = mc\left(1 - \frac{\kappa}{8\pi}\int \frac{\rho_0\,dv}{r}\right) + \frac{1}{2}\frac{m}{c}\left(1 + \frac{\kappa}{8\pi}\int \frac{\rho_0\,dv}{r}\right)q^2,$$

where m is the mass of the material point, q its velocity, c is the speed of light, ρ_0 the density of the surrounding matter, and κ the gravitational constant.

[25] Mentioned in this connection, incidentally, is V. Hoffman, a little-known Vienna mathematician, who, in the early 1900s, was developing a similar concept and to whose work Mach referred in the 6th edition of *Mechanics*.

[26] $\Box h_{\mu\nu} = 0 \quad (\mu,\nu \neq 4),$

$\Box h_{14} = -\kappa\rho_0\dot{x},$

. . . .

$\Box h_{44} = -\kappa\rho_0 c^2,$

$\ddot{\mathbf{r}} = -\frac{1}{2}\nabla g_{44} + \dot{\mathbf{g}} - [\dot{\mathbf{r}}, \operatorname{rot}\mathbf{g}],$

where $h_{\mu\nu} = g_{\mu\nu} - \eta_{\mu\nu}$, $\eta_{\mu\nu}$ being the metric tensor of pseudo-Euclidean space, $\mathbf{r}$ is the point's radius-vector, and $\mathbf{g}$ is vector (g_{14}, g_{24}, g_{34}).

REFERENCES

Barbour, Julian, and Pfister, Herbert, eds. (1995). *Mach's Principle: From Newton's Bucket to Quantum Gravity*. Boston: Birkhäuser.

Blackmore, John T. (1972). *Ernst Mach. His Work, Life and Influence*. Berkeley: University of California Press.

Blackmore, John, and Hentschel, Klaus, eds. (1985). *Ernst Mach als Aussenseiter: Machs Briefwechsel über Philosophie und Relativitätstheorie mit Persönlichkeiten seiner Zeit; Auszug aus dem letzten Notizbuch (Faksimile) von Ernst Mach*. Vienna: Wilhelm Braunmüller, Universitäts-Verlagsbuchhandlung.

Bunge, Mario (1966). "Mach's Critique of Newtonian Mechanics." *American Journal of Physics* 34: 585–596.

Cassirer, Ernst (1921). *Zur Einstein'schen Relativitätstheorie. Erkenntnistheoretische Betrachtungen*. Berlin: B. Cassirer.

Delokarov, K. Kh. (1979). "Einshtein i Mach." ["Einstein and Mach"] In *Einshtein i filosofskiye problemy fiziki XX veka. [Einstein and Philosophical Problems of Twentieth-Century Physics*] K. Delokarov, ed. Moscow: Nauka, pp. 484–503.

Dicke, Robert (1964). "The Many Faces of Mach." In *Gravitation and Relativity*. H.-Y. Chiu and W. F. Hoffman, eds. New York: Benjamin, pp. 121–141.

Einstein, Albert (1911). "Über den Einfluss der Schwerkraft auf die Ausbreitung des Lichtes." *Annalen der Physik* 35: 898–908.

—— (1912). "Gibt es eine Gravitationswirkung, die der elektrodynamischen Induktionswirkung analog ist?" *Vierteljahrsschrift für gerichtliche Medizin und öffentliches Sanitätswesen* 44: 37–40.

—— (1913). "Zum gegenwärtigen Stande des Gravitationsproblems." *Physikalische Zeitschrift* 14: 1249–1262.

—— (1914a). "Physikalische Grundlagen einer Gravitationstheorie." *Naturforschende Gesellschaft in Zürich. Vierteljahrsschrift* 58: 284–290.

—— (1914b). "Zur Theorie der Gravitation." *Naturforschende Gesellschaft in Zürich. Vierteljahrsschrift* 59. Part 2, *Sitzungsberichte*: iv–vi.

—— (1914c). "Zum Relativitäts-Problem." *Scientia* 15: 337–348.

—— (1914d). "Die formale Grundlage der allgemeinen Relativitätstheorie." *Könglich Preussische Akademie der Wissenschaften* (Berlin). *Sitzungsberichte*: 1030–1085.

—— (1916a). "Ernst Mach." *Physikalische Zeitschrift* 17: 101–104.

—— (1916b). "Die Grundlage der allgemeinen Relativitätstheorie." *Annalen der Physik* 49: 769–822.

—— (1917). "Kosmologische Betrachtungen zur allgemeinen Relativitätstheorie." *Königlich Preussische Akademie der Wissenschaften* (Berlin). *Sitzungsberichte*: 142–152.

—— (1918). "Prinzipielles zur allgemeinen Relativitätstheorie." *Annalen der Physik*: 241–244.

—— (1922). *The Meaning of Relativity*. Princeton: Princeton University Press.

—— (1954). "Notes on the Origin of the General Theory of Relativity." George A. Gibson Foundation Lecture, delivered at Glasgow, 20 June 1933. Reprinted in A. Einstein, *Ideas and Opinions*. New York: Crown Publications, pp. 285–290.

—— (1979). *Autobiographical Notes: A Centennial Edition*. P. A. Schilpp, trans. and ed. La Salle, Illinois: Open Court.

—— (1993). *The Collected Papers of Albert Einstein*. Vol. 5, *The Swiss Years: Correspondence, 1902–1914*. Martin J. Klein, et al., eds. Princeton, New Jersey: Princeton University Press.

Einstein, Albert, and Grossmann, Marcel (1913). *Entwurf einer verallgemeinerten Relativitätstheorie und Theorie der Gravitation*. Leipzig: Teubner.

Feyerabend, Paul (1980). "Zahar on Mach, Einstein and Modern Science." *British Journal for the Philosophy of Science* 31: 273–282.

—— (1984). "Mach's Theory of Research and its Relation to Einstein." *Studies in History and Philosophy of Science* 15: 1–22.

Frank, Philipp (1949). *Einstein. Sein Leben und seine Zeit*. Munich: P. List.

Einstein and Mach   87

Goenner, Hubert (1970). "Mach's Principle and Einstein's Theory of Gravitation."
In *Ernst Mach: Physicist and Philosopher*. Robert S. Cohen and Raymond J.
Seeger, eds. Dordrecht: Reidel, pp. 200–218.

Haller, Rudolf, and Stadler, Friedrich (1988). *Ernst Mach—Werk und Wirkung*.
Vienna: Holder-Pichler-Tempsky.

Heller, K. D. (1964). *Ernst Mach. Wegbereiter der modernen Physik (mit aus-
gewählten Kapiteln aus seinem Werk)*. Vienna and New York: Springer.

Herneck, Friedrich (1966). "Die Beziehungen zwischen Einstein und Mach,
dokumentarisch dargestellt." *Wissenschaftliche Zeitschrift der Friedrich-
Schiller-Universität* (Jena). *Mathematisch-naturwissenschaftliche Reihe* 15:
1–14.

— (1974). *Albert Einstein*. 2nd ed. Leipzig: Teubner.

— (1983). *Die heilige Neugier. Erinnerungen, Bildnisse, Aufsätze zur Geschichte
der Naturwissenschaften*. Berlin: Der Morgen.

— (1984a). *Wissenschaftsgeschichte: Vorträge und Abhandlungen*. Berlin:
Akademie-Verlag.

— (1984b). "Mach's Theory of Research and its Relation to Einstein." *Studies in
History and Philosophy of Science* 15: 1–22.

Hertz, Heinrich (1894). *Gesammelte Werke*. Vol. 3, *Die Prinzipien der Mechanik.
In neuem Zusammenhange dargestellt*. Philipp Lenard, ed. Leipzig: Johann
Ambrosius Barth.

Hiebert, E. (1970). "The Genesis of Mach's Early Atomism." In *Ernst Mach:
Physicist and Philosopher*. Robert S. Cohen and Raymond J. Seeger, eds.
Dordrecht: Reidel, pp. 79–106.

Holton, Gerald (1973). "Mach, Einstein, and the Search for Reality." In *Thematic
Origins of Scientific Thought. Kepler to Einstein*. Cambridge, Mass.: Harvard
University Press, pp. 219–259.

Hönl, Helmut (1966a). "Zur Geschichte des Machsche Prinzip." *Wissenschaftliche
Zeitschrift der Friedrich-Schiller-Universität* (Jena). *Matematisch-natur-
wissenschaftliche Reihe* 15: 25–36.

— (1966b). "Das Machsche Prinzip und seine Beziehung zur Gravitationstheorie
Einsteins." In *Entstehung, Entwicklung und Perspektiven der Einsteinschen
Gravitationstheorie*. H.-J. Treder, ed. Berlin: Akademie-Verlag, pp. 238–278.

Illy, József (1979). "Albert Einstein in Prague." *Isis* 70: 76–84.

Konoplyova, N. P. (1978). "Ob evolyutsii ponyatiya inertsii (N'yuton, Mah,
Einshtein.)" ["On the Evolution of the Notion of Inertia (Newton, Mach,
Einstein)"] In *Einshteinovskiy Sbornik 1975–76*. V. L. Ginzburg and U. I.
Frankfurt, eds. Moscow: Nauka, pp. 216–245.

Kuznetsov, B. G. (1967). "Einshtein i printsip Maha." ["Einstein and Mach's
Principle"] In *Einshteinovskiy Sbornik 1967*. I. Tamm, ed. Moscow: Nauka,
pp. 134–174.

Mach, Ernst (1872). *Die Geschichte und die Wurzel des Satzes von der Erhaltung
der Arbeit*. Prague: J. G. Calve.

— (1901). *Die Mechanik in ihrer Entwicklung. Historisch-kritisch dargestellt.* 4th ed. Leipzig: F. A. Brockhaus.

— (1905). *Erkenntnis und Irrtum. Skizzen zur Psychologie der Forschung.* Leipzig: Ambrosius Barth.

— (1907). *The Science of Mechanics. A Critical and Historical Account of its Development.* T. J. McCormack, trans. Chicago: Open Court. [Translation of Mach 1901]

— (1911). *History and Root of the Principle of the Conservation of Energy.* P. Jourdain, trans. Chicago: Open Court. [Translation of Mach 1872]

— (1976). *Knowledge and Error. Sketches on the Psychology of Enquiry.* T. J. McCormack, trans. Dordrecht: Reidel. [Translation of Mach 1905]

Narlikar, Jayant V. (1979). *Lectures on General Relativity and Cosmology.* London: Macmillan.

Neumann, C. (1870). *Die Prinzipien der Galilei–Newtonschen Theorie.* Leipzig: Teubner.

Pais, Abraham (1982). *"Subtle Is the Lord. . .": The Science and the Life of Albert Einstein.* Oxford: Clarendon Press.

Pyenson, Lewis (1974). *The Göttingen Reception of Einstein's General Theory of Relativity.* Ph.D. Dissertation. Baltimore: John Hopkins University.

Reinhardt, M. (1973). "Mach's Principle—A Critical Review." *Zeitschrift für Naturforschung* 28a: 529–537.

Sanchez, Ron J. M. (1981). *Relatividad especial, relatividad general (1905–1923): origenes, desarollo y recepción por la comunidad científica.* Barcelona: Univ. Autonoma de Barcelona.

Sciama, Dennis (1969). *The Physical Foundations of General Relativity.* Garden City, New York: Doubleday.

Seeliger, Hugo (1896). "Über das Newton'sche Gravitationsgesetz." *Münchner Akademie. Sitzungsbericht.*

Speziali, Pierre, trans. and ed. (1972). *Albert Einstein–Michele Besso. Correspondance, 1903–1955.* Paris: Hermann.

Stachel, John (1979). "Einstein and the Rigidly Rotating Disc." *General Relativity and Gravitation* 1: 30–45.

Stein, Howard (1977). "Some Philosophical Prehistory of General Relativity." In *Foundations of Space-Time Theories.* J. Earman, C. Glymour, and J. Stachel, eds. Minneapolis: University of Minnesota Press, pp. 3–49.

Vasiliev, A. V. (1923). Prostranstvo, vremya, dvizheniye: Istoricheskiye osnovy teorii otnositel'nosti. [*Space, Time, Motion: Historical Foundations of Relativity Theory*] Petrograd: Tovarishchestvo Obrazovanie.

Vizgin, Vladimir P. (1981). *Relyativistskaya teoriya gravitatsii: istochniki i formirovaniye, 1900–1915.* [*Relativistic Theory of Gravitation: Sources and Formation, 1900–1915*] Moscow: Nauka.

— (1985). "Formirovaniye tenzorno-geometricheskoy kontseptsii gravitatsii." ["Development of the Tensor-Geometrical Concept of Gravitation"] *Issledo-*

vaniya po istorii fiziki i mekhaniki. [*Studies in the History of Physics and Mechanics*] Moscow: Nauka, pp. 61–77.

Wheeler, John Archibald (1964). "Mach's Principle as a Boundary Condition for Einstein's Equations." In *Gravitation and Relativity*. H.-Y. Chiu and W. F. Hoffman, eds. New York: Benjamin, pp. 303–349.

Wolters, Gereon (1987). *Mach I, Mach II, Einstein und die Relativitätstheorie. Eine Fälschung und ihre Folgen*. Berlin and New York: Walter de Gruyter.

Zahar, Elie (1977). "Mach, Einstein and the Rise of Modern Science." *British Journal for the Philosophy of Science* 28: 195–213.

— (1981). "Second Thoughts About Machian Positivism: Reply to Feyerabend." *British Journal for the Philosophy of Science* 32: 267–276.

Zel'dovich, Yakov Borisovich, and Novikov, Igor Dmitrievich (1971). *Teoriya tyagoteniya i evolyutsiya zvyozd*. [*Gravitational Theory and Stellar Evolution*] Moscow: Nauka.

— (1975). *Struktura i evolyutsiya vselennoy*. [*The Structure and Evolution of the Universe*] Moscow: Nauka.

Hermann Weyl and Large Numbers in Relativistic Cosmology

Gennady Gorelik

1. Introduction

Since time immemorial philosophers, poets, and scientists have pondered the relationship between the micro- and the mega-world. The relevant scale, what counts as the micro- and the mega-, has always been determined by the scientific knowledge of the time. Since Newton, the scales of the largest and the smallest have extended by ten orders of magnitude in both directions. Equally strikingly, the meanings of 'micro' and 'mega' have changed in the historical development from the unification of celestial and terrestrial mechanics, to the physical study of stars by means of spectral analysis, to the micro-physical explanation of the baryon asymmetry of the universe.

It was only in the late 1910s, however, that the first physical fact was discovered that could provide a quantitative clue to the interconnection between the micro- and mega-worlds. It was a famous mathematician, Hermann Weyl, who made this discovery.

His discovery later gave rise to such different ideas as the hypothetical variation of the gravitational constant and the anthropic principle. More cautiously, it was referred to as "an unexplained empirical connection between meta-galactic parameters and micro-physical constants" (Zel'manov 1962, p. 496).

Although this link between the micro- and mega-worlds is regarded as an empirical fact, its recognition was intertwined with developments in advanced theoretical physics. Before turning to the circumstances of the discovery of this fact, let us look at its contemporary status, which clearly points to its empirical nature.

Yuri Balashov and Vladimir Vizgin, eds., *Einstein Studies in Russia.*
Einstein Studies, Vol. 10. Boston, Basel, Birkhäuser, pp. 91–106.

2. The Cosmological and Micro-physical Parameters

The universe as a whole is the largest physical object that can be described by a limited number of parameters. The basic mega-parameters are the average density of mass (or energy), $\rho \approx 10^{-31}$ g/cm^3, and the Hubble constant, $H \approx 2 \times 10^{-18}$ s^{-1}, which describes the rate of the expansion of the universe by relating the speed of the recession of galaxies to the intergalactic distance according to $v = Hr$. The gravitational constant G and the velocity of light c can also be attributed to mega-physics.

The characteristic parameters of micro-physics are the mass of an elementary particle m, the elementary charge e, the Planck constant $\hbar$, and the velocity of light c. The uncertainty of a few orders of magnitude in "the" mass of a particle does not really matter in the present context. For definiteness, we shall assume it to be the electron mass. Thus, mega-physics is represented by the quantities H, ρ, G, and c, while micro-physics is represented by m, e, $\hbar$, and c. Let us also put aside the question of whether one should use the coupling constant of the strong or weak interaction in place of e: in the 1920s, no such question existed.

Only dimensionless quantities, the values of which are independent of the units of measurement, can be regarded as being of genuine theoretical interest. Among such quantities, constructed from the parameters

$$H, \rho, G, m, e, \hbar, c \tag{1}$$

are three combinations whose values lie very close to the unusually large number 10^{40}. One of them is the ratio of the strengths of electromagnetic and gravitational interactions

$$Q_1 = e^2/Gm^2 = 4 \times 10^{42} \approx 10^{40}. \tag{2a}$$

Two other combinations can be obtained by taking the ratio between the distance and density scales characteristic of the micro- and macro-worlds: between the so-called radius of the universe, $R = c/H$, and the classical electron radius, $r_e = e^2/mc^2$, and also between ρ and m/r_e^3:

$$Q_2 = R/r_e = 3 \times 10^{40} \approx 10^{40}, \tag{2b}$$

$$Q_3 = mr_e^3/\rho = 3 \times 10^{40} \approx 10^{40}. \tag{2c}$$

The fact that these independent dimensionless quantities, Q_1, Q_2, and Q_3, are all so close to the large number 10^{40},

$$Q_1 \approx Q_2 \approx Q_3 \approx 10^{40} , \tag{3}$$

is called the *coincidence of large numbers*. This coincidence turns out to be a "purely" empirical fact that is not incorporated in any theory. (The quotation marks here are intended to remind the reader that it is only our enlightened age that permits us to regard such quantities as the universe's average density and the rate of its expansion as being empirical.)

Since the Hubble constant, H, appears in relations (2) and (3), one might think that the coincidence of large numbers came to be known only after 1929, when Hubble's law was discovered. This is not true: The coincidence came to the attention of physicists in the early 1920s and Arthur Eddington was one of its most active proponents. To clarify these circumstances, let us turn to the history of relativistic cosmology, with which the history of "large numbers" is closely connected.

3. The First Steps of Relativistic Cosmology

Relativistic cosmology was born in Einstein's 1917 paper, "Kosmologische Betrachtungen zur allgemeinen Relativitätstheorie," in which he suggested the first cosmological model based on general relativity (GR). Einstein's universe was described by two parameters, the radius of curvature R and the density of matter ρ. Einstein did not regard his model as a purely theoretical construct. For him, its most essential features, such as homogeneity, isotropy, constancy in time, and non-zero average density, were experimental and observational facts. Evidently, he knew nothing about the galactic structure of the universe (it was no more than a hypothesis then), nor about Vesto Slipher's discovery in 1913 of the enormous velocities of some galaxies (referred to, at that time, as "spiral nebulae"), which later developed into Hubble's law. Einstein supported his assumption about the static nature of the universe by referring to "the most important fact that we draw from experience as to the distribution of matter . . . that the relative velocities of the stars are very small as compared with the velocity of light" (1917, p. 148, trans. by W. Perrett and G. B. Jeffery). Einstein never specified which "experience" he had in mind. One thing is clear: this fact was of signal importance to Einstein. He mentioned it seven times in the eleven-page paper.

Einstein's acceptance of the static nature and the non-zero density of the universe forced him to generalize the field equations of GR by introducing the cosmological constant Λ:

$$R_{ik} - \tfrac{1}{2}g_{ik}R = \kappa T_{ik} - \Lambda g_{ik}, \tag{4}$$

which led him to assume a closed spatial geometry and the relation

$$1/R^2 = \kappa\rho/2 = \Lambda, \tag{5}$$

connecting the quantities R and ρ.

Einstein himself never attempted to estimate R and ρ on the basis of astronomical data: "Whether, from the standpoint of present astronomical knowledge, it [i.e., Einstein's cosmological model] is tenable, will not be here discussed" (1917, p. 152; trans. by W. Perrett and G. B. Jeffery).

Willem de Sitter was the first to do this, in the same year, 1917. In his third (and last) paper in a series that discussed Einstein's gravitation theory and its astronomical consequences, he turned to the cosmological problem (de Sitter 1917). Having described Einstein's cosmological model, he suggested another model known today as the de Sitter model, in which the matter density is zero while space, depending on the sign of Λ, can be either open or closed and the curvature radius is related to Λ:

$$3/R^2 = \Lambda. \tag{6}$$

De Sitter concluded his paper with an estimate of R based on the astronomical data.

The Einstein model allowed two types of test: by deriving geometrical consequences for the curvature (for example, for the behavior of light rays) and by obtaining the density from equation (5). De Sitter's model, on the other hand, permitted only geometrical considerations to be used, as it assumed $\rho = 0$.

De Sitter relied on much more extensive astronomical data than the single fact that the stars are at rest. He took into account the galactic structure of the universe, the size and mass of our Galaxy, and even the recently discovered shifts in the spectra of three galaxies ("spiral nebulae"). He used Schwarzschild's (1900) work, which compared astronomical data with assumptions about spatial curvature. Based, of course, on pre-relativistic considerations, Schwarzschild put a constraint on the curvature radius from below. For example, he noted that the light from the other side of the sun must be absorbed as it travels "around the world." (In a transparent, spatially-closed universe, any object can be observed in two opposite directions).

A constraint on the curvature radius from above was more interesting. In Einstein's cosmological model, this constraint is based on the estimate of the average matter density in the universe. De Sitter used the available data on the density of stars at the center of our galaxy and on its size (about 10^4 parsecs), and he took inter-galactic distances to be approximately equal to our galaxy's size. The result was

$$\rho > 10^{-27} \text{ g/cm}^3, \tag{7}$$

and, according to (5), the curvature radius is constrained by

$$R < 10^{27} \text{ cm.} \tag{8}$$

Assuming that space is closed and that the sun's reverse side is invisible, the estimate of R from below gave $R > 4 \times 10^{24}$ cm, while $R > 10^{25}$ cm follows from sufficiently small angular dimensions of the "spiral nebulae" and the hypothesis that their linear size is of the same order as that of our galaxy. Although de Sitter emphasized that these estimates were very crude, he believed that their agreement is remarkable and could not be expected a priori.

This is how the first quantitative cosmological parameters based on real observational data appeared. They claimed to describe the universe as a whole giving, for its radius

$$R \approx 10^{27} \text{ cm,} \tag{9}$$

and for the average density

$$\rho \approx 10^{-27} \text{ g/cm}^3. \tag{10}$$

Notably, despite all the difference between Einstein's model and present-day cosmology (today, the concept of the "radius of the universe" does not presuppose closed space), contemporary estimates of R and ρ differ from (9) and (10) only by a few orders of magnitude.

Thus, as early as in 1917, the coincidence of large numbers (2b) and (2c) could have been noticed. But no theoretical framework was available to account for it.

4. An Empirical Fact Discovered by a Mathematician

Hermann Weyl made a step towards discovering a "coincidence" of large numbers while working on his unified field theory (Weyl 1918). This takes us from observational astronomy to theoretical physics, or even to "purely" theoretical physics, as Weyl's theory was never developed beyond the state of a "draft theory," a mathematical construct that could not be tested by experiment. (This did not render it fruitless, for the notion of gauge symmetry, a key ingredient of contemporary physics, was first suggested by Weyl's theory; see Vizgin 1994.)

The notion of spatial-temporal standards, or scales, served as the starting point in Weyl's theory. In Einstein's general relativity, which Weyl sought to generalize, the space-time metric

$$ds^2 = g_{ik}dx^i dx^k$$

is brought into correspondence with experiment by means of definite standards of length associated with solid bodies or light signals. This makes it possible to compare the lengths of metric intervals measured at different points of space-time.

Weyl suggested that, in accordance with the principle of locality, the lengths should be strictly comparable only at individual space-time points. This led him to a geometry that, while generalizing Riemannian geometry, described the properties of space-time, not only by ten components of the metric tensor $g_{ik}(x)$, but also by four quantities $\varphi_i(x)$ that could play the role of the vector potential of the electromagnetic field.

Besides arbitrary coordinates, Weyl's theory also involved the so-called gauge, or scale, transformation

$$g'_{ik} = \lambda g_{ik}, \quad \varphi'_i = \varphi_i + \partial_i \lambda,$$

where $\lambda(x)$ is an arbitrary function of the space-time coordinates. This transformation was interpreted as a change of scale or the measuring standard (which, according to this theory, should be chosen at each point of space-time), since $ds'^2 = \lambda ds^2$.

Invariance under scale transformations was thus the pivoting point of Weyl's theory. This led Weyl to field equations that cannot be transformed into Einstein's equations in the limit $\varphi_i(x) = 0$ (the Lagrangian $R\sqrt{g}$, which leads to Einstein's equations, is not scale invariant; therefore Weyl adopted, as his Lagrangian, the square of the curvature, $R^i_{jkl}R^{jkl}_i\sqrt{g}$). To defend this feature of the theory, Weyl noted in 1918:

> [I]t is highly improbable that Einstein's equations for the gravitational field are strictly valid. First and foremost, it is improbable because the gravitational constant is out of place among other natural constants, so that the gravitational radii of the charge and the mass of the electron have values of quite different orders of magnitude than, for example, the radius of the electron itself (they are smaller than the latter—the first by 10^{20} and the second by 10^{40}). (Weyl 1918, p. 476)

Here Weyl repeats his own remark from an article written in 1917 entirely from the point of view of GR. Having found a solution for an electrically charged point mass (also known as the Reisner–Nordström solution), Weyl introduced the lengths $r_{gm} = Gm/c^2$ (gravitational radius of mass m) and $r_{ge} = e\sqrt{G}/c^2$ (gravitational radius of the electric charge e). Comparing these lengths with the "electron radius," $r_e = e^2/mc^2$, he noted that, for the electron, $r_{ge}/r_e = 10^{20}$ and that this solution could hardly be used to understand the physics of the atom, since the gravitational field's influence would be important, in the case of the electron, only at a distance of the order $r_{ge} \approx 10^{-33}$ cm (Weyl 1917, p. 145). In 1918, Weyl also mentioned that "in the most general case, for a world with [the curvature scalar] $R \neq 0$, one can obtain, by appropriately choosing an arbitrary unit, $R = \text{const} = \pm 1$" (Weyl 1918, p. 475).

It seems that further elaboration of these findings led Weyl, for the first time, to turn to the coincidence of large numbers in his article of 1919.

The lack of a physical interpretation of the procedure of "choosing an arbitrary unit of length" was, from the physical point of view, the Achilles' heel of Weyl's theory. (It was precisely this point that prompted Einstein's critical comment, which, however, failed to convince Weyl.) Physics had always attached importance to the notion of measurement standards, and the standards and scales employed had invariably been presupposed to be as stable as possible. Physics had not recognized arbitrary spatio-temporal changes of scale. True, there was a certain arbitrariness in choosing the standards, but it had a discrete, or global, character. Having introduced an arbitrariness in calibrating the scales and being convinced that this represented a true realization of the principle of locality (as distinct from Einstein's approach, which was "half-hearted and inconsistent"), Weyl had to eliminate this arbitrariness one way or the other, so that his theory could be experimentally tested.

Weyl failed to find a convincing solution to this problem, but in his search for it he stumbled upon the values of length that could be regarded as fundamental characteristic scales. In 1918, he returned to comparing the quantities r_{ge}, r_{gm}, and r_e, his "official" goal being to demonstrate that Einstein's theory of gravitation could not serve as the basis of atomic physics.

In his 1919 paper, Weyl suggested a new variant of a unified theory that generalized GR and was based on a geometry that he introduced (i.e., the *Weyl geometry*). In this new theory, the gravitational Lagrangian remained the square of the curvature. The Maxwellian part of it, however, was no longer emerging in a natural way but was added "by hand." Through artful manipulations, Weyl managed to bring this Lagrangian into a quasi-Einsteinian form. This led him to a variational principle that included the Einsteinian component (with the cosmological term), the Maxwellian component, and a non-Maxwellian component, the square of the vector potential $\varphi_i \varphi^i$ ("the simplest expression found in the Mie theory" (Weyl 1919, p. 122), a theory that claimed to provide a unified field description of the electron and the electromagnetic field).

This "spoiled" both the Maxwellian and Einsteinian equations in the new version of the theory. However, according to Weyl, there was no contradiction with experiments (having to do, in the first place, with electromagnetism), since the non-Maxwellian term entered the equations with a very small factor of the order $1/R^2$, where R is the radius of the universe. (It seems that Weyl assumed a connection between the value of the cosmological constant and the radius of the universe, $\Lambda \approx 1/R^2$, figuring in Einstein's cosmological model.) By following this path, Weyl arrived at a unit of charge, the gravitational radius of which, $G^{1/2}e/c^2$, has the same order of magnitude as the radius of the universe. Noting that, in his work, the unit of electricity and the unit of action both have "cosmic values," Weyl emphasized: "*The 'cosmological' term that Einstein first added to his theory is a natural consequence of our original principles*" (Weyl 1919, p. 124, italic in the original).

In this way, Weyl clearly demonstrated, in 1919, his cosmological tendency, which had been absent in his previous works, and supplemented the microscopic quantities r_e, r_{ge}, and r_{gm} with a megascopic quantity, the radius of the universe R. It is in this paper, in discussing the "problem of matter," that Weyl, probably for the first time, pointed to

the fact that, for the electron, there are dimensionless numbers the order of magnitude of which differs from unity by a great degree. Such is the relation between the radius of an electron and the gravitational radius of its mass, which is equal to a magnitude of the 10^{40} order. The ratio of the electron radius to the radius of the universe may be of the same order. (Weyl 1919, p. 129)

What he has in mind here is the relation

$$\frac{r_e}{r_{gm}} \left(= \frac{e^2/mc^2}{Gm/c^2} = \frac{e^2}{Gm^2} \right) \approx \frac{R}{r_e}, \qquad (11)$$

that is, the "coincidence of large numbers." Weyl's essay does not mention any specific numerical value for R, nor is there any reference to works that mention it. But Equation (11) suggests that Weyl adopted de Sitter's estimation $R \approx 10^{27}$ cm. In 1923, in the same context, Weyl put the radius of the universe at about 10^9 light years ($\approx 10^{27}$cm) (Weyl 1923, p. 323).

It is hard to tell if it was actually Weyl who, in that article (Weyl 1919), discovered, for the *first time*, the "coincidence of large numbers." The apropos tone of his remark suggests that this fact could already have been known to Weyl and was referred to as something curious. In any case, this fact could not originate before de Sitter made his astronomical estimates of R in 1917.

Weyl's 1919 paper may produce an impression that relation (11) *follows* from his theory. However, neither at that time, nor later, did his theory reach the stage at which it could be compared with observational data. Relation (11) was just *consistent* with Weyl's ideas. Indeed, some of his claims seem dubious; for example, the specific value for the radius of the universe R adopted in (11) was obtained by de Sitter in the framework of Einstein's cosmology and it was unclear whether this result could be imported into Weyl's theory.

Weyl's transition from his original Lagrangian, which was quadratic in the curvature and produced a field equation of the *fourth* order, to the *second*-order equations and his observations regarding a possible relationship between the electron and the universe are inferences that appear even more precarious. From a mathematical point of view, Weyl's manipulations were never properly justified (see Bergmann 1942, Pauli 1958). The most important unresolved problem was to effect the correspondence between the new theory and Newtonian gravitation, and to explain how "half" of the initial conditions (the second and third derivatives at time zero) in the Newtonian limit could be ignored. This explains why Dirac, who turned to Weyl's geometry fifty years later, avoided this difficulty by introducing an additional scalar field; he used it to construct a scale-invariant Lagrangian that was linear in the curvature and that led, consequently, to equations of the second order (Dirac 1973).

Despite the circumstances mentioned above, Equation (11) reinforced by Weyl's ideas was perceived as suggesting a unified approach to both micro-physics and cosmology. A. S. Eddington, the most enthusiastic adherent of this approach, believed that the "coincidence of large numbers" did follow from Weyl's theory. In his *Space, Time and Gravitation*, the introduction to which is dated May 1920, he described the relationship between electricity and gravitation wholly in terms of Weyl's unified theory:

> [Weyl's] theory suggests a mode of attacking the problems of how the electric charge of an electron is held together; at least it gives an explanation of why the gravitational force is so extremely weak compared with the electric force. It will be remembered that associated with the mass of the sun is a certain length, called the gravitational radius, which is equal to 1.5 kilometers. In the same way the gravitational mass or radius of an electron is 7×10^{-56} cms. Its electrical properties are similarly associated with a length 2×10^{-13} cms., which is called the electrical radius. The latter is generally supposed to correspond to the electron's actual dimensions. The theory suggests that the ratio of the gravitational to the electrical radius, 3×10^{42}, ought to be of the same order as the ratio of the latter to the radius of curvature of the world. This would require the radius of space to be of the order 6×10^{29} cms., or 2×10^{11} parsecs, which though somewhat larger than the provisional estimates made by de Sitter, is within the realm of possibility. (Eddington 1920, pp. 178–179)

Later, when it became clear that Weyl's ideas failed to produce a viable theory, Eddington turned to other theoretical constructs. Still, his central motivation had always been the search for a unity of micro- and mega-physics. Equation (11) frequently figured in his books and articles and was probably the least speculative of his arguments.

The meaning of R in relationships such as (11) changed radically with the emergence of Friedmann's and Lemaître's evolutionary models and especially after Hubble's discovery of the redshift law in 1929. Redshifts were immediately interpreted as a result of the universe's expansion. Consequently, R (the radius of the universe) could no longer be firmly associated with a closed cosmological model, although "aesthetic" preferences for this model survived for a long time. The value of the Hubble constant H obtained from observations determines the characteristic cosmological distance $R = c/H$, although a precise geometrical meaning of this quantity can only be specified within a particular cosmological model.

The "coincidence of large numbers" retained its importance, since the Hubble radius, $R = c/H$, proved to be close to the Einsteinian radius of the universe estimated by de Sitter. But the meaning of R changed: it was now a function of time $R(t)$, not a constant of nature. This undermined the belief in an underlying connection between the mega- and microscopic parameters. As a result, the "coincidence of large numbers" became an empirical relation and ceased to depend on purely theoretical speculations. At the same time, it became even more of an enigma. Indeed, two large numbers coincide, one of which, according to the theory, is constant (e^2/Gm^2), while the other depends on time ($R(t)/r$)

$$Q_1 = Q_2(t). \qquad (12)$$

Two totally different approaches were suggested for interpreting the "coincidence of large numbers" understood in this way. The first, initiated by Dirac (1937), attempted to find a new physical theory in which the value of $Q_1 = e^2/Gm^2$ would be dependent on time $Q_1 = Q_1(t)$, so that the equality $Q_1(t) = Q_2(t)$ would always hold. Since Dirac associated the dependence of Q_1 on time with the gravitational constant, $G = G(t)$, one would expect a new gravitational theory to be closely connected with cosmology, as the gravitational constant became bound up with the cosmological parameters H and ρ. Although this project led to some further interesting physical ideas, by itself, it failed to produce a viable physical theory.

The second interpretation of coincidence (12), which emerged two decades after the first one, does not presuppose the variation of physical constants with time. Relation (12) is regarded as an equation that determines a certain moment of time, or, more exactly, a certain epoch, namely the present cosmological epoch. The manner of this determination resurrects (in a new form) the anthropocentric approach to the universe and leads to challenging questions.

5. New Anthropocentrism in Cosmology

The anthropocentric approach to the "coincidence of large numbers" originated in two papers published by Robert Dicke in 1957 and 1961. Dicke's theoretical ideas were far from consistent. He expressed doubts that GR had a reliable experimental basis (Dicke 1957a). He believed that the equivalence principle was invalid for "weak interactions" (he used the term to refer both to the gravitational force and to Fermi's concept of weak interaction). Dicke suggested, as a consequence of his approach, that the constants of these interactions depended on time and space. He tried to formulate a new theory of gravitation (Dicke 1957b) that he regarded as a manifestation of electromagnetism with variable permeability of the vacuum.

Dicke thought that the "coincidence of large numbers" was sure evidence that the gravitational constant changed with time. He offered, in passing, his explanation for the large values of coinciding numbers. Being complex physical structures, human observers could not have evolved within the time characteristic of the atomic scale. The epoch of man should be described by a time that is both large and random with respect to that scale (see Dicke 1957a, p. 356). In discussing the variability of physical constants, Dicke insisted that "the age of the Universe 'today' is not fortuitous. It is biologically determined." He argued that, if the fine-structure constant had been much less or much larger than its contemporary value, the stars (the luminosity of which is highly sensitive to the value of this constant) would still have been, or would already have been, so cold that human existence would have been impossible (Dicke 1957b, p. 375).

In 1961, Dicke put forward a more detailed anthropocentric explanation of the "coincidence of large numbers": Since the elements heavier than hydrogen were indispensable to life ("to create physicists we need carbon") and were formed inside the stars, the epoch of humankind could be determined by the life span of main-sequence stars. This life span could approximately be expressed in terms of the physical constants

$$ T \approx \left(\frac{m_p}{m}\right)^{5/2} \left(\frac{e^2}{\hbar c}\right)^3 \left(\frac{Gm_p^2}{\hbar c}\right)^{-1} \left(\frac{\hbar}{m_p c^2}\right), $$

which, with the accuracy of $(e^2/\hbar c)\,(m/m_p)^{1/2} \approx 10^{-4}$, represents the coincidence of large numbers (Dicke 1961).

Whereas the anthropic approach to the coincidence of large numbers is regarded today as an alternative to the hypothesis of the variability of physical constants, Dicke never separated these two ideas. Furthermore, at the very same time (i.e., around 1961), he was working on the scalar-tensor theory of gravitation in which the gravitational "constant" was a scalar field.

It is tempting to trace the currently popular anthropocentrism in physics and cosmology to Boltzmann's fluctuation cosmological hypothesis. Back in the 1930s, Matvei Bronstein and Lev Landau offered a distinctively anthropic description (in order to refute it), according to which large cosmological fluctuations were a necessary prerequisite of the existence of observers, that is, a *sine qua non* of human existence (see Gorelik and Frenkel 1994, §5.5).

Despite all the arguments that endow the physical properties of the universe with anthropocentric interpretations, the anthropic principle belongs today to metaphysics (in the original sense of the term), rather than to physics. The principle adds a genetic connection to the link between the observer and the object of observation. If the object is the universe itself, it should be possible for the observer to emerge within it; that is to say, "our universe is what it is because we were able to appear in it."

This should not be taken to mean that the anthropic principle has no place in physics. It provides an independent reason to postulate a super-universe consisting of numerous separate domains. Our universe, which proved itself capable of creating human observers, is one of them (see, e.g., Okun 1991, 1996).

6. Conclusion. A Mathematician in Physics

Mathematician Hermann Weyl found himself a place in the history of fundamental theoretical physics in the twentieth century. His interest in physical problems, his concrete results, fruitful ideas, and outstanding books on general relativity and quantum mechanics set him apart from the mathematicians of his generation. One might say that, in this respect, he inherited the role of Henri Poincaré.

The difference between mathematics and physics, despite their close and mutually beneficial interaction, is indeed great. What Weyl wrote about our knowledge of the physical world was a graphic illustration of a *mathematician's view* of such knowledge.

One of the earliest images of science's omnipotence—"Give me a firm point, and I will turn the world around"—invites one to view mathematics as the lever and to saddle theoretical physics with the task of looking for the Archimedean point. The search has to proceed in a realm that cannot, in principle, be mathematized, the realm of physical reality. It demands considerable physical intuition, possible only in those who have extensive experience of this reality. A mathematician who spent his life creating various levers will find physical reality—with its apparently chaotic collection of phenomena, facts, properties, and its mathematical eclecticism—to be rather alien. This impression persists even if the mathematician is able to find the notion of a lever that can be used to "turn the world around."

This alienation and resulting frustration can be easily perceived in a dialogue of two apostles of relativism recounted by Weyl (1924). The ideas of Mach discussed in the dialogue were instrumental in formulating the general theory of relativity. So far, however, no one has managed to transform Mach's principle from a verbal into a mathematical form. Cases in which an illuminating idea burns to the ashes are not infrequent in the history of physics: the light from burning often helps one to make some steps forward, which would otherwise have to be made in complete darkness. In the history of mathematics, more often than not, it is fortuitous and unnecessary parts of the original idea that are destroyed by the cleansing fire; a mathematical structure of beauty comes out of it to take its proper place.

Several years divide Weyl's two papers, (1924) and (1931). In fact, they belong to two different epochs in theoretical physics. In 1924, the hope of creating a unified geometrical field theory ran high. It was stimulated by Weyl's own work, which resulted in the first such theory suggested in 1918. By the early 1930s, the unsuccessful program of unified theories had withered under the impact of quantum theory. Nobody was more resolute in proclaiming the end of the epoch than Weyl himself when, in 1931, he called the project of unified theories "geometrical trinkets," allowing the community of physicists to choose a proper funeral rite. One can even suspect Weyl of nursing a grudge against physics for its rejection of the goal to which he had pointed, namely, the idea of locality, more consistently than it was done in Einstein's general relativity.

Cosmology was another significant element in Weyl's papers of 1924 and 1931. Both give one a good sense of the highly speculative nature of relativistic cosmology in the pre-Friedmann era. The earlier of the two papers should be dated to this stage as well, since nobody, including Einstein, paid any attention to Friedmann's achievements. It is much more amazing that, in 1931, Weyl still failed to notice a new stage in the development of relativistic cosmology, after Lemaître had published his works and Hubble had made his discovery, and when the expansion of the universe had already become an observational fact. As a mathematician, Weyl could afford to ignore developments in astrophysics. Yet, there was a more profound reason for his neglect.

It was Weyl the mathematician who tried, in 1919, to connect his unified theory with physical reality. He was the first to pay attention to the empirical fact of the "coincidence of large numbers," which he regarded as providing a connection between cosmology and micro-physics. The manner in which Weyl argued for this connection lacked mathematical (and even physical) rigor; he was driven by his desire to implement the unification project. He abandoned the project, yet the result he obtained survived and acquired a life of its own. The majestic idea about possible links between cosmology and micro-physics fascinated Eddington and drew Dirac far away from physics. No wonder that Weyl, the philosopher and mathematician who had given birth to this idea, was too much involved with it to pay attention to a non-static cosmology that was not part of his unified picture of the world.

Non-static cosmology is, in fact, absent from Weyl's later book (Weyl 1949). This can hardly be explained by the empirically unsatisfactory state of cosmology at that time (Hubble's estimate of the age of the universe was in conflict with geo- and astrophysical data). More likely than not, this absence can be explained by Weyl's general philosophical and mathematical approach to the physical representation of the world. The coincidence of large numbers, discredited by Eddington's ambitious yet barren constructs, received a totally different treatment in Weyl's works. He denied that gravitation was a fundamental property and tried to explain it as a certain kind of statistical effect of a large number of particles in the universe, in line with Mach's principle. This attempt remains nothing more than a fact of Weyl's biography.

There is no need to regret this, however. Under more or less similar circumstances, Boltzmann had the following to say (about his fluctuation cosmology):

> Certainly no one will think that this and similar [cosmological] speculations are important discoveries. No one will agree with the ancient philosophers that they are science's ultimate aim. Yet, does one have the right to ridicule them as totally devoid of any significance? Probably they are extending our horizons, making our thinking more flexible, and contributing to our knowledge of reality. (Boltzmann 1898, p. 90).

REFERENCES

Bergmann, Peter G. (1942). *Introduction to the Theory of Relativity.* New York: Prentice-Hall.

Boltzmann, Ludwig (1898). *Vorlesungen über Gastheorie.* Part 2, *Theorie Van der Waals'; Gase mit zusammengesetzten Molekülen; Gasdissociation; Schlussbemerkungen.* Leipzig: Johann Ambrosius Barth.

De Sitter, Willem (1917). "On Einstein's Theory of Gravitation and Its Astronomical Consequences." *Royal Astronomical Society. Monthly Notices* 78: 3–28.

Dicke, Robert (1957a). "Principle of Equivalence and the Weak Interactions." *Reviews of Modern Physics* 29: 355–362.

— (1957b). "Gravitation without a Principle of Equivalence." *Reviews of Modern Physics* 29: 363–376.

— (1961). "Dirac's Cosmology and Mach's Principle." *Nature* 192: 440–441.

Dirac, Paul (1937). "The Cosmological Constants." *Nature* 139: 323.

— (1973). "Long-Range Forces and Broken Symmetries." *Proceedings of the Royal Society of London* 333A: 403.

Eddington, Arthur Stanley (1920). *Space, Time and Gravitation: An Outline of the General Relativity Theory.* Cambridge: Cambridge University Press.

Einstein, Albert (1917). "Kosmologische Betrachtungen zur allgemeinen Relativitätstheorie." *Königlich Preussiche Akademie der Wissenschaften* (Berlin). *Sitzungsberichte*: 142–152.

Gorelik, Gennady Efimovich, and Frenkel, Viktor Yakovlevich (1994). *Matvei Petrovich Bronstein and Soviet Theoretical Physics in the Thirties.* Boston: Birkhäuser.

Okun, Lev Borisovich (1991). "Fundamental Physical Constants." *Uspekhi fizicheskikh nauk* 161: 177–194.

— (1996). "Fundamental Constants of Nature." Preprint, LANL, hep-ph 9612249.

Pauli, Wolfgang (1958). *Theory of Relativity.* G. Field, trans. New York: Pergamon Press.

Schwarzchild, Karl (1900). "Über das zulässige Krümmungsmaass des Raumes." *Vierteljahrsschrift der Astronomischen Gesellschaft* 35: 337–347.

Vizgin, Vladimir (1994). *Unified Field Theories in the First Third of the 20th Century.* Basel and Boston: Birkhäuser.

Weyl, Hermann (1917). "Zur Gravitationstheorie." *Annalen der Physik* 54: 117–145.

— (1918). "Gravitation und Elekrizität." *Königlich Preussische Akademie der Wissenschaften* (Berlin). *Sitzungsberichte:* 465–478.

— (1919). "Eine neue Erweiterung der Relativitätstheorie." *Annalen der Physik* 59: 101–133.

— (1923). *Raum-Zeit-Materie. Vorlesungen über allgemeine Relativitätstheorie,* 5th ed. Berlin: Julius Springer.

— (1924). "Massenträgheit und Kosmos. Ein Dialog." *Die Naturwissenschaften* 12: 197–204.

— (1931). "Geometrie und Physik." *Die Naturwissenschaften* 19: 49–58.

— (1949). *Philosophy of Mathematics and Natural Science.* Olaf Helmer, trans. Princeton, New Jersey: Princeton University Press.

Zel'manov, A. L. (1962). "Kosmologiya" ["Cosmology"]. In *Fizicheskiy entsiklopedicheskiy slovar'* [*Encyclopedia of Physics*]. Moscow: Sovetskaya Entsiklopediya 2: 491–501.

Laws of Physics and the Universe

Yuri Balashov

1. Introduction

Are the laws of nature real? Do they belong to the world or merely reflect the way we speak about it? And if they are real, what sort of entity are they?

These questions have been intensely debated by philosophers. Modern cosmology, however, has given such questions a new twist by introducing a unique perspective on physical reality, the perspective which I shall call the *cosmological point of view*. In this perspective, the universe as a whole presents itself as a single individual entity that undergoes a radical change with time. Laws of physics, on the other hand, have both local and global significance. They characterize how things behave locally. But they also characterize the entire universe. This suggests an interesting connection between the universe as a whole and what laws of physics hold in this universe. From the cosmological point of view, these two totalities, the laws of physics and the universe, may be related. But how exactly? Are the laws "inscribed" in the fabric of the universe or do they in some sense "precede" it in the order of being? If the latter, what is a "medium," over and above the physical universe, in which physical laws are "written"? If the former, are they but a consequence of the universe's very existence? And if so, how could the laws of physics survive the dramatic change the physical state of the universe underwent in the course of time?

In this paper, I argue that questions of this sort have played a significant role in the history of twentieth-century cosmology. They were, in particular, critically involved in the battle between the big bang and steady-state theories in 1948–65. As is well known, the steady-state cosmological model lost this battle. But a concern of the proponents of that model about the status of physical laws in a changing universe has survived the model itself. To set a case study in the steady-state theory in a relevant context, let

Yuri Balashov and Vladimir Vizgin, eds., *Einstein Studies in Russia.*
Einstein Studies, Vol. 10. Boston, Basel, Birkhäuser, pp. 107–148.

me first indicate what sort of implications the cosmological perspective on laws may have in contemporary evolutionary cosmology.

2. Cosmology and History: Laws of Nature in the Evolutionary Perspective

The concept of physical law has traditionally been regarded as an essentially *atemporal* notion. The very idea of lawfulness seems to presuppose the independence of the *nomic* (lawful, pertaining to laws) characteristics of objects and systems of time. All physics, however, is performed on a stage that, taken as a whole, undergoes change. The time of cosmology is, in fact, the time of *history*. Because of its global character, cosmological evolution comprises everything that exists. And if "everything that exists" involves physical laws, should they be *a priori* excluded from the evolutionary perspective?

The idea that laws of nature may not represent absolutely immutable aspects of reality but undergo change is highly controversial. Yet it has been entertained at one time or other by an appreciable number of scientists and philosophers, although few of them have gone beyond mere conjecture and bothered to give a sufficiently clear account of what they meant by evolving laws. Among the advocates of the mutability of laws, one finds such diverse thinkers as Charles Sanders Peirce and Alfred North Whitehead. They were led to the idea by rather different considerations, but their common ground is found, as one can expect, in the broadly evolutionary worldview, which has made its way into various scientific disciplines in the last two centuries.

Peirce developed his views on the matter in a number of works written towards the end of the nineteenth century. His main concern in thinking about laws and lawhood was primarily epistemological. The idea that laws may evolve was a direct consequence of his belief that laws and regularities in nature require special explanation and that no explanation of laws is possible except a historical one:

> The only possible way of accounting for the laws of nature and for uniformity in general is to suppose them results of evolution. This supposes them not to be absolute, not to be obeyed precisely. It makes an element of indeterminacy, spontaneity, or absolute chance in nature. . . . Law ought more than anything else to be supposed a result of evolution. (Peirce 1956, pp. 162–163)

Whitehead, on the other hand, based his reflections on ontological rather than epistemological grounds. In his discussion of this topic in *Adventures of Ideas*, Whitehead (1933, p. 142) distinguished among what he believed to be the four prevalent contemporary doctrines concerning the laws of nature: the "doctrine of law as immanent," the "doctrine of law as imposed," the "doctrine of law as observed order of succession," and the "doctrine of law as conventional interpretation." Although Whitehead did not make it clear in *Adventures of Ideas*, his sympathies rested with the "doctrine of law as immanent," which is consistent with the all-embracing evolutionary view of nature, *process philosophy*, created and developed by him during the later stages of his philosophical career. Immanence of laws means, in this doctrine, their inherence in real properties and mutual relations of things. Common patterns of such relations exhibit themselves as laws of nature. One consequence of such an interpretation is that,

> since the laws of nature depend on the individual characters of the things constituting nature, as the things change, then correspondingly the laws will change. Thus the modern evolutionary view of the physical universe should conceive of the laws of nature as evolving concurrently with the things constituting the environment. Thus the conception of the Universe as evolving subject to fixed eternal laws regulating all behaviour should be abandoned. (Whitehead 1933, p. 143)

Organic and even social analogies played a major role in Whitehead's view of physical nature. As he observed in *Science and the Modern World*, "the laws of physics are the laws declaring how the entities mutually react among themselves" (Whitehead 1967, p. 106). Relations among fundamental physical entities are partially determined by their "environment," and if the latter undergoes a drastic enough modification, it is only natural to suppose that properties and relations of basic constituents and, hence, the laws of nature may also change.

> The assumption that no modification of these laws is to be looked for in environments, which have been observed to hold, is very unsafe. The physical entities may be modified in very essential ways, so far as these laws are concerned. . . . According to this theory the evolution of laws of nature is concurrent with the evolution of enduring pattern. For the general state of the universe, as it now is, partly determines the very essence of the entities whose modes of functioning these laws express. The general principle is that in a new environment there is an evolution of the old entities into new forms. (Whitehead 1967, pp. 106–107)

It should be noted that neither Peirce nor Whitehead, in their philosophical reflections on physical laws, referred to any particular scientific evidence of the day or explained how the idea of the evolution of laws could square with the available evidence favoring their essential immutability. This is not to say that these great philosophers were not familiar with the state of contemporary science. They most certainly were, and in the case of Whitehead, one can speak of a rather close, indeed first-hand, acquaintance with the most recent physical theories. The point is rather that Peirce's and Whitehead's conjectures concerning the evolution of natural laws were rooted in their respective *philosophical* cosmologies, not so much in their particular scientific beliefs.

Philosophical cosmologies of the evolutionary kind developed at the end of the nineteenth and the beginning of the twentieth centuries by such philosophers as Peirce, Henri Bergson, Whitehead, Samuel Alexander, Pierre Teilhard de Chardin, Jan Smuts and others were, however, themselves inspired, to a considerable extent, by contemporary scientific developments and cannot be adequately understood in abstraction from them. The philosophical style characteristic of the above mentioned thinkers more easily gives rise to broad extrapolation than rigorous scientific reasoning, which is always constrained by concrete empirical evidence. In this particular case, the organic model of development was taken to be the paradigm of any evolutionary process, including the realm of non-living physical matter. Nothing in the physics of the day suggested that such an extrapolation should be taken seriously. Subsequent progress of physical sciences, too, did not give much support to a global evolutionist perspective on physics. It was not until the 1970s, when the stage was set for a remarkable synthesis of fundamental physics and cosmology of the early universe, that physics began to assume a truly historical dimension. The idea that the temporal career of the universe may include not only the history of matter but also the history of its basic properties, which figure in the laws, is largely a product of this interplay between particle theory and cosmology in their joint effort to probe the physics of the very early universe.

This physics now includes symmetry breaking phase transitions in unified gauge theories, the transitions apparently changing the nomic properties of certain elementary particles, for example, the masses of the intermediate vector bosons responsible for weak interactions. Cosmological considerations suggest that such transitions may have occurred in the real history of the universe, as it cooled down from its initial hot state. We live in a low-energy epoch ($T \approx 3K$) with a broken symmetry between electromagnetic and weak interactions. Because of this, the photon is massless whereas the masses of $W^\pm$ and Z^0 bosons are not zero. But one has only to warm up the cosmic substratum to approximately 10^2 GeV (i.e., 10^{15} K), and the broken symmetry will be restored, and the masses of all bosons will vanish. In the history of the universe, the inverse process of separation of the electromagnetic and weak forces may have occurred at about 10^{-10} s after the big bang, and a similar process of separation of the strong force may have been operative at a still earlier epoch ($t \approx 10^{-35}$ s) corresponding to the energy 10^{14} GeV (i.e., about 10^{27} K). The scenario features, in Steven Weinberg's words, a "parallel between the history of the universe and its logical structure" (Weinberg 1977, p. 149). This expression is a bit misleading, for the universe does not have a "logical" structure. What Weinberg has in mind is its *nomic* structure. The idea is that if the material structure of the universe undergoes dramatic enough change, its nomic structure may not escape being influenced by this change.

Some physicists recognized the importance of a historical perspective on physics quite early in the process. Thus at the celebration of the hundredth anniversary of MIT in 1961, Richard Feynman conjectured that physics may develop a historical outlook and become deeply concerned with the "studies of astronomical history and cosmology": "There is at least the possibility that the laws of physics change with time, and if [so] . . . it is very likely that physics is enwrapped in the cosmological problem" (the quotation kindly provided by Sam Schweber, personal communication).

Feynman reiterated the same conjecture in his famous lectures:

> There is another *kind* of problem in the sister sciences [i.e., in biology, geology, and astronomy] which does not exist in physics; we might call it, for lack of a better term, the historical question. How did it get that way? . . . There is no historical question being studied in physics at the present time. We do not have a question, "Here are the laws of physics, how did they get that way?" We do not imagine, at the moment, that the laws of physics are somehow changing with time, that they were different in the past than they are at present. Of course, they *may* be, and the moment we find they are, the historical question of physics will be wrapped up with the rest of the history of the universe, and then the physicist will be talking about the same problems of astronomers, geologists, and biologists. (Feynman et al. 1963, pp. 3–9)

Biological analogies were used by other authors to illustrate the basic idea behind the historical view on physics. Yoichiro Nambu put the matter thus:

> In a more serious vein, one could ask whether the laws of physics are intimately bound up with the evolution of the universe, influenced not only by the initial conditions, but also by the subsequent evolutionary processes

themselves. In a way I am suggesting biological evolution as a possible model for physical evolution.

One would have to be more specific, however, so let me entertain the idea that the term "generation" means more than just an analogy. Is it at all possible that the generations of quarks and leptons have "evolved" one after another in some sense, that each generation is "born," so to speak, at the corresponding energy (or length) scale of an expanding universe, its properties being influenced, but not necessarily deterministically fixed, by what already exists?

Biological evolution is made possible by the vast degrees of freedom residing in complex molecules. If translated to particle physics, this might again bring back the compositeness issue. Are lower mass generations more complex than the higher ones? This is hardly likely although the opposite might be true. So what I should mean would be that the constants like mass are really dynamical quantities that were selected, with some degrees of chanciness, from among other possibilities in the course of the universal evolution. (Nambu 1985, pp. 108–109)

Walter Thirring recently took a similar position with regard to the laws of physics and their hierarchical order. The laws of an upper level, he maintained, may not be completely determined by the laws of a lower level and may, in a sense, present a "purely accidental fact" when looked at from the lower level, because the upper-level laws depend not only on the lower-level ones, but also on the particular historical circumstances. In this way, "the hierarchy of laws has evolved together with the evolution of the universe. The newly created laws did not exist at the beginning as laws, but only as possibilities" (Thirring, as quoted in Schweber 1997, p. 185).

The physicist and historian of physics Silvan Schweber takes the fact that "the notion of a natural selection of physical laws is being discussed in the most respectable fora of physics, astrophysics and cosmology" as an indication of the "changing metaphysics of physics" (Schweber 1997, p. 186).[1]

Enthusiastic as these views are, a word of caution is in order, because what is manifestly lacking in them is a serious analysis of the coherence and plausibility of the idea of nomic evolution.[2] But the fact that the idea is making its way into physics is notable. On the other hand, the fact is hardly surprising. If the laws of nature are real, they belong to the furniture of the world. From the cosmological point of view, the holding of particular laws in the universe can be considered as its basic property characterizing the kind of being the universe is. And if most, perhaps even all, of its other global properties undergo change in the process of evolution, should the laws be necessarily regarded as an exception?

One party's *modus ponens*, however, is another party's *modus tollens*. The above reasoning can be turned on its head. If there is, indeed, an interdependence between the structural properties of the universe and physical laws acting in it, then one might claim that the *permanence* of the latter requires, for its justification, the constancy of the former. This idea was instrumental in the emergence and development of the steady-state theory of the expanding universe (SST) put forward in 1948 by three Cambridge physicists, Hermann Bondi, Thomas Gold, and Fred Hoyle, as a principal alternative to the big bang cosmology.

3. Steady-State Cosmology: How a Scientific Failure Can Be Valuable

According to SST, the expanding universe, instead of evolving from the hot big bang, is stationary on the large scale. The dilution of matter due to the cosmic expansion is compensated for by the creation of new matter, and any other global process operative in the universe is regarded as being self-perpetuating. All evolutionary effects are thus merely local and no distinction between past, present and future can be made for the universe at large.

The 1964–65 discovery of the microwave background radiation, soon afterwards identified by the majority of cosmologists as relic of the hot big bang, effected a crushing blow to SST. As early as the 1950s, after recalibration of extragalactic distances, the pressing time-scale problem that afflicted the standard cosmology for more than two decades[3] had been taken off the agenda. The prevalent opinion was that there no longer existed any necessity for SST. One rarely reads about SST any more in textbooks.[4] In this view, SST looks like an awkward accident in the history of modern cosmology.

The case of SST raises an important question of the value of scientific failure, important from the points of view of both historiography and philosophy of science. It should be noted that even in its best days SST had far more opponents than advocates. It was often regarded as a good example of what a scientific theory should *not* be (see, e.g., Bunge 1962). But SST was the real mainspring of cosmology in the 1950s. To be convinced of this, one need only have a cursory look at the cosmological literature of the time. The mere presence of SST forced many astronomers and astrophysicists to invest a considerable amount of effort in observational and theoretical work for the purpose of refuting it. Its advocates, however, fought their case with persistence and ingenuity. It can be seen, in retrospect, how much benefit cosmology on the whole has gained from this controversy. At least two remarkable achievements, the theory of stellar nucleosynthesis and the development of radio astronomy, were directly stimulated by SST. In its initial stage, the idea of stellar nucleosynthesis showed a way to account for the observable abundance of chemical elements without recourse to a hot state in the remote past of the universe. That prompted Hoyle to work on this program. By an irony of fate, this contribution was later to become a part of the rival big bang cosmology. First counts of radio sources, on the contrary, gave strong promise of disproving SST. It is not at all obvious that these and some other achievements would have been made so rapidly if there had not been an SST to be defeated.

It proved, however, to be not so easy to defeat SST. The theory "died" several times, but invariably came back to a new life. This is not surprising. Cosmologically significant astronomical data being scarce and contradictory in those days, the general desire to refute SST quickly often led to hasty conclusions that were abandoned later on. In such circumstances debates moved to essentially theoretical issues. The parties were forced to resort to foundational arguments, and this is what makes the history of SST philosophically interesting.

The competition between SST and standard relativistic models in the 1950s was no less important for molding cosmology, as a scientific discipline with its particular methods, than were the discussions provoked by the ideas of Edward Milne, Arthur Eddington, and Paul Dirac in the 1930s.[5] In both cases disagreement about conceptual issues grew into controversies about foundational principles of science in general. It is still more important for the purpose of the present essay that a concern about the way in which the laws of physics relate to the universe as a whole played a vital role in the origin and development of steady-state cosmology. Various aspects of the relationship between the global properties of the universe and physical laws were explored both by the advocates of SST and by their opponents. Some of the questions they raised in this debate add new dimensions to the general concept of a law of nature.

My reconstruction of the early history of SST focuses on such questions.[6] I begin by recapitulating the two original versions of this theory (Bondi and Gold 1948; Hoyle 1948, 1960). First of all, however, I want to identify their common conceptual precursor. This precursor, I think, is to be found in a certain scientific school, the advocates of which were the first to argue explicitly that the methodology used in a science dealing with large-scale evolutionary processes cannot be independent of the particular resulting pattern of the historical unfolding of our actual world. This tradition is known as *uniformitarianism.*

4. Uniformitarianism as a Methodological Principle

Both cosmology and geology belong to what Whewell dubbed "palaetio-logical" sciences, which are concerned with events that happened in the remote past. The general problem of uniformity is common to all such sciences, though the particular form it takes depends on the context. No satisfactory explanation of the present state of the Earth, or of the universe, can be attained without inquiring into the former's geological, or the latter's cosmological, past. But obviously one has no direct observational access to either. Invoking hypotheses about the past is thus unavoidable in palaetiological sciences. Different schools of thought, however, hold divergent views about what kinds of hypotheses are admissible here.

The basic tenet of the uniformitarian school was that, unless the past of a global system under study is in some important way similar to its present, the freedom involved in hypothesizing about the former is so great that no genuine science of the system at hand is possible. A historically relevant interpretation of the basic uniformitarian requirement admits at least two senses of "similarity" sometimes blended together. In the weak sense, one can require that the *kinds* of processes at work in the geological past of the Earth be the same as at present. Put differently, this weak uniformitarian principle implies the temporal uniformity of natural *laws.* The strong uniformitarian assumption goes further and demands that not only the kinds of processes, but their *intensities* be the same in the past as at present.

It should be noticed that the weak principle of uniformitarianism was shared not only by the historical uniformitarians but also by many of their rivals (Rudwick 1971). It can reasonably be argued that the constancy of laws is an indispensable assumption of the scientific method in general, since no simple generalization of experience is possible without it. The weak uniformitarian thesis was in fact primarily directed against invoking non-scientific causes of past events explicitly violating the laws of nature. By itself, this thesis does not entail a particular, *non-developmental* geological scenario. A directional theory of geology (like that of the gradually cooling Earth) can be compatible with weak uniformitarianism, provided no currently unobservable *kinds* of processes are introduced to account for the present state of affairs. This by no means rules out a possible change in the *strength* with which the processes operate in various epochs.

Some geologists, and most notably James Hutton and Charles Lyell, believed this was not enough to make geology a science. They insisted that the particular *intensities* of processes, and not only the laws of their operations, should be the same throughout the entire geological history of the Earth. This does not exclude small-scale spatial and temporal fluctuations, and the latter were in fact used by Lyell for explaining climatic changes and the details of the fossil record. The overall large-scale picture, however, must be essentially *steady-state*. No systematic evolutionary effect (like the Earth's cooling) was allowed in it.

Strong uniformitarianism and the related steady-state pattern have been refuted by subsequent observations that have proved that some drastic changes did occur in the geological history of our planet. As to weak uniformitarianism, it has, in fact, lost its geological identity and become an implicit methodological presupposition of all natural sciences dealing with evolutionary phenomena. Although it poses some restrictions on theorizing, it is generally held that no particular picture of phenomena follows from weak uniformitarianism. One can easily see why this is the case in geology. The whole geological scene is nothing but a local superstructure over the basic level of the physico-chemical laws. The evolution of the "scene" can proceed against the unchanging background of the underlying laws.

The situation becomes more ambiguous in cosmology where the "scene" is not a local superstructure built above a more fundamental level, but an all-embracing totality coextensive with the realm of the most fundamental physical laws.

5. Cosmology Versus Local Physics

Philosophical considerations were centrally involved in the version of SST presented by Bondi and Gold (henceforth SST-I). Their seminal paper (1948) begins with an extensive methodological introduction. The history of SST is usually traced back to this work. The philosophy of the steady-state project, however, was already contained in the review of cosmology published by Bondi some four months earlier (Bondi 1948). Although there was no mention of the steady-state hypothesis in it, the ground was completely prepared for its introduction in the next issue of *Monthly Notices*. I shall follow both papers in my account of the philosophical foundations of SST-I.

Like his famous predecessor Milne, Bondi argued that, because of the uniqueness of its subject, cosmology is very different from "local physics." Hence, not all procedures employed in the latter may be entirely appropriate for the former. In local situations, one can always distinguish between laws and their particular instances. The laws reflect inherent, unchanging, and reproducible features of phenomena, whereas the laws' instances are normally taken to be accidental, contingent, and, generally speaking, irreproducible. Indeed, it is not possible to reproduce all conditions of a particular local experiment or observation, for a scientist does not have control over the time and place of their occurrence. It is, according to Bondi, a fundamental assumption of physical science that, while the "accidental" characteristics of the phenomena under study can obviously be affected by their temporal and spatial location (as well as by the entire collection of initial and boundary conditions), what is regarded as "inherent," or lawlike, cannot be so affected. Otherwise, no coherent physical explanation of the phenomena and processes could be attained.

For example, in any local branch of mechanics, such as ballistics (Bondi 1948, p. 105), actual motions can be infinitely varied by their initial conditions, including times and places of particular occurrences. But the law according to which the trajectories of all such motions are (approximately) conic sections is supposed to survive all the changing circumstances. Furthermore, it is usually taken for granted that "the law of motion does not only cover all the cases corresponding to the various initial conditions but all these cases are supposed to have a real or potential existence." In this sense, "the law of motion is neither too wide nor too narrow; it covers all existing and possible cases and no others" (Bondi 1948, p. 105).[7]

Does the "ballistic attitude" apply to cosmology? Not at all obviously. "The distinction between impossible and possible, but 'accidentally' not realized states, becomes absurd when we have to deal with something as fundamentally unique as the universe" (Bondi 1948, p. 106). In what way can this "fundamental uniqueness" manifest itself in cosmological theorizing? First of all, it can blur the demarcation line between the laws of nature and their particular instances. The universe is something more than just one particular instance of natural laws; indeed, the "instance" in question is coextensive with the laws themselves. With equal reason may the latter be regarded as a consequence of the universe's very existence.

One corollary of these considerations is this: the demarcation between what is "intrinsic" and what is "accidental," if it can be drawn at all for the universe as a whole, is not bound to coincide with that typical of local situations. In other words, there are reasons to doubt that observations of various features of the universe will tend naturally and automatically to sort themselves out into those pertaining to the "intrinsic" and those relating to the "accidental," as normally happens with observations performed in local physics. It is not so clear in advance where to draw a line separating these classes. Take, for example, two parameters, the constant of gravitation and the Hubble "constant." The former is usually held to be "inherent" and the latter "accidental." A measurement of the Hubble parameter, however, gives as unique a result as the determination of the gravitational constant (cf.: Bondi and Gold 1948, p. 252). And furthermore, there are theories in which the gravitational constant itself becomes "accidental," by virtue of its hypothetical dependence on the cosmological epoch.

The upshot is that as soon as one steps into the cosmological arena, the general arguments supporting a particular division of physical experience into "intrinsic" and "accidental" subcategories are no longer available. One should draw this division anew, based on some additional considerations. Big bang cosmology, as Bondi and Gold noted, chose the most straightforward way, that of a direct extrapolation of concepts, laws, and "demarcation principles" of the local physics to the cosmological scale, in the conviction that no problems will arise with this approach. In particular, a tacit agreement has been made to the effect that the global structure of the universe (say, the density and velocity distribution of matter) has no influence over the local physical laws. But this is not self-evident. One could recall Mach's principle stating that some local physical properties may be subject to the dynamic influence of distant matter in the universe. Any dependence of this sort, argued Bondi and Gold, could impede the appropriate interpretation of observations of distant objects so essential to cosmology.

According to the authors of SST-I, any possibility of such an influence must be precluded from the start. For this, a very special cosmology is needed, one that would postulate equality and indistinguishability of all parts and all stages of the physical history of the cosmos. Any large enough space-time fragment of the universe should be a fair sample of the whole.

By adopting the cosmological principle, the big bang theory made a first but insufficient step in this direction. The cosmological principle postulated the large-scale homogeneity and isotropy of the universe and ensured a uniform description of all parts of the universe at each moment of time, but not for the entire duration of its evolution. Any cosmological theory contemplating local laws in a universe undergoing changes must make, as Bondi later stressed, "definite assumptions about the effect of these changes on the laws of physics. Even the statement that there are no such effects is evidently an assumption, in fact a highly arbitrary assumption" (Bondi 1957, p. 197). Indeed, it is not at all clear that the laws of physics discovered here and now, at a later cosmic epoch, would be suitable for dealing with the early evolutionary stages, as depicted by the big bang cosmology, when the matter of the universe is supposed to have been in a rather different physical state.

One could adopt another strategy and posit a possible explicit dependence of the laws on the changing physical structure of the universe, as was done by Dirac (1937), or (somewhat anachronistically, but to the point) by Carl Brans and Robert Dicke in their scalar-tensor theory of gravity (1961). Generally speaking, many possibilities arise at this point, this freedom being, according to the steady staters, a defect, rather than an advantage, of the received methodology. For, again, there is only one universe, and its evolution is perhaps a unique cosmic event. It would not be unreasonable to require that the scientific picture of this event be also "unique" so as to cover "all existing and possible cases and no others" (Bondi 1948, p. 105).

The most radical way to avoid these problems, as well as arbitrary assumptions concerning possible effects of the changing cosmological environment on the physical laws, is to exclude such effects altogether, by extending the cosmological principle. Bondi and Gold's *perfect cosmological principle* (PCP) requires the large-scale structure of the universe to be not only uniform in space but also constant in time. "We do not claim that this principle must be true," Bondi and Gold observed, "but we say that if it does not hold, one's choice of the variability of the physical laws becomes so wide that cosmology is no longer a science. One can then no longer use laboratory physics without relying on some arbitrary principle for their extrapolation" (Bondi and Gold 1948, p. 255).

The uniformitarian leitmotif is clearly recognizable in this claim. The situation is unlike that in geology, however, as the distinction between weak and strong versions of the uniformitarian assumption disappears in steady-state cosmology. The whole point of Bondi and Gold is that once we let the "intensities" of physical processes in the past of the universe be drastically different from what they are at present, no guarantee can be given for the stability of physical laws themselves across the entire evolutionary track.

The universe is, of course, under no obligation to live up to PCP. But according to Bondi and Gold, no science of cosmology is possible in a universe that does not satisfy this principle. In other words, cosmology is only possible in a steady-state universe. Consequently, we either have to abandon any hope of building a viable cosmology or to explore the opportunities provided by the steady-state picture, as long as its consequences do not conflict with observations. It is only rational to try moving ahead rather than simply standing still, and Bondi and Gold proceeded to derive consequences from PCP.

Before following them in these derivations, I want to reflect a bit more on the idea of the interdependence of local laws and the large-scale structure of the universe presupposed in Bondi and Gold's "prolegomena" to SST.

6. The Origin of Inertia and the "Interaction Principle"

First of all, one has the impression that a shift of meaning occurs throughout the discussion of the issue "cosmology versus local physics" in Bondi 1948, Bondi and Gold 1948, and also in later works (Bondi 1957, 1960). The argument starts with the locally ascertainable distinction between the laws of nature and their particular instances. Such a distinction, then, is supposed to vanish or become blurred with respect to the entire universe. This seems to imply something like a "law of the universe" in the first place. Even if this hypothetical law collapses with its only instance, becoming thus "degenerate," it still has to possess some distinctive features of a law. Otherwise there is no reason to call it by that name. No examples of this type of law are given, though, and one wonders if there can be any. This does not impair the main argument, since the latter essentially hinges on quite a different usage of the term "law," to which the discussion eventually switches.

As a matter of fact, it is the familiar laws of local physics that, according to Bondi and Gold, may be non-uniformly affected by the structure of the universe, unless one adopts the perfect cosmological principle making such an influence uniform and hence imperceptible. "As the physical laws cannot be assumed to be independent of the structure of the universe, and as conversely the structure of the universe depends upon the physical laws, it follows that there may be a stable position," Bondi and Gold remark (1948, p. 254). Clearly they mean the local physical laws acting *in* the universe, and not some hypothetical law *of* the universe. It would be appropriate to assume, then, that in the steady-state universe satisfying PCP, the action of a local law may, in the general case, consist of two components: (a) an intrinsic local "source" and (b) a uniform global cosmological "contribution." Of course, there may be no such contribution at all. But even if there is, PCP guarantees the universal validity of the same local laws in the range of the whole universe and at any moment of time.

But this still leaves it unclear how the presumed *two-way* interaction between the laws of nature and the material content of the universe is to be conceived. Bondi and Gold frequently quote Mach's principle in this connection. This principle is ambiguous and has nearly as many interpretations as there are interpreters (see, e.g., Barbour and Pfister 1995). One particular attempt to conceptualize Mach's principle is worth dwelling upon, however, as it was put forward in 1953 by a young convert to the steady-state cosmology, Dennis Sciama (1953, 1957). He definitely derived inspiration not only from the physics of SST but also from its philosophy and, specifically, from the idea of a possible influence the universe as a whole may have on the local properties of matter.

The original thesis known as "Mach's principle" is often expressed as a requirement to reinterpret the dynamical theory of mechanics in purely relational fashion, so that kinematically equivalent motions be also dynamically equivalent. Since the former can only be defined in terms of irreducible relations among moving bodies, the dynamical properties of all bodies and, in particular, their *inertia* must be understood as arising out of their interaction with the rest of the matter in the universe.

Sciama (1953) suggested a tentative sketch of a theory of gravitation incorporating Mach's principle and based on a reasonable assumption that the total gravitational field induced by the matter of the universe should vanish in the rest-frame of any given body. As a first approximation, Sciama assumed this field to be derivable from a vector potential in Minkowski space. He considered a smoothed-out homogeneous and isotropic model of the universe of density ρ expanding in accordance with Hubble's law $\mathbf{v}_H = H\mathbf{r}$, neglected relativistic effects, and restricted the bulk of the matter inducing gravitation/inertia on a test particle, residing at the origin of the coordinate system, to the spherical volume V_H of radius c/H. A natural state of rest at each point in such a model universe is defined by the isotropic distribution of the redshifts of distant galaxies.

Now suppose the test particle moves with a rectilinear velocity $-\mathbf{v}(t)$ relative to the universe and to another body of mass M, which is at rest with respect to the universe. One straightforward way to describe the dynamics of the test particle is to determine its acceleration in the rest-frame of the universe by means of Newton's laws of motion and gravitation

$$-\frac{d\mathbf{v}}{dt} = -G\frac{M}{r^2}\frac{\mathbf{r}}{r}, \tag{1}$$

where $\mathbf{r}$ is the distance to the other body.

Alternatively, one could work in the rest-frame of the test particle, in which the system consisting of the universe and the body moves with velocity $\mathbf{v}(t)$ and has acceleration $-d\mathbf{v}/dt$. In this reference frame, as in any other, the gravitational field on a particle is, on Sciama's assumption, generated[8] by the scalar and vector potentials of the universe, Φ_U and $\mathbf{A}_U$, and of the body, Φ_M and $\mathbf{A}_M$,

$$\mathbf{E} = -\nabla\Phi - \frac{1}{c}\frac{\partial \mathbf{A}}{\partial t}, \qquad \mathbf{H} = \nabla \times \mathbf{A}, \qquad (2)$$

where $\Phi = \Phi_U + \Phi_M$ and $\mathbf{A} = \mathbf{A}_U + \mathbf{A}_M$. Provided $v/c \ll 1$,

$$\Phi_U \approx -\int_{V_H} \frac{\rho}{|\mathbf{r}|} dV = -\frac{2\pi\rho c^2}{H^2}, \qquad \mathbf{A}_U \approx -\int_{V_H} \frac{v\rho}{c|\mathbf{r}|} dV = -\frac{\Phi}{c}\mathbf{v}(t),$$

$$(3)$$

$$\Phi_M = -\frac{M}{r}, \qquad \mathbf{A}_M = -\frac{\Phi_M}{c}\mathbf{v}.$$

This gives

$$\mathbf{E} = -\frac{M}{r^3}\mathbf{r} - \frac{M}{rc^2}\frac{\partial \mathbf{v}}{\partial t} - \frac{\Phi}{c^2}\frac{\partial \mathbf{v}}{\partial t}, \qquad \mathbf{H} = 0. \qquad (4)$$

The total field $\mathbf{E}$ induced by the system universe + body at the test particle should be zero in the particle's rest frame. This requirement, after some simplifications, is expressed thus:

$$\frac{M}{r^2} = -\frac{\Phi_U + \Phi_M}{c^2}\frac{\partial \mathbf{v}}{\partial t}. \qquad (5)$$

By the relationist assumption, the two descriptions of the situation, (1) and (5), must be equivalent. Hence, provided $\Phi_M \ll \Phi_U$,

$$\frac{2\pi G\rho}{H^2} = 1,$$

or better, given the approximate character of the above derivation of (4),

$$\frac{G\rho}{H^2} \approx 1. \qquad (6)$$

In this relation, the "interaction" principle finds its manifestation: the gravitational constant G, which supposedly represents a *nomic* feature of the universe, is coupled to the density of matter in the universe (ρ) and its global velocity pattern (H), which are, presumably, of a purely "factual," non-nomic nature. It can be shown that the main contribution to G in Sciama's theory comes from very distant matter inaccessible to observation. In this sense, local measurements—in fact, the whole structure of local physics—give us information about the structure of the universe as a whole; the former indeed carry an "imprint" of the latter.

Another consequence of (6) that has a direct implication for steady-state cosmology is that, unless the universe is stationary on the large scale, the gravitational constant must be changing with cosmic time. Thus, because of the influence of the universe on the local nomic properties of matter, the latter would not survive unchanged in a changing universe.

Sciama has also shown that the case of uniform rotation can be treated similarly. His considerations outlined above do not, however, constitute a viable theory of gravity. On the one hand, they are limited to the simplest cases of uniform rotation and rectilinear accelerated motion and incapable of describing, in the same relational fashion, cases of arbitrary motion. For this purpose, a tensor rather than a vector potential is needed. Secondly, a theory should be relativistic. Sciama promised to develop a theory satisfying these desiderata in a subsequent paper, to which he referred as "paper II" throughout his (1953) article. "Paper II," however, was never published. One reason was probably empirical. Equation (6) relates three observable quantities, G, ρ, and H. Given the value of H, as it was estimated at that time, and even the value reduced more than thrice after the recalibration of extra-galactic distances in the late 1950s, the average density of matter in the universe should, according to (6), be of the order $\rho \approx 10^{-27}$ g/cm^3, which is too far off the mark (by a factor of 1000).

Nonetheless, Sciama's work on Mach's principle stimulated further developments of a similar kind,[9] and it was clearly supposed to be an exemplification of the philosophy of the steady-state project. Bondi and Gold, however, refer to Mach's principle as if it were just one manifestation of a *general interaction* principle, the latter equally pertaining to all local physical laws. It remains to be seen what such interaction could mean from a physical point of view and whether it could imply anything more than the existence of a more general law subsuming the hypothetical interaction under a more comprehensive principle that is itself unaffected by the state of the universe as a whole (see Balashov 1992). Here I put these questions aside and return to the exposition of SST-I.

7. The Perfect Cosmological Principle

The entire content of SST-I was expected to be deducible from the perfect cosmological principle (PCP) postulating the large-scale uniformity of the universe in space and time. Bondi and Gold considered it of crucial importance to stress that PCP and the cosmological principle (CP) of the big bang theory (assuming the universe to be uniform in space but changing in time) differed not only in their formulations but also in their status in the corresponding theories. According to Bondi and Gold, CP is no more than an auxiliary hypothesis needed to derive a particular empirically adequate cosmological model from the field equations of general relativity. Should a conflict arise between a model and the astronomical data, CP could well be replaced with a more complicated assumption without abandoning the framework of the big bang theory. In SST-I, on the contrary, PCP was supposed to be an essential element. SST-I and PCP stand or fall together, for, as Bondi and Gold invariably stressed, the scientific value of cosmology derives from the strong uniformitarian assumptions inherent in PCP.

On this view, the "rank" of PCP is higher than that of ordinary physical laws, for the very *raison d'être* of ordinary physical laws hinges on the validity of PCP. As Bondi and Gold noted in this connection,

> we regard the principle as of such fundamental importance that we shall be willing if necessary to reject theoretical extrapolations from experimental results if they conflict with the perfect cosmological principle even if the theories concerned are generally accepted. (Bondi and Gold 1948, p. 255)

The main target here was the principle of conservation of matter and energy. To satisfy PCP in an expanding universe, the creation-of-matter hypothesis must be introduced, in order to keep the density of cosmic matter constant. The rate of creation required for it[10] turns out to be too low to be discoverable by any observational effects. As the universe expands, new matter is created, leading thereby to local evolutionary phenomena, like the formation of new galaxies and stars. There is, however, no large-scale evolution in this picture. Any sufficiently large fragment of space contains objects at all stages of their development. No global feature of the universe, such as the mean density of matter, the integral luminosity or the spectral distribution of radiation, is subject to a systematic temporal change in SST. In other words, any large-scale process operative in the steady-state universe should be self-perpetuating.

This feature of the theory can be illustrated by the mechanism of galaxy formation elaborated by Sciama (1955) in the framework of SST. New galaxies were supposed to form permanently in the wake of the old ones. The latter moving through space served as attractors of intergalactic matter including its newly created fraction. The subsequent separation of the daughter and mother galaxies provided for the continuous rejuvenation of the cosmic population, keeping its average age constant. Unlike the big bang cosmology, SST allowed no unique "catastrophic" event associated with the *original* formation of galaxies in the remote past. Still more importantly, Sciama's mechanism was intended to account for "the *actual* distribution of matter in the universe entirely in terms of the general laws and constants of nature" (Sciama 1955, p. 3). The requirement that the properties of the self-perpetuating system of galaxies be independent of time determines, in Sciama's theory, these properties *uniquely*, without introducing any free parameters. The key point here is the following chain of relations: the mass of a parent galaxy determines the degree of compression of intergalactic matter (including its newly created fraction) which, in turn, determines (through appropriate considerations of gravitational instability) the characteristic mass of a stable daughter configuration. This chain is then closed by the requirement of self-consistency, namely, that the mass of a daughter galaxy determined in this way be equal to that of a parent galaxy. As Sciama claimed, in complete conformity with the philosophy of the steady-state project, "we have here the first example of an actual property of the universe being calculated from general principles, without the intervention of any arbitrary initial conditions" (Sciama 1959, p. 191).

Let us now return to the creation of matter hypothesis. To ensure the steady state of the universe, Bondi and Gold had thus sacrificed the conservation laws. From the cosmological point of view, which Bondi (1957, 1960) later expounded, this was not a deadly sin. Although creation events constitute anomalies contradicting the conservation principle, these anomalies are too small to be manifested in any observable effects. In actuality, the conflict is only between the creation hypothesis and the simplest theoretical generalization (viz., the laws of exact conservation of matter and energy) of multitudinous experimental facts testifying that energy is conserved with great accuracy. Inference from experience to theory should not, however, ignore the cosmological point of view. The cosmological case is not just one instance of local physical laws. The latter have their locus in the expanding universe. Any statement of their form is at the same time a statement of the global properties of the unique physical whole, the universe. The creation process required by PCP implies violation of the *exact* conservation principle thereby substantially reducing its simplicity, but this is "more than counterbalanced by the gain in simplicity" in the resulting cosmological model (Bondi 1957, p. 196).

> When observations indicated that matter was at least very nearly conserved it seemed simplest (and therefore most scientific) to assume that the conservation was absolute. But when a wider field is surveyed then it is seen that this apparently simple assumption leads to the great complications discussed in connexion with the formulation of the perfect cosmological principle. The

principle resulting in greatest overall simplicity is then seen to be not the principle of conservation of matter but the perfect cosmological principle with its consequence of continual creation. From this point of view continual creation is the simplest and hence the most scientific extrapolation from the observations. (Bondi 1960, p. 144)

This argument suggests a highly inductivist interpretation of the fundamental conservation principles of physics. Such an inductivism occurs in a theory that places strong emphasis on the hypothetico-deductivist methodology. Whether or not one is willing to accept this argument, there is nothing impossible or even peculiar in this combination which is entirely consistent in its own right. One should certainly agree with John North (1965, p. 210) that it was naive to reject (as many physicists in fact did) an empirically successful theory (which SST was in the early 1950s) for this reason alone, that "some inviolable Principle of the Conservation of Energy" is violated in it.

8. The Perfect Cosmological Principle and General Relativity

Yet PCP is definitely in conflict with the field equations of general relativity, for the latter's mathematical formalism requires strict conservation of energy. Therefore, Bondi and Gold could not employ the available theory of gravity in deriving their model. Remarkably, no theory of gravity at all was needed for that. Bondi and Gold (1948, p. 260) proceeded from the generic Robertson–Walker metric for homogeneous and isotropic models, which was shown to be obtainable independently of any particular dynamical theory,

$$ds^2 = c^2 dt^2 - R^2(t)(dr^2 + r^2 d\theta^2 + r^2 \sin^2\theta d^2)(1 + kr^2/4)^{-2}, \qquad (7)$$

where $(r, \theta,)$ are constant coordinates of a fundamental particle partaking in the cosmic expansion, $k = -1, 0, 1$ is the parameter defining the geometry of a particular model, and $R(t)$ is an arbitrary function of time usually called the scale factor in the relativistic models.

The steady-state model can be formally derived from (7) in the following way (see, e.g., Bondi 1960, pp. 145–146). The square of the radius of curvature of the $(r, \theta,)$ space, k/R^2, is responsible for certain observable effects (for example, the number of galaxies observable in the unit proper volume of space) and, hence, according to PCP, must be constant. Since obviously $R(t)$ is not constant (otherwise there would be no redshifts in the spectra of distant galaxies), this gives $k = 0$. The Hubble parameter H is also an observable quantity as it accounts for the receding of galaxies. From $H = \dot{R}/R = const.$ it follows that $R(t) = \exp(Ht)$. Thus the metric of the stationary universe is

$$ds^2 = c^2dt^2 - (dr^2 + r^2d\theta^2 + r^2\sin^2\theta d\varphi^2) \exp(2Ht), \qquad (8)$$

which formally reproduces one of the early de Sitter solutions, as expressed by Lemaître and Robertson (see North 1965, p. 112).

Whether this formal similarity has any physical meaning depends on SST's relation to the received field theory of gravitation. Bondi and Gold discuss this problem in detail. Because of the violation of conservation principles, it is not possible to incorporate the steady-state model into general relativity. It may, however, be possible to proceed the other way round and to derive the proper theory of gravity from SST-I, as a consequence of PCP (Bondi and Gold 1948, p. 270). From the cosmological point of view, this procedure may be legitimate. For PCP has priority over any particular physical law, and the methodology of SST-I assumes that implying laws from cosmological considerations is at least as fundamentally important as a general formulation of the laws.

In this sense, argued Bondi and Gold, the theory of gravity underlying the big bang cosmology is not entirely satisfactory. Locally, it proclaims an equality of all reference frames. At the cosmological level, however, the equality is violated owing to adoption of the *Weyl postulate*, which is nothing but a mathematical corollary of the cosmological principle. According to the Weyl postulate, the world lines of the fundamental cosmic units partaking in the general expansion of the universe are geodesics orthogonal to the spatial hypersurfaces $t = const$. The existence of such hypersurfaces and hence of t, the "cosmic time," is due to the uniformity of the universe in the smoothed-out model. Consequently, at each point of space-time there exists a time-like vector associated with the state of motion of a fundamental cosmic unit and an observer moving with it is privileged in the sense that she sees a strictly isotropic expansion picture.

The geometrical structure of the expanding universe thus naturally gives rise to a preferred vector field. This field, however, plays no role in the general formulation of the received gravitation theory. Because of this, wrote Bondi and Gold, the latter becomes too wide: "It covers a far greater range of possibilities than actually exist" (Bondi and Gold 1948, p. 268). An additional postulate (viz., that of Weyl) is then invoked to narrow down this range. "To us this narrowing-down of the theory in its final form seems to be utterly unsatisfactory, these restrictions should enter the theory at the beginning and not at the end" (Bondi and Gold 1948, pp. 268–269). There is, in other words, no reason to require a complete symmetry of the laws of nature while assuming that their most important application, corresponding to the unique structure of the universe, is signally asymmetrical.

SST, Bondi and Gold held (1948, p. 266), has an important advantage over relativistic cosmology in that it attributes a direct physical and not only a geometrical meaning to the field of privileged vectors, imposed by the uniform expansion of the universe, by identifying these vectors with velocities of the newly created particles of matter. Because of its universal significance, the vector field defined in this way should play as essential a role in the general formulation of the gravitation theory as the tensor field.

This idea, which can also be traced back to Bondi's "prolegomena" (1948), is most unusual and looks, at first sight, entirely opportunistic. A cosmological point of view, however, suggests a rationale for it. What Bondi and Gold seem to be implying is a certain *symmetry principle* requiring that the symmetric properties of the (smoothed-out) cosmological model match the dynamical symmetries of the underlying theory. In particular, the latter should not possess extra symmetries over and above those required to support the model of the actual universe. Such a requirement could be related to a somewhat similar principle operative in the methodology of local space-time theories. As stated by John Earman, this Symmetry Principle demands that every *dynamical* symmetry of such a theory be its *space-time* symmetry, and vice versa. Behind this principle "lies the realization that laws of motion cannot be written on thin air alone but require the support of various space-time structures. The symmetry principles then provide standards for judging when the laws and the space-time structure are appropriate to one another" (Earman 1989, p. 46). For example, the space-time appropriate to Newtonian mechanics should contain (Newton's own view notwithstanding) no less and no more symmetries than are needed to "bring out" the rotational and translational (i.e., Galilean) symmetries of the classical laws of motion.

Bondi and Gold's "cosmological symmetry principle" stems, in effect, from a further idea that the space-time at hand is not a generic space-time providing a suitable structure for the infinite class of local motions, but a very *particular* space-time of the cosmological model describing the global geometry of the *actual* universe. In other words, the space-time in which we live and do physics is the space-time associated with a single *model*, not with the general theory; hence its particular significance:

> While there is no logical argument against theories which are too wide, it is generally agreed that such a theory is unsatisfactory and likely to be mislead-

ing in physics. If then we postulate that a theory of relativity should not be too wide, then any such theory makes certain demands on the structure of the universe. This is a most interesting point, in view of the distinction we had previously drawn between laws of motion and actual motion. While a theory of relativity as such only makes statements about the validity of the laws of motion, its cosmological implications will to some extent make statements about the actual motion of matter in the universe. (Bondi 1948, p. 106)

This disparity characteristic of generic relativistic models is, in Bondi's view, unsatisfactory precisely because the theory at hand (i.e., general relativity) turns out to possess a "superfluous symmetry" that is not manifest in the symmetries of the real universe: "although according to general relativity the laws of nature are immutable, it cannot in its current form provide a universe which is both homogeneous and stationary" (Bondi 1948, p. 107).

In order to deprive the existing theory of gravity of its "superfluity," this theory should be substantially modified. Bondi and Gold promised to present in another paper a formulation of a field theory immune from the above objections (Bondi and Gold 1948, p. 270). This idea, however, was eventually completely abandoned. Later on, Bondi gave reasons for that: "We feel that, as the assumption that the universe is in a steady state leads to observable consequences without any field theory formulation, no advantage is gained by tackling now the obscure and highly ambiguous problem such a formulation presents" (Bondi, in Stoops 1958, p. 78).

This later remark, however, looks more like a post factum rationalization than a real rationale for not pursuing the field theory project. As a matter of fact, a gravitation theory formulation satisfying, in many respects, the above requirements and leading to a steady-state model of the expanding universe already existed at the time Bondi and Gold's original paper came out, and awaited its appearance in the next issue of *Monthly Notices* (Hoyle 1948). Moreover, this theory had been criticized by Bondi and Gold in their paper before it was actually published.[11]

9. Hoyle's Theory

The author of SST-II was apparently less concerned with philosophical problems. He posed a physical question instead: where did the observable matter of the universe come from? There are two main alternatives: either it had been created all at once in the remote past, or it was, and now continues to be, created as the universe expands. The one-time "catastrophic" creation-in-the-past implicit in the big bang cosmology was,

Hoyle maintained (1948, p. 372), "against the spirit of scientific inquiry," for the theory, in fact, deals only with the already created matter and does not consider the process of creation itself.

Pushing the awkward question of creation of the material content of the universe back to the past is, one could say, quite similar to invoking suitable catastrophic events in the geological history of the Earth in order to account for its currently observable features. In geology this approach has been severely criticized and rejected by the uniformitarians. In cosmology, almost the same sort of criticism was presented by the authors of the steady-state theory, who suggested that severe constraints should be placed on cosmological speculation about the remote and inaccessible past of the universe by postulating that the processes that have occurred in the past are basically the same as those going on in the universe now. The most important process of this kind is the continuous creation of matter. Hoyle insisted furthermore that, because of its fundamental importance, this process should be explained and not simply postulated, as was done in SST-I. Contrary to the "philosophical" approach of the latter, he suggested a mathematical account of the creation process, by way of a modification of the field equations of general relativity.

This is reminiscent of the first steps taken in relativistic cosmology. In 1917 Einstein modified his field equations of gravitation, by introducing the famous Λ-term, in order to obtain the static model of the universe (Einstein 1917):

$$R_{\mu\nu} - \frac{1}{2}Rg_{\mu\nu} + \Lambda_{\mu\nu} = -\frac{8\pi G}{c^4}T_{\mu\nu}. \tag{9}$$

Hoyle aimed to justify a stationary picture. By invoking the Einsteinian precedent, he too introduced an additional symmetrical tensor term into the equations of general relativity

$$R_{\mu\nu} - \frac{1}{2}Rg_{\mu\nu} + C_{\mu\nu} = -\frac{8\pi G}{c^4}T_{\mu\nu}, \tag{10}$$

where

$$C_{\mu\nu} \equiv C_{\mu;\nu} = \frac{\partial C_\mu}{\partial x^\nu} - \Gamma^\alpha_{\mu\nu}C_\alpha \tag{11}$$

and

$$C_\mu = \frac{3c}{a}(1,0,0,0), \quad a=const. \tag{12}$$

The construction of $C_{\mu\nu}$ requires the introduction of a special reference frame as follows (Hoyle 1948, pp. 375–376). The universe is assumed to be uniform and satisfy the Weyl postulate. In such a universe, a family of geodesics always pass through a single point O in space-time. The choice of a particular geodesic passing through O is then fixed by providing three "space" coordinates x_1, x_2, x_3 of any other point P on it, whereas a suitable choice of the length along a geodesic from O to P gives the "time" coordinate. "Suitable" means that the measure of time is common to all geodesics; that is, the hypersurfaces $t = const.$ are, at each point, orthogonal to the geodesics. Given the postulated uniformity, together with the above assumptions, the generic metric of the model is

$$ds^2 = c^2 dt^2 - R^2(t) h_{ij} dx_i dx_j, \quad i,j = 1,2,3,$$

where the curvature of space is constant. The additional assumption that this curvature be everywhere zero (suggested by the similarity of the implied stationary universe model with the de Sitter model) leads to further simplification

$$ds^2 = c^2 dt^2 - R^2(t)(dx_1^2 + dx_2^2 + dx_3^2). \tag{13}$$

The vector $C_\mu = 3c/a(1,0,0,0)$ directed at each point along a geodesic gives rise, via (11), to the symmetrical tensor $C_{\mu\nu}$, which can be shown, given (10) and (13), to have the non-vanishing components

$$C_{ij} = -\frac{3 R \dot{R} \delta_{ij}}{ac}, \quad i,j = 1,2,3.$$

In the modified field equations, $C_{\mu\nu}$ plays a role similar to that of the Λ-term in the de Sitter model, except that, unlike in the latter, where $\Lambda g_{00} \neq 0$, $C_{00} = 0$. It is precisely this difference that allows one to obtain the desired model of the steady-state universe, which is *both* de Sitter-like and non-empty.

The vector C_μ, which is parallel to a geodesic at each point of the homogeneous and isotropically expanding universe (thus satisfying the Weyl postulate), represents precisely the vector field that Bondi and Gold expected should play as fundamental a role in the general formulation of the gravitation theory as the tensor field.

Under the normal assumption that the only non-vanishing component of $T_{\mu\nu}$ is $T_{00} = \rho c^2$, a solution of (10)

$$ds^2 = c^2 dt^2 - (dx_1^2 + dx_2^2 + dx_3^2) \exp(2ct/a) \tag{14}$$

is of a de Sitter type and gives the metric of the stationary universe (Hoyle 1948, pp. 375–377). Of course, the proper density of matter in SST-II, unlike that in the de Sitter model, is a constant non-zero quantity; it is given by

$$\rho = \frac{3c^2}{8\pi Ga^2}. \tag{15}$$

It can be shown that the vector field C_μ is responsible for the creation of matter process. From Eq. (10) we have

$$(C^{\mu\nu})_{;\nu} = -\frac{8\pi G}{c^4}(T^{\mu\nu})_{;\nu}. \tag{16}$$

Since $(C^{0\nu})_{;\nu} \neq 0$, a continuous creation of matter and energy uniformly occurs.

The only free parameter in Hoyle's theory is a, and it can be adjusted to fit the actual redshift data. By (15), these data uniquely determine the value of the matter density, which makes SST-II more specific than SST-I, where no constraints are imposed on ρ. On the other hand, the steady-state metric (14) is not the only possible solution of (10). In this respect, Hoyle's 1948 theory is less stringent than SST-I.

Although Hoyle's approach was much less philosophical and more mathematical than Bondi and Gold's, the idea of cosmological uniqueness, which played so essential a role in deducing the model of the universe from the perfect cosmological principle in SST-I, had also found a very clear manifestation in Hoyle's theory. To appreciate this fact, one need only look at the modified field equations of gravity (10). These equations represent a *general* relation between physical quantities $g_{\mu\nu}$ and $T_{\mu\nu}$. Incorporated in this *nomic* relation, however, is another quantity, $C_{\mu\nu}$, having, it would seem, a purely *factual* significance, as it is constructed from the vector field C_μ, which has its origin in the features of a particular *model* of the universe. Thus, in order to derive this model from the modified field theory of gravity, one has first to ground the theory itself in the model at hand. What legitimizes creating such a "centaur," in which nomic and apparently non-nomic features are blended together in a single relation, is, again, the idea that for the universe as a whole, the distinction between the general and the particular fades away. Equation (10) describes the physics of the world in a general way. Such a description, one might think, should, by its very nature, be devoid of any *concrete* parameters. The concrete parameters of the whole universe, however, have a status different from that of the particular initial and boundary conditions attendant to a description of a local situation. General physics described by (10) has its locus in the *concrete* universe, not in some abstract set of possibilities, of which our universe is just one instance. In this view, the incorporation of the vector field C_μ into the structure of a general physical theory, which Hoyle's modified equations of gravity were supposed to represent, may be legitimate.

Nonetheless, neither Bondi and Gold nor Hoyle himself were entirely comfortable with the centaur-like equation (10). A major question looming in the background of this discontent was that of *violated covariance*. Indeed, the introduction of the "creation field" $C_{\mu\nu}$ presupposes the existence of the preferred vector C_μ at any point whose definition makes appeal to a particular coordinate system. As a result, the field equations of gravity (10) become non-covariant.

10. Covariance and Cosmological Expansion

Hoyle's discussion of this problem in his paper (1960) brings out, once again, the distinctive features of cosmological theorizing that played such an important role in the development of SST. Hoyle, in effect, makes two different points there. He argues, first, that the violation of general covariance manifested in (10) is not a defect but rather an advantage of his theory. The reasoning behind this claim can be reconstructed as follows. It is the phenomenon of global and uniform cosmological expansion that introduces a preferred direction at any space-time point. The question is whether this state of affairs should be written into the very structure of the general laws of nature. One might think that it shouldn't, for the state of affairs at hand constitutes an accidental characteristic of the universe and thus should be rooted not in the laws, but rather in the appropriate boundary conditions. This was the approach adopted in the big bang cosmology, an approach that, according to the steady-state advocates, does not seem to square properly with the "point of view of the whole."

The alternative is then to attribute more than merely factual significance to the existence of the preferred vector field in the expanding universe. In that view, the field of preferred directions in space-time should somehow originate in the laws of nature themselves. But—and this is the crucial point —"no theory working entirely in terms of covariant laws (i.e., without reference to a special coordinate system) could explain the development of preferred directions in space-time" (Hoyle 1960, p. 258). Indeed the ideal goal for a cosmologist who wishes to retain general covariance, or maximal symmetry, of the fundamental laws of nature, but, at the same time, aspires to explain the violation of this symmetry at the level of the cosmological application of such laws, would be to demonstrate the following result: "Given any arbitrary distribution of matter and motions, and arbitrary values of the metric tensor and its derivatives, on a space-like surface, prove that the universe must ultimately evolve to its present isotropic, homogeneous state" (Hoyle 1960, p. 258). But this is unattainable, for "it is difficult to see how this result could ever follow from covariant laws, since neither the laws nor the initial situation single out any particular coordinate system for especial preference" (Hoyle 1960, p. 258).

As is clear from these remarks, Hoyle's purpose was to represent, in conformity with the philosophy of the steady-state project, the violation of covariance involved in Eq. (10) as a natural requirement of the cosmological point of view. If the existence of preferred directions possesses a nomic status, in virtue of being incorporated into the field equations (10), one may reasonably pursue the ideal of obtaining a homogeneous and isotropic universe from *arbitrary* initial and boundary conditions.

Having stated this much, by way of defending his approach in (1948), Hoyle, however, shifts emphasis rather sharply, by noting that the ideal specified above may in fact be futile and not worth pursuing, for "all that really needs proving is the stability of a large scale isotropic homogeneous distribution":

> According to this second point of view, the notion of starting with an arbitrary distribution of matter could well be an invalid concept, because the [steady-state] universe need never have been in any state other than one of homogeneity and isotropy however far back in time we go. (Hoyle 1960, p. 258)

Relaxing the requirements on a viable cosmological theory in this way opens a possibility for the rehabilitation of general covariance. Hoyle then outlines a new and covariant formulation of the laws of creation of matter.

The key point in this formulation (Hoyle 1960, pp. 259ff) was to relate the creation field not to the structure of space-time of the universe, but to the already existing matter. The sought-for creation field φ is now assumed to be a scalar with a source proportional to the density of the existing mass:

$$\Box\varphi = \kappa\rho, \tag{17}$$

where $\Box$ is the operator of covariant differentiation (twice) and subsequent contraction by two indices, that is:

$$g^{\mu\nu}\varphi_{;\mu;\nu} = \kappa\rho. \tag{18}$$

The tensor $C_{\mu\nu}$ is now related to the material field rather than to the preferred direction in space-time:

$$C_{\mu\nu} = \frac{\partial^2\varphi}{\partial x^\mu \partial x^\nu} - \Gamma^\alpha_{\mu\nu}\frac{\partial\varphi}{\partial x^\alpha}. \tag{19}$$

Substituting (19) into (10) and assuming that φ is a function of time only (in virtue of the homogeneity and isotropy of the universe) gives the solution $R = \exp(Ht)$, similar to that obtained in (Hoyle 1948). The precise phenomenological law for the creation field follows from (14), (18), $T^{\mu\nu} = \rho v^\mu v^\nu$, and $v^\mu = c(1,0,0,0)$:

$$\varphi = \frac{3Hc^4}{8\pi G}t, \quad \kappa = 3c^2. \tag{20}$$

This law, Hoyle maintained, should eventually be explained by microphysics.

Put in this form, Hoyle's theory confronts a serious empirical objection. According to (18), the rate of creation of new matter is proportional to the density of the already existing matter, which, in the real universe, is distributed rather non-uniformly. Hence most of the new matter must be created within the existing galaxies and in the interior of stars, which contradicts the available observational data. In the subsequent versions of the creation scalar field theory, the character of the creation of new matter was indeed determined by super-massive objects, such as galactic nuclei and quasars. In general, Hoyle's 1960 article marks a transition from the early SST-II to its later modifications in the work of Hoyle and Narlikar (1964, 1966). This program is still running (Hoyle 1989, 1990; Narlikar 1989, 1991, 1993), but it has become less and less adequate from the empirical point of view, especially after the detection of the microwave background radiation and in light of the progress of big bang cosmology in the recent two decades.[12]

I will not deal in this essay with these later developments of SST-II. Instead, I want to return to the formative years of steady-state cosmology and discuss another problem that gave a headache to all the steady-state defenders but was handled by them in rather different ways. This is the problem of conservation laws.

11. Steady-State Cosmology and Conservation Principles

As we saw in Section 7, Bondi and Gold argued that the simplicity of exact conservation principles could be traded off for the simplicity of the overall steady-state cosmological pattern. They also offered another argument to the same effect, by raising a question about the *meaning* of the notion of conservation of matter and energy in an *infinite* universe. It makes no sense, they argued, in dealing with an infinite universe, to speak of any conservation principle without specifying the type of volume in which the quantity at hand is supposed to be conserved. Relativistic models normally specify the coordinate volume of space partaking in the cosmic expansion and defined in terms of co-moving coordinates. The density of matter is assumed to be constant in this particular kind of volume. In the steady-state model, however, matter and energy are also conserved, but in a different type of volume, namely, "in the part of the universe observable with a telescope of given power, that is, in the part within any fixed distance from the observer. In this sense, matter is conserved in any constant proper volume of space" (Bondi 1960, p. 144).

Which type of volume, "co-moving" or "proper," is, then, more fundamental to cosmology? No obvious answer is available, according to Bondi and Gold. But certain empirical considerations could be brought up in favor of the "proper" volume: an observer keeping the resolution of her instrument constant will always see, in the stationary universe, a finite and constant amount of matter. Upon transcending the boundary defined by the power of resolution, matter becomes invisible. If this process were not compensated for by the continual creation of new matter within the limits of observation, the conservation principle would *not* hold for *observable* matter. The observable region of the universe can be taken to be more fundamental than any other type of volume in the infinite universe in the sense that "different observers might agree in ostensively defining it" (North 1965, p. 209). "It may well be considered more correct," Bondi concludes rather paradoxically, "to speak of conservation of mass in the steady-state model rather than in relativity, since proper volume is more fundamental than coordinate volume" (Bondi 1960, p. 144).

Hoyle took up this idea in his (1960) paper:

> For a universe of infinite volume (as in the steady-state theory) energy conservation for the whole universe is a meaningless notion. Conservation of energy must be considered in relation to a box of finite volume. Two important cases evidently arise: the box can have a fixed proper volume, or its walls can expand as the universe expands. When there is creation of matter, the

conservation equations require energy to be conserved in a box of fixed proper volume. When there is no creation of matter, the conservation equations require energy to be constant in a box that expands with the universe. (Hoyle 1960, p. 257)

This argument strikes one as unconvincing, at the very least, for it presupposes a purely operational definition of energy. Such a definition would carry no weight at all in local circumstances, where it is always possible to apply conservation principles to a system of objects or a configuration of fields, rather than to a particular volume of space. A cosmological perspective, however, suggests new ways of looking at familiar notions, and the steady-state theorists did not hesitate to exploit the potential inherent in this perspective.

In 1951, McCrea suggested another, in fact more inventive, way of dealing with conservation of energy in SST-II. Instead of looking at Hoyle's equations (10)

$$R_{\mu\nu} - \frac{1}{2}Rg_{\mu\nu} + C_{\mu\nu} = -\frac{8\pi G}{c^4}T_{\mu\nu}$$

as a modification of Einstein's theory of gravity, McCrea proposed to view (10) as a *redefinition* of a conserved entity. If one takes

$$T'_{\mu\nu} = T_{\mu\nu} + \frac{c^4}{8\pi G}C_{\mu\nu} \qquad (21)$$

to be the energy-momentum tensor in a new sense, all the cosmological results of Hoyle's theory are obtained "without any modification of Einstein's equations" (McCrea 1951, p. 563).

The legitimacy of such a redefinition of $T_{\mu\nu}$, according to McCrea, is due to the fact that Einstein's equations by themselves do not suffice to determine $g_{\mu\nu}$, for the general theory of relativity does not decide what form the energy-momentum tensor takes. A particular expression for $T_{\mu\nu}$ is normally borrowed from classical theory, or from special relativity, on the assumption that such expressions are the correct limits of the (general) relativistic case. In other words, the importation of a certain form of $T_{\mu\nu}$ into the framework of general relativity bears on the relation of correspondence between an old and a new theory. Since such relations are far from being clear and unambiguous, the classical limit of a sought-for relativistic expression may not determine the latter uniquely. "Relativistic investigations," McCrea observes, "may disclose phenomena in relativity theory that have no classical analogs" (McCrea 1951, p. 564). One particular phenomenon of this kind is the physical significance of the absolute value of pressure and its contribution to *gravitational mass*.

This notion was suggested in a work by Edmund Whittaker (1935) in which he offered an extension of Gauss's theorem to general relativity. He showed that the surface integral of what he termed the "gravitational force" over an arbitrary closed surface is equal to the volume integral

$$I = \frac{8\pi G}{c^2}\int\int\int(T_0^0 - \frac{1}{2}T)\sqrt{-g}\,dx^1dx^2dx^3,$$

which prompted Whittaker to interpret the quantity

$$\sigma = 2(T_0^0 - \frac{1}{2}T) = T_0^0 - T_1^1 - T_2^2 - T_3^3$$

as the density of "gravitational mass" in a distribution of matter and energy. In the case of

$$T^{\mu\nu} = \left(\frac{p}{c^2} + \rho\right)v^\mu v^\nu - \frac{p}{c^2}g^{\mu\nu},$$

this reduces to

$$\sigma = \rho - \frac{3p}{c^2}. \tag{22}$$

Equation (22) shows that, in a general relativistic case, the absolute value of pressure p effectively contributes to "gravitational mass" σ.

Consequently, instead of modifying the field equations of gravity in order to get the steady-state model (as Hoyle did in his 1948 paper) with the consequence that energy is not conserved, one can assume that conservation equations $T^{\mu\nu}_{\;\;;\nu} = 0$ hold and then derive the form of $T_{\mu\nu}$ appropriate to the steady-state model (14),

$$ds^2 = c^2dt^2 - (dx_1^2 + dx_2^2 + dx_3^2)\exp(2ct/a), \tag{23}$$

from the original Einstein equations

$$R_{\mu\nu} - \frac{1}{2}Rg_{\mu\nu} = -\frac{8\pi G}{c^4}T_{\mu\nu}. \tag{24}$$

This being done, one has to provide a reasonable physical interpretation for $T_{\mu\nu}$ so obtained.

Substituting (23) into (24) and taking into account[13]

$$T^{\mu\nu} = \left(\frac{p}{c^2} + \rho\right)v^\mu v^\nu - \frac{p}{c^2}g^{\mu\nu} \tag{25}$$

gives

$$\rho = \frac{3c^2}{8\pi Ga^2}, \qquad p = -\frac{3c^4}{8\pi Ga^2}. \tag{26}$$

Thus, the steady-state model results from the Einstein equations (24) if one admits the existence of a uniform negative pressure (or "stress") $p = -3c^4/8\pi Ga^2$ pervading the universe. Since the gradient of this stress is everywhere zero, it does not give rise to any directly observable mechanical effects. Its contribution to cosmology, however, is due to the fact that its absolute value contributes to the gravitational mass (22). The effect of the negative uniform pressure is, in a sense, similar to the effect of the Λ-term in the de Sitter model. The negative pressure performs a *positive* work on the expansion of the universe, which transforms into the mass-energy of newly created matter. The creation of matter proceeds, on this account, in conformity with conservation principles. Hoyle's tensor $C_{\mu\nu}$ represents a reservoir of *negative* energy (in the form of the negative pressure driving the exponential expansion of the universe) whose *rarefaction*, due to the universal expansion, supplies, in effect, *new* energy in the form of created matter, which, in turn, keeps the density of normal matter and of the energy reservoir constant.[14]

Hoyle later (1960) took McCrea's interpretation into account and represented the energy-momentum tensor in the form

$$T^{\mu\nu} = T^{\mu\nu}_{(g)} + T^{\mu\nu}_{(e)} + T^{\mu\nu}_{(c)} + T^{\mu\nu}_{(n)},$$

putting the "creation field" (c) on the same footing with more familiar physical fields: gravitational (g), electromagnetic (e), and "nuclear" (n). The gravitational and electromagnetic components of $T^{\mu\nu}$ were given by their standard expressions. The "nuclear" component was, of course, unavailable at the time, whereas the "creation component" was related to φ via (18), (19), and (10):

$$T^{\mu\nu}_{(c)} = \varphi^{;\mu\nu}_{;} = g^{\mu\lambda}g^{\nu\sigma}\left(\frac{\partial^2\varphi}{\partial x^\lambda \partial x^\sigma} - \Gamma^\alpha_{\lambda\sigma}\frac{\partial\varphi}{\partial x^\alpha}\right). \tag{27}$$

Responding to George Lemaître's objection at the 1958 Solvay Congress, Hoyle said: "I would not agree that the steady-state theory violates conservation. It changes the nature of the quantity that is conserved, but the whole history of the conservation laws of physics shows repeated changes of the conserved quantities" (Hoyle 1958, p. 80).

It must be clear from the above that SST was not a single theory. Its two main versions were based on rather different foundations. Whereas SST-II was, in essence, a mathematical hypothesis, SST-I constituted a rare example of a scientific theory developed from explicitly philosophical arguments. Whereas SST-I, because of its logical inflexibility, was difficult to develop further, SST-II turned into a chain of modifications continuing up to the present. To be sure, most of the consequences of SST-I and SST-II were basically the same, and their authors, together with some other converts, formed a single front in their struggle with the common big bang rival in the 1950s. Hoyle, as mentioned earlier, never had any sympathy with Bondi and Gold's "philosophical" approach. Nonetheless, he fully shared with them the idea that cosmology imposes an unusual perspective on physical laws:

> I take the view that the laws of physics are not what people think they are. What we count as the laws are a combination of the true laws together with a cosmological influence. There are long-range interactions. When you look at a book on particle physics and look at the masses of the fundamental particles, if you believe in the canonical view of physics, then all that is a part of basic physics. I don't believe it. There *is* a basic physics. But in my way of looking at things, I don't have to assume that the various peculiar aspects of physics—particular masses, etc.—depend wholly on the basic laws. They are also a product of the way the universe actually is. What we actually see in the laboratory is a product of two things: long-range cosmological influence and the laws, which are very very much more elegant and symmetrical than particle physicists believe. (Hoyle, in Lightman and Brawer 1990, p. 65)

12. Steady-State Theory and Observations

Following the fate of SST is not my purpose here. (See note 4 for references to available historical accounts.) But a brief comment on the relationship between SST and the astronomical observations that finally overthrew it is in order. Much of the observational work in astronomy in the 1950s was stimulated by a desire to refute SST. The task, however, proved difficult. The theory was put to various tests including the redshift-magnitude data, counts of radio sources, and the theory's ability to explain the origin and abundance of the chemical elements in the universe. At first, none of these produced conclusive results, while some led to contradictory observational reports, much to the advantage of SST. Real problems for SST started to pile up in the early 1960s. First the discrepancy between different surveys of radio sources had been removed and their divergence from the prediction of SST had been firmly established. Then it was shown that the theory could not account for the considerable amount of helium actually observed in the universe and also for the perceptible presence of deuterium. At about the same time, a staunch steady-stater, Sciama, together with his student Martin Rees, performed a careful analysis of the number-redshift diagram for quasars discovered by that time and concluded that it "rules out the steady-state model of the universe" (Sciama and Rees 1966, p. 1283).

Formally, SST can be regarded as having been falsified by these observational data, if only barely so. Various modifications of SST-II were proposed in the early 1960s to explain away the mounting negative evidence. Retrospectively, they all look manifestly *ad hoc*.

Bondi and Gold took the failure of the uniformitarian cosmology at face value. They also seem to have remained true to their methodological principle that a non-uniformitarian "catastrophic" cosmology cannot be a science. Neither Bondi nor Gold joined the big bang mainstream. Their cosmological activity stopped in the early 1960s.

The final blow to SST was the discovery of the microwave background. It was, of course, no less a surprise to the steady-staters than to the rest of the cosmological community. Bondi recalls:

> I did not remain active in the field for long—the last paper I wrote on cosmology was a joint study on the radio number counts, published in 1955—but from early on I kept challenging adherents of evolving models. I told them that if the universe had ever been in a state very different from what it is today, they should please show me some fossils of that earlier age. At that time, there was no answer at all to this challenge. I began to suspect that the amount of helium might be very important as a potential fossil. . . . I must confess that the three degree radiation did not cross my mind. (Bondi 1982, pp. 60–61)

13. Conclusion

Uniformitarian cosmology lost the empirical and methodological battles. As in geology, its "developmental" rival proved its capacity to be a science. But what happened to the underlying philosophy of "cosmological uniqueness"? As we have seen, this philosophy gave rise to rather atypical views regarding the relations:

- between theory and model in the description of the physical universe;
- between laws and boundary (initial) conditions; and, in general,
- between the nomic and non-nomic features of the world.

It was the central claim of the philosophy of the steady-state project that a cleavage between these factors becomes the more problematic the more seriously one takes into account the cosmological point of view. In what sense are the accidental features of the universe *as a whole* different from the nomic characteristics of its content? What can prevent one from incorporating such "universal accidental" features into the formulation of general laws of nature, given that they both have the very same "jurisdiction"? If such incorporation is indeed possible and legitimate, what consequences may result from this procedure for both the accidental and the nomic properties? Can't the former somehow partake of the "nomicity" of the latter, and the latter of the "accidentality" of the former? Can't it be, in other words, that the allegedly accidental features of the universe are not really completely "accidental," whereas the nomic characteristics of its contents are not really that "necessary"?

Eugene Wigner once drew attention to the "three categories which physics used to describe the world and its events[:] . . . initial conditions, laws of nature, and symmetries."

> The initial conditions describe the structure as it now exists. And about this, physics does say virtually nothing. . . . There is one exception to that which I will admit; no, two exceptions: first, and this we often forget to mention, that all electrons have the same charge, that all electrons have the same mass. And the same applies for protons. In other words, there is some part of the structure of the world which has a high regularity. But these are about the only "initial conditions" regularities. Physics assumes, in fact, that the other initial conditions, the present structures of the world, are as irregular as conceivable except for what one can really view, and see, and experience. (Wigner 1980, p. 14)

The hybrid notion of "initial-conditions regularities" alludes to the sort of issues mentioned above. Although most of these issues were introduced into cosmological theorizing by the proponents of a failed theory, they have a more general significance. As was indicated earlier, they can be meaningfully posed and examined in the context of contemporary evolutionary cosmology. That the cosmic evolution of physical matter could, in principle, carry with it the evolution of its basic properties and, perhaps, even of natural laws grounded in them was regarded as a highly unpalatable consequence of a cosmological theory by the advocates of the steady-state cosmology. A desire to avoid it was a driving force behind the steady-state project. But what was viewed in the 1950s as an unacceptable implication of a cosmological point of view is now regarded by some as a real possibility.

It should not escape one's attention that although the steady state proponents raised an important problem of interaction between the nomic and the structural features of the universe, they had an easy and straightforward way of avoiding any serious discussion of the nature of the supposed interaction. To save the constancy and uniformity of laws, they postulated the steady state of the universe. The whole issue of how one should conceive of such an interaction was brought up only for a moment, in order to be immediately side-stepped, by ensuring that whatever interaction there might be between laws and objects thereby governed, one should not really be bothered by it inasmuch as one lives and does cosmology in a stationary universe.

To be sure, some of the steady state theorists came closer than others to an attempt at conceptualizing possible forms of "interaction." Sciama's work on Mach's principle, as we saw, was clearly stimulated by the methodology of steady-state cosmology. This work showed how the project of a purely relational dynamics in a cosmological setting naturally gives rise to the idea of the dependence of the gravitational "constant" (G) on the average density of matter in the universe (ρ) and its velocity distribution (H). Sciama concluded that making ρ and H constant (as in the steady-state model) allows one to avoid undesirable physical consequences of changing G. Hoyle's initial steady-state theory (1948) can be looked upon as a mathematically perspicuous way of expressing the idea of the supposed interdependence of the nomic and non-nomic properties of the universe by incorporating the features of a particular cosmological model (namely, the vector field defined by a family of geodesics in the homogeneous and isotropically expanding universe) into the modified field theory of gravitation. Later developments of Hoyle' theory, however, signal a retreat from this radical position through replacing the dependence of general features of a theory on properties of a single space-time model derived from it by a less contentious dependence of the former on the hypothetical properties of cosmic matter. In both cases, since the universe is supposed to be (and always to have been) in a steady state, the dependence of the relevant sort does not infringe upon the immutability of factors figuring in the equations of a theory.

The steady staters were in a fortunate position. On the one hand, they could be praised for drawing attention to a problem important both physically and philosophically, that of the status of physical laws from a cosmological point of view. On the other hand, they cannot be held responsible for not providing a detailed account that would render the idea of possible dependence of laws on the material structure of the universe intelligible. The authors of SST could pursue their goals without bothering to give such an account.

The contemporary big bang theory of cosmology gives rise to an entirely different situation. Here the idea of a global cosmological evolution is presupposed from the beginning and cannot be made an issue. All other ideas, including that of a possible dependence of laws on the properties of the universe as a whole, must square with this global evolutionary perspective. Anyone taking the possibility of such dependence seriously has to say much more about it than the steady staters did. Despite a growing enthusiasm with the idea of the evolution of laws, it is unclear if the idea is tenable or even coherent. But it is certainly an idea worthy of an analysis. Laws of nature are, after all, part and parcel of the universe. The steady staters were among the first to make this much clear.

Acknowledgments. I wish to thank Michael J. Crowe, James T. Cushing, George Gale, Helge Kragh, and Ernan McMullin for many helpful comments on earlier drafts. Versions of this paper were presented at the Seven Pines Symposium on "Issues in Modern Cosmology" (Lewis, Wisconsin, May 2000) and the Summer School on "Cosmology and Philosophy" (Santander, Spain, July 2000). My thanks to the organizers who made these possible and to the audiences for stimulating discussions. Sections 3–5 and 7–12 of this article are based on substantial parts of my paper previously published in *Studies in History and Philosophy of Modern Physics* 25B (1994): 933–958 under the title "Uniformitarianism in Cosmology: Background and Philosophical Implications of the Steady-State Theory." This material is used here with kind permission from Elsevier Science.

NOTES

[1] The idea of a natural selection of physical laws is explored, e.g., in Smolin 1997.

[2] Henri Poincaré was one of the first to offer such an analysis in one of his last philosophical essays (Poincaré 1911). He concluded that the essential immutability of the fundamental laws is a necessary precondition of the entire scientific enterprise: to deny the first is simply to undermine the second. For a critical examination of Poincaré's arguments, see Balashov 1992.

[3] That is, the age of objects in the universe was estimated to be greater than the age of the universe as derived from the original Hubble constant in the most popular relativistic models.

[4] Historical accounts of the development of SST can be found in North 1965, Merleau-Ponty 1965; Brush 1992, 1993; Balashov 1994, 1998; Kragh 1996; Gale and Urani 1999. Personal recollections of the main protagonists are contained in Terzian and Bilson 1982. See also Bondi 1988, 1990, 1993.

[5] See Gale 1992; Gale and Urani 1993, 1999; Urani and Gale 1994; Gale and Shanks 1996.

[6] The best available account of the history of SST (Kragh 1996) downplays the significance of these questions. I hope to provide reasons for my disagreement with this stance below. See also Balashov 1994, 1998.

[7] This statement is somewhat misleading. One wonders what other cases can there be except existing and possible ones.

[8] In a way precisely the same in which the components of the electromagnetic field are generated by their vector and scalar potentials in the theory of electrodynamics. Hence the similarity of notation below, whose symbols, of course, should not be confused with the familiar electromagnetic quantities.

[9] In particular, the Brans–Dicke scalar-tensor theory of gravity (Brans and Dicke 1961).

[10] In SST-I, this rate corresponded to the emergence of 1 hydrogen atom per 1 m^3 every 3×10^5 years.

[11] Why did three physicists working together came up with two different theories? The story behind the publication of Bondi–Gold's and Hoyle's versions of SST is interesting in its own right. Slightly diverging recollections of Bondi, Gold, and Hoyle in Terzian and Bilson (1982) throw some light on it. Bondi and Gold had read Hoyle's work in manuscript. This prompted them to finish quickly their own paper which they submitted to the *Monthly Notices* two weeks earlier than Hoyle. As a result, it came out first containing, not surprisingly, a critique of Hoyle's views to be published four months later. For more detail, see Balashov 1994 and Kragh 1996, Ch. 4.

[12] The interested reader is referred to Ch. 7 of Kragh 1996.

[13] This form of $T_{\mu\nu}$, rather than the normal idealized expression $T^{\mu\nu} = \rho v^\mu v^\nu$, is needed precisely because, on McCrea's supposition, one may expect a cosmologically significant contribution from the absolute value of p.

[14] One can easily see, in retrospect, many features of modern inflationary scenarios in this picture. This fact prompted some steady-state defenders to view the inflationary theory as a revival of the steady-state cosmology. See Hoyle 1989, Narlikar 1988, pp. 221ff. One can even register, somewhat whiggishly, a further interesting anticipation of later discoveries in McCrea's comments that the origin of the negative pressure is to be found in the quantum theory of fields (McCrea 1951, pp. 573–74).

REFERENCES

Balashov, Yuri V. (1992). "On the Evolution of Natural Laws." *The British Journal for the Philosophy of Science* 43: 343–370.
— (1994). "Uniformitarianism in Cosmology: Background and Philosophical Implications of the Steady-State Theory." *Studies in History and Philosophy of Science* 25: 933–958.
— (1998). "Two Theories of the Universe." *Studies in History and Philosophy of Modern Physics* 29: 141–149.
Barbour, Julian, and Pfister, Herbert, eds. (1995). *Mach's Principle: From Newton's Bucket to Quantum Gravity.* Boston: Birkhäuser.
Bertotti, B., Balbinot, R., Bergia, S., and Messina, A., eds. (1990). *Modern Cosmology in Retrospect.* Cambridge: Cambridge University Press.
Bondi, Hermann (1948). "Review of Cosmology." *Monthly Notices of the Royal Astronomical Society* 108: 104–120.
— (1957). "Some Philosophical Problems in Cosmology." In *British Philosophy in the Mid-Century.* A. Mace, ed. London: George Allen and Unwin, pp. 195–201.
— (1960). *Cosmology.* 2nd ed. Cambridge: Cambridge University Press.
— (1982). "Steady State Origins: Comments I." In Terzian and Bilson, pp. 58–61.
— (1988). "Steady-State Cosmology." *The Quarterly Journal of the Royal Astronomical Society* 29: 65–67.
— (1990). "The Cosmological Scene 1945–1952." In Bertotti et al., pp. 189–196.
— (1993). "Origins of Steady State Theory." In *Encyclopedia of Cosmology: Historical, Philosophical, and Scientific Foundations of Modern Cosmology.* Norriss Hetherington, ed. New York and London: Garland Publishing House, pp. 475–478.
Bondi, Hermann, and Gold, Thomas (1948). "The Steady-State Theory of the Expanding Universe." *Monthly Notices of the Royal Astronomical Society* 108: 252–270.
Brans, Carl, and Dicke, Robert H. (1961). "Mach's Principle and a Relativistic Theory of Gravitation." *Physical Review* 124: 925–935.
Brush, Stephen G. (1992). "How Cosmology Became a Science." *Scientific American* 267: 62–70.
— (1993). "Prediction and Theory Evaluation: Cosmic Microwaves and the Revival of the Big Bang." *Perspectives on Science* 1: 565–602.
Bunge, Mario (1962). "Cosmology and Magic." *The Monist* 47: 116–141.
Dirac, Paul (1937). "The Cosmological Constants." *Nature* 139: 323–324.
Earman, John (1989). *World Enough and Space-Time: Absolute Versus Relational Theories of Space and Time.* Cambridge, Mass.: The MIT Press.
Einstein, Albert (1917). "Kosmologische Betrachtungen zur allgemeinen Relativitätstheorie." *Königlich Preussiche Akademie der Wissenschaften* (Berlin). *Sitzungsberichte*: 142–152.

Feynman, Richard P., Leighton, Robert B., and Sands, Matthew (1963). *The Feynman Lectures on Physics.* Vol. 1. Reading, Mass.: Addison-Wesley.

Gale, George (1992). "Rationalist Programmes in Early Modern Cosmology." *The Astronomy Quarterly* 8: 4–21.

Gale, George, and Shanks, Niall (1996). "Methodology and the Birth of Modern Cosmological Inquiry." *Studies in History and Philosophy of Modern Physics* 27B: 279–296.

Gale, George, and Urani, John (1993). "Philosophical Midwifery and the Birth-pangs of Modern Cosmology." *American Journal of Physics* 61: 66–73.

— (1999). "Milne, Bondi and the 'Second Way' to Cosmology." In *The Expanding Worlds of General Relativity.* H. Goenner, J. Renn, J. Ritter, and T. Sauer, eds. Boston; Basel; Berlin: Birkhäuser, pp. 343–375.

Gold, Thomas (1982). "Steady State Origins: Comments II." In Terzian and Bilson, pp. 62–65.

Hoyle, Fred (1948). "A New Model for the Expanding Universe." *Monthly Notices of the Royal Astronomical Society* 108: 372–382.

— (1960). "A Covariant Formulation of the Law of Creation of Matter." *Monthly Notices of the Royal Astronomical Society* 120: 256–262.

— (1989). "The Steady-State Theory Revived." *Comments on Astrophysics* 13: 81–86.

— (1990). "An Assessment of the Evidence Against the Steady-State Theory." In Bertotti et al., pp. 221–231.

Hoyle, Fred, and Narlikar, Jayant V. (1964). "A New Theory of Gravitation." *Proceedings of the Royal Society of London* A278: 465–478.

— (1966). "A Radical Departure from the Steady State Concept in Cosmology." *Proceedings of the Royal Society of London* A290: 162–176.

Kragh, Helge (1996). *Cosmology and Controversy: The Historial Development of Two Theories of the Universe.* Princeton: Princeton University Press.

Lightman, Alan P., and Brawer, Roberta. (1990). *Origins: The Lives and Worlds of Modern Cosmologists.* Cambridge, Mass.: Harvard University Press.

McCrea, William H. (1951). "Relativity Theory and the Creation of Matter." *Proceedings of the Royal Society of London* 206A: 562–575.

Merleau-Ponty, Jacques (1965). *Cosmologie du XX Siècle.* Paris: Gallimard.

Nambu, Yoichiro (1985). "Directions of Particle Physics." *Progress of Theoretical Physics. Supplement* 85: 104–110.

Narlikar, Jayant V. (1988). *The Primeval Universe.* Oxford and New York: Oxford University Press.

— (1989). "Did the Universe Originate in a Big Bang?" In *Cosmic Perspectives.* S. K. Biswas, D. C. V. Mallik, and S. K. Vishveshwara, eds. Cambridge: Cambridge University Press, pp. 109–120.

— (1991). "What if the Big Bang Didn't Happen? (Big Bang vs. Steady State Theory)." *The New Scientist* 129 (Mar 2): 48–51.

— (1993). *Introduction to Cosmology.* Cambridge: Cambridge University Press.

North, John (1965). *The Measure of the Universe. A History of Modern Cosmology.* Oxford: Clarendon Press.

Peirce, Charles Sanders (1956). *Chance, Love, and Logic: Philosophical Essays.* New York: George Braziller, Inc.

Poincaré, Henri (1911). "L'Évolution des Lois." *Scientia* 9: 275–292.

Rudwick, Martin (1971). "Uniformity and Progression: Reflections on the Structure of Geological Theory in the Age of Lyell." In *Perspectives in the History of Science and Technology.* D. H. D. Roller, ed. Norman: University of Oklahoma Press, pp. 209–227.

Schweber, Silvan S. (1997). "The Metaphysics of Physics at the End of a Heroic Age." In *Experimental Metaphysics.* Robert S. Cohen, Michael Horne, and John Stachel, eds. Dordrecht: Kluwer, pp. 171–198.

Sciama, Dennis W. (1953). "On the Origin of Inertia." *Monthly Notices of the Royal Astronomical Society* 113: 34–42.

— (1955). "On the Formation of Galaxies in a Steady State Universe." *Monthly Notices of the Royal Astronomical Society* 115: 3–14.

— (1957). "Inertia." *Scientific American* 196: 99–109.

— (1959). *The Unity of the Universe.* Garden City: Doubleday & Company, Inc.

Sciama, Dennis W., and Rees, Martin J. (1966). "Cosmological Significance of the Relation between Redshift and Flux Density for Quasars." *Nature* 211: 1283.

Smolin, Lee (1997). *The Life of the Cosmos.* New York: Oxford University Press.

Stoops, R., ed. (1958). *La Structure et l'Évolution de l'Univers.* Bruxelles: Institut International de Physique Solvay.

Terzian, Y. and Bilson, E. M., eds. (1982). *Cosmology and Astrophysics: Essays in Honor of Thomas Gold.* Ithaca; London: Cornell University Press.

Urani, John, and Gale, George (1994). "E. A. Milne and the Origins of Modern Cosmology: an Ubiquitous Presence." In *The Attraction of Gravitation: New Studies in the History of General Relativity.* John Earman, Michel Janssen, and John D. Norton, eds. Boston: Birkhäuser, pp. 390–419.

Weinberg, Steven (1977). *The First Three Minutes: A Modern View of the Origin of the Universe.* New York: Basic Books.

Whitehead, Alfred North (1933). *Adventures of Ideas.* Cambridge: Cambridge University Press.

— (1967). *Science and the Modern World.* New York: The Free Press.

Whittaker, Edmund T. (1935). "On Gauss' Theorem and the Concept of Mass in General Relativity." *Proceedings of the Royal Society of London* 149A: 384–395.

Wigner, Eugene (1980). "The Role and Value of the Symmetry Principles and Einstein's Contribution to Their Recognition." In *Symmetries in Science.* B. Gruber and R. S. Millman, eds. New York: Plenum Press and Urbana: Illinois Academy of Science.

Vsevolod Frederiks, Pioneer of Relativism and Liquid Crystal Physics

Vladimir Vizgin and Viktor Frenkel

1. Introduction

The dramatic age in which he lived left its imprint on Vsevolod Frederiks's personal fate. An offspring of a noble family prominent in the Russian political and cultural scene, he dedicated his life and talent to advances in physics in the Soviet Union. Having started as an experimentalist, he can clearly be identified as a physicist-theoretician. Moreover, his studies of relativity theory give reasons to regard him as a pure theoretician. He was a highly professional physicist who combined experimental skills with a profound interest in fundamental problems of theoretical physics.

His interest in relativity theory was not induced by the then current fashion but by the fact that life had brought him together with David Hilbert, when the latter was working on general relativity (GTR), and later with Aleksandr Friedmann. Today, when the theory of gravitation and relativistic cosmology are in the forefront of theoretical physics, Frederiks's activity as one of the leading proponents of GRT in the 1920s is especially notable. It adds a new dimension to our understanding of his personality and work.

The other area of his interests was the physics of liquid crystals. Together with a small group of associates, Frederiks made a fundamental contribution to the study of liquid crystals, which has since advanced to the front line in experimental physics and technology.

2. Biographical Essay[1]

Vsevolod Frederiks was born on 13 April 1885 in Warsaw into the family of a high-placed government functionary. His mother was a writer of

Yuri Balashov and Vladimir Vizgin, eds., *Einstein Studies in Russia.*
Einstein Studies, Vol. 10. Boston, Basel, Birkhäuser, pp. 149–180.

children's stories, known under the pen-name of Vsevolodskaya (probably derived from her son's name, just as her stories were probably first told to him). Soon after his birth, Vsevolod's father was moved to Tobolsk in Siberia where Vsevolod finished the first class of a grammar school. In the early 1890s, his father was appointed vice-governor of Nizhni Novgorod and the family moved to the Volga. Vsevolod continued his studies in a boys' high school located far outside the city and better known as the famous Alexander Institute for Nobility. The building has survived the tearing-down zeal of the twentieth century and can still be seen in its place not far from the Nizhni Novgorod Kremlin. Its walls carry memorial plaques telling anyone who cares to look that at one time Ilya Ul'yanov (Lenin's father), Miliy Balakirev (a well-known composer), and German Lopatin (the first translator of Marx's *The Capital* into Russian) taught there. By a strange coincidence, a neighboring building housed the first Russian Circle of Lovers of Physics and Astronomy set up in 1881.

The city archives have a vast collection of documents related to this educational establishment, including papers covering the period the future physicist spent under its roof. They contain detailed information about progress in studies and diligence, the pupils' behavior and, oddly enough, the time they spent on written exams. Frequently, themes of compositions or mathematical problems offered at exams were also registered.

One has to admit that Vsevolod Frederiks never appeared at the top of the list, yet he was a "very diligent pupil who displayed a great interest in all subjects." He was one of the best in mathematics and the natural sciences. At the exams he was the fifth or sixth to finish his assignment, obviously not because he was not sharp enough to come up the first. More likely than not this was due to his balanced and placid nature, noted by all who knew him in his later years. Only once his name was mentioned among "breachers of the peace" though his crime was a petty one: he was noticed at an evening performance, something that was forbidden.

Great prospects opened up before him upon graduation. One of them was a career in state service like his father's. Like his elder sister Natalie, who became a mathematician, and his younger brother Vladimir, a biologist, he preferred science.

Moscow or St. Petersburg Universities would have been natural choices for a young man living in one of the two capitals. But since Vsevolod had to leave his home to continue education anyway, his parents chose Switzerland. It is hard to explain this choice since Germany was the Mecca of young and aspiring Russian boys. It seems that Switzerland's favorable geographical situation attracted the family: after all, he would be able to travel easily to Germany, France, and Italy to improve his education.

Vsevolod used this opportunity to study in Paris for one semester and attend Paul Langevin's lectures; he also went to Germany. It remains unclear, however, why the family selected Geneva University while Zurich had a more famous university and the still more famous Federal Polytechnical School.

Of all the lecturers at the University of Geneva there was only one who outlived his time: Charles Guye (1866–1942) whose wide-ranging physical interests embraced both experiment (measuring the mass of fast-moving electrons, molecular and solid state physics) and theory (the theory of relativity and statistical thermodynamics).[2] In 1908–1909, under Guye's guidance, Frederiks completed a work on intrinsic viscosity in solid bodies. Guye started his research on this subject precisely when the Russian student began his university course. As early as 1914, the fourth edition of the first volume of Orest Khvolson's famous course (it included theories of gases, liquids, and solid bodies) (Khvolson 1914) offered a detailed discussion of the achievements of Guye's school.[3] At the initial stage efforts were made to establish the coefficient of internal friction of different materials at temperatures higher than $0°$ Celsius. Frederiks was the first whom his teacher entrusted to extend the temperature range. Geneva, in fact, is a city with a glorious "cryogenic past." It was in this city that Raoul Pictet liquefied oxygen back in 1877. It seems that in the late 1900s, researchers at the University of Geneva had obtained a significant amount of liquid nitrogen. Frederiks, to whose thesis Khvolson referred in his course, had proved reliably that, in certain materials, the coefficient of intrinsic viscosity dropped sharply when the temperature reached $-196°$ C. (In some cases the picture was even more complicated.)

In his thesis, which he defended on 22 November 1909 and published as a booklet in 1910 (Frederiks 1910), he cited his detailed calculations of vibrations of a thread made of the material under study and described the equipment used. He also summed up the results of changes of the internal friction coefficient for aluminum, magnesium, iron, gold, and quartz within a wide temperature range.

Besides Guye, Frederiks mentioned a German physicist Woldemar Voigt (1850–1919) as his predecessor. He was the author of a two-volume theoretical course and an experimentalist who worked, among other things, on elasticity of solids. It was no wonder then that having defended the thesis and received the doctorate from the University of Geneva, Frederiks decided to continue his research at Göttingen University, where Voigt worked at the Physical Institute. He was destined to spend more than eight years there, from 1910 to 1918. In the autumn of 1911 he was admitted to the Department of Theoretical Physics of the Physical Institute at Göttingen University as a *Hilfsassistent*. He retained this post until World War I (see Appendix 1). He was working under Voigt on the optics of metals and the physics of crystals. In 1915, he and his scientific advisor published their research on the piezo-electric effect in quartz. All in all, Frederiks published about ten articles in the leading German scientific journals. He devoted much attention to the lab practice in optics, heat, and mechanics he headed at the university and supervised a number of doctorates. Voigt highly valued his young Russian assistant and even invited the young man to accompany him on his American lecture tour on the eve of World War I.

It was in Göttingen that Frederiks first met Hilbert. This probably happened while he was still working with Voigt. As the war broke out he failed to leave Germany, was dismissed from the university and interned. According to international law he became a "civilian prisoner-of-war." Voigt, who had been defeated in his attempts to preserve Frederiks as his assistant, found a possibility for him to conduct experiments in his laboratory. At this point Hilbert significantly helped Frederiks: he hired him as his "private assistant" whom he financed from his pocket (see Appendix 1). These efforts allowed the Russian researcher to survive and continue his work. Probably at this period he plunged into general relativity which interested Hilbert (see below for details).

In summer 1918 Frederiks returned to Russia and came to Moscow. On 21 July 1918 he informed Hilbert that he had safely reached Moscow and that what the German newspapers were writing about Russia was not true (see Appendix 2). In Moscow Frederiks found a job at the Institute of Physics and Biophysics, the first physical institute set up in the Soviet Union with academician Pyëtr Lazarev as its director.

He knew some of the St. Petersburg physicists, including Dmitri Rozhdestvenskiy and Abram Ioffe, from the pre-revolutionary times. It seems that their range of interests was much closer to his (optics, solids, theoretical physics; the latter was practically non-existent in Moscow in the early 1920s). In February 1919 he moved to Petrograd. There he began working at the State Institute of Optics (SIO) where he remained till the end of 1923. He was an active member of the Atomic Commission headed by Rozhdestvenskiy, and in 1919 he was appointed assistant professor at Petrograd University. He was one of the first who brought the latest news on GRT into Petrograd, which was walled off from the West. There he made friends with his colleagues at the institute and the university, especially V. Bursian, Yu. Krutkov, V. Pavlov and V. Fock, then a very young man. For some years Frederiks lectured on optics, electrodynamics, the theory of relativity, and other areas of theoretical physics.

Immediately after the revolution the circle of physicists in Petrograd was very narrow and concentrated at SIO, closely connected with the University and the Institute of Physics and Technology, which equally closely cooperated with the Department of Physics and Mechanics of the Polytechnical Institute. As time went by, Frederiks frequented seminars at the Institute of Physics and Technology (IPT) and very soon (starting in 1923) transferred his experimental activity there. On 1 September 1923 he entered the Polytechnical Institute as a lecturer. N. Semenov signed his reference presented to the Institute's Council, which said that "Frederiks exhibited a rare combination of splendid experimental and profound theoretical achievements." Besides teaching a lecture course, Frederiks headed a Special Physical Laboratory of the Department of Physics and Mechanics, where students practiced.

This union was sealed when Frederiks moved from downtown Leningrad to Lesnoye, a suburb. The family made its home within walking distance from IPT and the Polytechnical. While maintaining his friendly relations with colleagues from the Optical Institute and the University, Frederiks made many friends in Lesnoye as well. He was especially close with N. Andreyev, V. Kondratiev, N. Semenov, D. Talmud, and Ya. Frenkel.

He was an exceptionally hard worker. His research and teaching commitments load included IPT, the Polytechnical, and the University, where he headed the local branch of the Russian Society for Physics and Chemistry and supervised the research his associates were conducting at the University's Institute of Physics. In addition, he performed regular consultations in the Optical Institute. As if this were not enough, he lectured at the Hertsen Teachers College (between 1920 and the early 1930s). Since 1932, he was professor at the Military Academy of Communications. During this time he also wrote several books and extended his interests to the history of science, of which more below. This was done not for the love of money or thirst for honor: his sense of duty pushed him to fill in as many gaps as possible in the situation of the 1920s and 1930s, when physicists were still a rarity in the Soviet Union.

We have every reason to classify GRT and liquid crystal physics as fundamental areas. Starting in the late 1920s, Frederiks spent increasingly more time on experiments with liquid crystals. At the same time he paid much attention to applied problems as well: beginning in 1926, together with Bursian, Frederiks devoted much of his time to the physical foundations of geological sensing (in particular, to electrical sensing), a field that was rapidly developing at the time. Together they elaborated methods and equipment for developing oil reservoirs, copper mines, and mines of other metals. Frederiks was a scientific advisor in some of the Leningrad and Moscow research establishments engaged in the same problem and at the Leningrad University and the Institute of Physics and Technology. In 1929 he summarized his research in a book (Frederiks 1929). He never shunned expeditions to the Urals or Central Asia where he lived in tents, rode on horseback, and in general displayed what his younger colleagues called "geophysical tenacity."

His rich teaching experience quite naturally expressed itself in a number of university textbooks, for example, *Electrodynamics and Introduction to the Theory of Light* (Frederiks 1934) based on his course at Leningrad University. It was ready by the fall of 1933. In the introduction, the author promised that the second volume would deal with what he, in the old-fashioned way, called the "the electrodynamics of bodies in motion."

Soon after his text on electrodynamics, *A Course of General Physics* (Afanasiev and Frederiks 1935) appeared. It was edited by Frederiks and his University colleague, Professor A. Afanasiev, and was published jointly by the University and the Polytechnical. It was approved as a standard textbook for freshmen. Some of the chapters were contributed by physicists from the IPT and the Polytechnical: A. Alexandrov, N. Dobronravov, I. Kikoin, D. Nasledov, Yu. Khariton, and E. Shtrauf. Each of these names speaks for itself.[4] The idiosyncracies of the authors made it quite difficult to put their contributions together to produce a seamless yet highly individual whole. It was an instant success: the second edition of the book appeared just before the Great Patriotic War (Dobronravov, Nasledov, Khariton, and Shtrauf 1941) with no mention of Frederiks on its title page. During and immediately after the war it was the main textbook of the Department of Physics and Mechanics, other departments, and other higher-education institutions. Later it was replaced with the well-known courses by A. Timoreva and S. Frish, G. Landsberg, and others.

Frederiks displayed a great interest in the history of physics: he studied the works by the founders of optics, such as Newton, Huygens, and Fresnel. He did even more than this: armed with sound knowledge of the subject and European languages, he also edited and commented upon the works of Huygens (Huygens 1935)[5] and Fresnel (Fresnel 1928) published in the series *Classics of Science*. Below we shall discuss his historical work on relativity theory.

3. The Theory of Relativity

Along with Frenkel, Friedmann, Khvolson, and Sergei Vavilov, Frederiks greatly contributed to promoting the theory of relativity in the USSR in the twenties and thirties. There is no doubt that, as a private assistant to Hilbert in Göttingen,[6] he was well familiar with Einstein's and Hilbert's articles on GRT. It is all the more probable because, during this period, Hilbert was deeply involved in the work on GRT and the unified field theory (see Vizgin 1981, 1994). Hilbert greatly contributed to GRT in his works of 1915 to 1917: He derived the general covariant equations of the gravitational field from the variation principle with scalar curvature as a Lagrangian, established a connection between the energy-momentum tensor and variation of action for the gravitational potentials, formulated an analogue of the second Noether theorem, probed into the problem of conservation of energy-momentum in GRT, and so on (Hilbert 1915, 1917). These works are marked by an axiomatic approach to GRT and unified field theories and intensified interest in abstract mathematical methods and structures: variational calculus, theory of groups and their invariants, differential geometry, and other mathematical tools. No wonder that, once Frederiks came back home, he started to actively promote the ideas of GRT, as exposed by Einstein and Hilbert.

Later Fock recalled that Frederiks and Friedmann had been the first in Petrograd to bring GRT to the attention of the academic community. Fock highly valued Frederiks's physical intuition and his understanding of the theory's fundamental problems. To illustrate, he cited the following example: "To my question: 'How should one formulate the law of motion of many bodies in general relativity?' he promptly answered: 'Motion of bodies is determined by the motion of singularities of the metric tensor.' This was many years before the equations of motion of a system of bodies were formulated in Einstein's works and my works of 1938–1939" (Fock 1966, p. 399). Another example of this kind cited by Fock relates to quantum theory and the fundamental nature of the problem of observation in micro-physics. Even before wave mechanics and the principle of uncertainty were formulated, Frederiks highly assessed de Broglie's concept of matter waves. One can get a good idea of his lectures on relativity theory delivered in the early 1920s at Petrograd University from a vast survey published in the second issue of *Uspekhi Fizicheskikh Nauk* in 1921 (Frederiks 1921). It was the first detailed and consistent exposition of the fundamentals of GRT in the Soviet scientific literature. In a short historical survey he mentioned Hilbert's works of 1915–1917 (Hilbert 1915, 1917), H. Lorentz's works of 1915–1917 (Lorentz 1916), and H. Weyl's book (Weyl 1918), along with Einstein's works. He emphasized the fundamental status of the new theory ("the theory formulated by Einstein embraces all of physics"), the high level and the original nature of mathematics it used,[7] and the profoundly physical and non-deductive method of building up the theory of relativity. Before turning to the theory itself, Frederiks discussed the relationship between geometry and physics. He then formulated the theory's basic propositions and discussed them in a succinct yet meaningful manner.

Thus his discussion of the Riemannian nature of the metrical structure of space-time,

$$ds^2 = a_{ik} dx_i dx_k,$$

was accompanied by a remark that "an experiment alone can decide what the a_{ik} functions are." The second and fourth propositions combine to formulate the general principle of relativity and its mathematical expression, the principle of general covariance. The third proposition is the tensor-geometrical conception of gravitation, according to which the components of the metric tensor of space-time are identified with gravitational potentials.

In his further exposition (in deriving the equations of the gravitational field) Frederiks followed Hilbert rather than Einstein. He greatly simplified the former's constructs and calculations. He based his derivation on the generally covariant variational principle with the Lagrangian, or the "world function" $H = K + L$, where K is the scalar curvature of the Riemannian four-dimensional space-time and L is the Langrangian's "material part," which coincides with Maxwell's Lagrangian in the case when there is no other matter besides an electromagnetic field, or with the Lagrangian of the Mie theory. (This was also the case in Hilbert's work. He believed, in conformity with the Mie theory, that all matter was reducible to the electromagnetic field.) Variation for the metric provides a gravitational equation while variations for "material" variables (such as electromagnetic potentials) provide the equations of motion of "matter," in particular, the equations of the electromagnetic field.

He then discussed certain applications of the field equations of gravitation, in particular, Schwarzschild's solution and the three classical effects of GRT: the shift in the perihelion of Mercury, the bending of light rays in the presence of large bodies, and the gravitational redshift. Frederiks noted that the bending of light rays at the edge of the solar disc observed in 1919 by two British expeditions headed by Eddington was a convincing experimental support of GRT.

Friedmann came back to Petrograd from Perm (a city in the Urals) in 1920. Frederiks, who had contacts with Hilbert and who "imported" GRT from Göttingen, could interest Friedmann with a new and mathematically sophisticated theory and help him get oriented in the relativistic literature. This was the time when they began working together on *The Basics of Relativity*, the first issue of which appeared in 1924 (Frederiks and Friedmann 1924). This was by no means everything that Frederiks did to promote relativistic ideas in the USSR. He was active in translating and editing foreign publications on the theory of relativity, co-authored, with Matvei Bronstein, a large article on it for the *Encyclopedia of Technology*, and repeatedly wrote articles for memorable dates related to the theory.

One of the most popular expositions of relativity theory, by an astronomer Erwin Freundlich (who had greatly contributed to its experimental corroboration), appeared under Frederiks's editorship and with his foreword (Frederiks 1924). In 1933, Frederiks published his translation of August Kopff's book (1921) on GRT and wrote an introduction to it (Frederiks 1933). In a collection of classical works on relativity theory, *The Principle of Relativity*, the editors, Frederiks and Dmitri Ivanenko, gathered under one cover the best works by Lorentz, Henri Poincaré, Einstein, and Hermann Minkowski on the special theory of relativity and Einstein's key works on GRT. As distinct from a similar collection published in Germany, the Soviet edition contained Poincaré's fundamental article of 1906. The book also included biographies of the classics of relativism and important historical commentaries by Frederiks and Ivanenko. In fact, this was one of the first significant works on the history of relativity (Frederiks and Ivanenko 1935). One should not forget that Frederiks belonged to the generation of physicists who were contemporaries of the pioneers of relativity and who learned about latest developments from German physical periodicals in Geneva and Göttingen.

In their joint article for the *Encyclopedia of Technology*, Frederiks and Bronstein (1931) discussed the foundations and history of relativity theory. In contrast to Frederiks's earlier survey of 1921, the two authors followed the logic of Einstein, not of Hilbert, especially in deriving the equations of gravity. Much space was given to cosmology, in particular, to one of its main achievements, Friedmann's non-stationary models and their further independent development by Lemaître.

They offered an extremely pessimistic view on cosmology's prospects—mainly because Friedmann's and Lemaître's estimate of the age of the Universe differed by four orders from that accepted by the stellar evolution theory current at that time. They wrote: "One has to recognize that the situation of the cosmological theory is extremely unfavorable" (Frederiks and Bronstein 1931, p. 364). The prospects for a unified geometrical field theory seemed to them bleak enough. They offered a concise discussion of what happened in this field within the recent decade and looked at the theories suggested by Weyl, Eddington, and Kaluza and at Einstein's theoretical constructions, such as the theory with an absolute, or distant, parallelism. Their criticism led them to conclude that "the program of the unified field theory will probably turn out to be unrealizable" (Frederiks and Bronstein 1931, p. 366). This may have been due mostly to Bronstein's position, who was more skeptical of the unified geometrized field theories (see below on Frederiks's later stand on this issue).

Jubilees of prominent physicists and memorial dates in physics have always provided an opportunity to look at the foundations of scientific knowledge, its problems, and the dissemination of scientific ideas in society. Frederiks frequently commented, on such occasions, on the theory of relativity. In 1927 a special issue of *Uspekhi* was devoted to the 200th anniversary of Newton's death. In his article, "Foundations of Newton's Mechanics and the Principle of Relativity," Frederiks (1927a) compared the classical ideas of space, time, mass, and force with the corresponding concepts in GRT and emphasized, not so much the gap between them created by relativism, as a profound similarity of the original principles of classical mechanics and GRT: "One should not oppose Newton's mechanics and the relativity principle; the latter is nothing but a natural and logical extension of the ideas of the former. This has become possible nowadays due to the enlargement of mathematical knowledge" (Frederiks 1927a, p. 84). According to Frederiks, the Newtonian idea that "geometry is based on mechanical practice and is nothing more than a chapter in general mechanics" was the core of the relativistic approach to space and time. At that time Frederiks had not yet abandoned the program of unified geometrized field theories. While pointing out a possibility of a radical revision of the ideas of space, time, and field created by quantum mechanics, he concluded: "I believe there is no need for a radical change in our views of things. What Newton began and Einstein continued can be brought to its logical end. The attempts to incorporate an electromagnetic field into the four-dimensional geometry by Weyl–Mie, Eddington, Einstein himself, and others are the first and shaky steps in this direction. One can only hope that the task successfully dealt with at the diametrically opposite end by quantum mechanics can be fulfilled in this way too" (Frederiks 1927a, p. 85).

Eight years later, in his article on the thirtieth anniversary of relativity theory Frederiks would return to the theory's past, the central propositions of STR and GTR, and the main directions along which they were developing (Frederiks 1935). While giving due credit to Lorentz and Poincaré in preparing the ground for SR, Frederiks insisted that it was Einstein who offered a truly relativistic interpretation of the electrodynamics of moving bodies that resulted in SR.[8]

In this article Frederiks gave much attention to the three main trends in GRT: the synthesis of the ideas of GRT and quantum mechanics, cosmology, and the geometrical unification of gravitational and electromagnetic fields. His views on the last of these three trends never changed since 1931 (i.e., since the *Encyclopedia* article written together with Bronstein), even though his critical considerations of the unified geometrized field theories became more pointed.[9] At the same time he acknowledged the program's positive significance.[10]

The attempts to combine the relativistic and quantum ideas (including five-dimensional models) made through the mid-1930s were failures in his eyes. He insisted that "every evidence with which quantum mechanics supplied us, concerning the spatial-temporal properties of matter in elementary phenomena, convinces us that a revision of the physics of space-time is again in order" (Frederiks 1935, p. 13). It seems that his forecast is coming true as the quantum theory of gravitation is being elaborated.

In his article he concentrates mostly on Friedmann's non-stationary cosmological model: "The solution suggested by Friedmann is interesting, because the radius of the Universe in it depends on time and is always increasing" (Frederiks 1935, p. 14). While being aware that the available observational data did not support any of the models (including the non-stationary one), Frederiks wrote that relativistic cosmology rested on a firm foundation: "Yet one can conclude that, as physicists will be amassing astrophysical observational material, they will reach a point where the theory of relativity will be able to solve similar problems [experimental, or observational, in support of cosmological models]. In this respect its failure is a temporary one" (Frederiks 1935, p. 15). This forecast proved to be correct. A more pessimistic account of cosmology's future expressed in the joint article of 1931 was due to Bronstein's overestimation of the divergence between the theory's prediction of the age of stars and the astronomical data.

Frederiks undertook special theoretical studies in GRT and unified geometrized field theories. He wrote six articles on various aspects of these theories, four of them jointly with young theoreticians G. Mandel, A. Izakson, and A. Shekhter. All of these were written between 1926 and 1928.

In his popular articles and surveys of that period Frederiks always pointed out three main trends in the development of GTR: (1) cosmology, (2) the unified geometrized theory of gravitation and electromagnetism, and

(3) the unification of GRT with quantum theory. The greater part of his work was, to a certain degree and in various ways, connected with these trends. At the same time, being immersed in an atmosphere of sharp discussions of the theory's physical interpretation and being himself one of its interpreters as a lecturer, Frederiks was drawn to the fundamental problems of a methodological nature: the interpretation of inertia in GRT and the related problems of rotation and the equivalence of the coordinate systems associated correspondingly with the Earth and the sky, the discussion of the behavior of the classical electron in a homogeneous gravitation field, and so on. These issues, which may look deceptively simple, are still under debate (see, e.g., Ginzburg 1969, 1975).

Frederiks and his coauthors never achieved results comparable to those of Friedmann in relativistic cosmology, or of Fock (on the geometrical aspects of wave quantum-mechanical equations and on the derivation of the equations of motion from the field equations of gravitation), or of Bronstein (on the quantization of gravitation). Still, these works moved with time, appeared in major physical journals, such as *Zeitschrift für Physik*, and in general reflected the state of relativistic research in the 1920s. Let us discuss them in more detail.

3.1 TWO WORKS ON THE INERTIAL FORCES, PARALLAXES, AND ABERRATION IN GRT

The methodological issues regarding the status of inertial forces and rotating coordinate systems in GRT continued to be debated into the twenties (Kopff 1921, Eddington 1923). Back in 1918, H. Thirring had established that the rotation of a hollow sphere generates a force having two components that could be interpreted as the centrifugal force and the Coriolis force (Thirring 1918). In his article Frederiks (1925) demonstrated that the Thirring effect could be applied to interpreting similar forces that arise on the Earth's surface as it turns around its axis, if we consider the Earth to be stationary and the stars as rotating about it. To do this, Frederiks had to redefine the rotating coordinate system so as to exclude superluminal velocities of very distant objects. He demonstrated that this redefinition in no way distorted the picture of the rotation of the firmament. In this way he interpreted inertial forces as due to a kinematic effect while the mass of the distant stars, he believed, ensured the metrical structure of space-time, including the "flat metric structure" which he used in his considerations.

In their article, Frederiks and Shekhter (1928) discussed the rotation of the firmament and correspondingly the effects of parallax and aberration in the cosmological models with non-zero curvature. These were the models of Einstein, de Sitter, and Friedmann. "The results of these calculations," the authors concluded, "are totally in line with the accepted astronomical ideas" (Frederiks and Shekhter 1928, p. 484).

3.2 TWO WORKS ON THE ELECTRON IN GRT

In an article published in 1926 Frederiks and Mandel discussed the result obtained within special relativity by Max Born in 1909 and by Arnold Sommerfeld in 1910, namely, that a relativistic electron moving hyperbolically (that is, along a straight line with constant acceleration) does not radiate. Frederiks and Mandel interpreted this result in line with GRT by introducing a uniformly accelerated system of reference in which the electron is at rest and by regarding it, in accordance with the equivalence principle, in a constant and homogeneous gravitational field. They concluded that in this case radiation appears only in a non-stationary field. The discussion which revived in the 1960s (Ginzburg 1969, 1975) testified that the question of the radiation of a hyperbolically moving electron is not trivial.

Frederiks and Izakson's article (1926b) discussed the problems related to unified field theories. The authors relied on Einstein's unified field theory of 1919 with the field equations

$$-\kappa E_{ik} = K_{ik} - (\tfrac{1}{4}) g_{ik} K,$$

where K_{ik} is the Ricci tensor, K the scalar curvature, g_{ik} the metric tensor, κ the gravitational constant, and E_{ik} the energy-momentum tensor of the electromagnetic field. Einstein demonstrated that these equations, together with Maxwell's equations, allowed an arbitrary spherically symmetrical distribution of charge, which could be interpreted as an extended electron of an indeterminate size (Vizgin 1994, pp. 163–167). This ambiguity forced Einstein to abandon this theory. Together with his co-author, Frederiks tried to supplement the energy-momentum tensor of the electromagnetic field E_{ik} with a certain material energy-momentum tensor M_{ik} that would correspond to a physical model of the electron (such as the hydrodynamic model or the ideal-gas model) to remove the ambiguity and discuss all possible values of the size of the electron. However, the idea of a classically extended electron ran into difficulties of relativistic and quantum nature and was not developed in this direction. It should be noted that in 1925–26 Einstein made an attempt to return to his 1919 theory but dropped his efforts in favour of five-dimensional models (Vizgin 1994, pp. 212–214).

3.3 TWO WORKS ON UNIFIED GEOMETRIZED FIELD THEORIES

The first of the two articles on unified geometrized field theories (Frederiks and Izakson1926a) dealt with Einstein's affine-metric theory of 1925. It introduced, as the basic variables, the components of a non-symmetrical affine connection Γ^l_{ik}, as well as the components of a non-symmetrical metric tensor, the symmetrical part of which was interpreted as gravitation while its anti-symmetrical part as electromagnetic field (Einstein 1925). Since his student days in Göttingen, Frederiks was always very much taken by the idea of a geometrical unification of the gravitational and electromagnetic fields. It seems that he infected Friedmann with this passion, who undertook, first on his own and later in collaboration with Jan Schouten, to develop a version of a unified field theory based on a particular geometry of the affine connection. It should be noted that Frederiks and Izakson started their work with establishing a link between Einstein's affine-metrical theory and Schouten's classification of possible four-dimensional generalizations of Riemannian geometry. They proceeded to discuss, using Schouten's terms, the specific and particular cases of Einstein's theory. Just like their predecessor, they failed to advance the theory's physical content. Einstein himself abandoned his theory shortly. But later, Einstein, Schrödinger, and others returned to the affine-metric theories on many occasions. One such occasion was an attempt to fit the conception of spin into the gravitation theory (Schrödinger 1963; Ivanenko, Pronin, and Sardanashvili 1985).

The second work appeared in one of the first Soviet publications on quantum mechanics (Frederiks 1927b). It dealt primarily with the synthesis of GRT, electrodynamics, and quantum theory based on the five-dimensional approach. It also provided a survey of previous attempts due to Fock (1926) and O. Klein (1926). In their classical works they tied the five-dimensional schemes of joining gravitation and electromagnetism with quantum theory. It was precisely Frederiks who introduced Fock to the idea of five dimensions: he drew his attention to a typewritten copy of Mandel's work intended for *Zeitschrift für Physik* (Mandel 1926). Totally unaware of Kaluza's (1921) pioneering five-dimensional synthesis of gravitation and electromagnetism and of Klein's work (1926), Mandel independently elaborated a five-dimensional theory. Fock identified the equations of the charged particle's motion with the equations of a geodesic in the five-dimensional space-time possessing a Riemannian metric of a special type. As a result he obtained a relativistic generalization of Schrödinger's equation as a five dimensional wave equation (the Klein–Gordon–Fock equation). In his article Fock acknowledged: "The idea was prompted by a conversation with Prof. Frederiks. I am grateful to him for valuable suggestions" (Fock 1926, p. 226). In his own paper, Frederiks summarized the results obtained by Fock and Klein and demonstrated that their approaches were very close. Following, to some extent, V. Bursian's footsteps (who had discussed the interconnection between five-dimensionality and quantum mechanics with Ehrenfest), he compared the wave solutions to the five-dimensional wave equations with de Broglie's waves. His judgment was a sober one. He wrote:

> We all can see that the five-dimensional space discussed by Fock and Klein has distinctive features. We cannot speak of the "unified" field geometry yet, but these works seem to be an important and interesting stage on the way towards it. . . . This stage is especially interesting because of its connection with quantum theory. The relativity theory has failed to approach it so far. (Frederiks 1927b, p. 89)

It is important to note that the recent decades have witnessed an upsurge of interest in the multidimensional analogies of the Kaluza–Klein scheme in connection with the gauge projects of unifying four fundamental interactions (see, e.g., Friedmann and van Nieuwenhuizen 1985).

To sum up, Frederiks was one of the few who actively introduced GRT in the Soviet Union in the 1920s. There is no doubt that his influence on Friedmann and Fock, two physicists who made important contributions to cosmology and the relativistic gravitation theory, was immense. After all, he was the author of the first surveys, popular articles, and encyclopedic entries that dealt with GRT and a co-author of the first (regrettably unfinished) textbook on the theory.

4. Liquid Crystals

Anisotropic liquids (AL), or liquid crystals, were discovered three years after Frederiks was born. This discovery was reported by Friedrich Rheinitzer, professor of the Botanical Institute in Gratz, in a letter posted, on 14 March 1888, to Otto Lehmann, a Dresden professor well known for his works in crystallography (see Sonin 1983). Some of the physicists were attracted to the unusual and amazing optical properties of liquid crystals (LC) and the very paradox of substances that combined two apparently opposite features, anisotropy (crystals) and the absence of shape (liquid).[11] Voigt, one of whose works Frederiks quoted, was among them. At an earlier period, when he was studying abroad, liquid crystals were a purely academic subject: few were tempted to plunge into it and, consequently, publications were few and wide apart. One can surmise that Frederiks's interest in LC was heated up by a consideration that a thin film of AL that was believed to consist of extended molecules possessing a dipole moment resembled, to a certain extent, a piezo-electric crystal. (Back in the 1910s, while in Germany, he devoted much of his time to piezo-electricity.)

In any case, a report of the research work done in 1923–24, submitted by the Institute of Physics and Technology (Frederiks's first year in it), indicates that he was studying "anisotropic liquids in the magnetic field" (Archives of IPT, record group 1, inventory 1, file 23, p. 3). The report for 1925 mentions that he engaged in investigating the orientation of AL molecules in thin films and tubes; according to the report for 1926, he delivered a paper on AL at the Institute's seminar.

The first publication on this subject appeared in the proceedings of the Fifth Congress of Russian Physicists held in December 1926 in Moscow. It was written by Frederiks and Repieva and was entitled "On the Theory of Anisotropic Liquids and Some New Observations of Them" (Frederiks and Repieva 1927a). It seems that their joint paper served as the basis for an article they sent in February 1927 to *Zeitschrift für Physik* and *JRPCS* (*Journal of the Russian Physico-Chemical Society*) (Frederiks and Repieva 1927b).

In 1935, at the request of academician Ioffe, the Institute's Director, Frederiks, like all his colleagues, submitted a summary of his research activity. It formed the basis of the section "Studies of the Liquid Crystal State of Matter" of Ioffe's report to the Academy's session in March 1936 (Ioffe 1936). Frederiks's concise summary presents both a rationale for his research and its assessment from the point of view of the work done over more than a decade. The discussion below is based on this summary. Here is what he wrote:

> Studies of liquid crystals are important as (a) studies of one of the phases of matter; (b) studies of a substance fairly widespread in biology; (c) being closely connected with technological problems where putting a fine layer of a liquid (such as a lubricant) on a rigid body is of signal importance; (d) a methodological issue: while sharing some of the properties with common liquids, these crystals allow to study phenomena undetectable in common liquids: motion and so on. (Ioffe 1936)

Let us have a closer look at his joint article with Repieva (Frederiks and Repieva 1927b). It deals with the effect of a magnetic field on AL and is based on W. Kast's experiments (1924), which produced an experimental dependence of the value $\Delta\varepsilon = \varepsilon_H - \varepsilon_0$ on the magnetic field (where ε_H is the dielectric constant of AL in the field of strength H and ε_0 in the absence of the field). The article contains an experimental and a theoretical part.

The experimental methods employed in (Frederiks and Repieva 1927b) became classical and were repeatedly used in his works and those of his colleagues. A sample of AL was put between a slide and a convex glass so that it was thicker along the sides than at the center. This makes the forces that push the molecules together different at different points: they are considerable at the center while the regions far from the slide's surface are less affected by them. The tray was placed in an electric magnet with intensity of up to 30 kG (kilogauss). The behavior of AL was observed in a monochromatic light ($\lambda = 0.546$ μ, the green line of mercury was used). There were standard polarizers on both sides of the tray. The tray itself was placed in an oven that was powerful enough to produce AL (the "meso-phases" of matter).

One of these experiments studied a nematic with the optical axis normal to the slide's plane. In the absence of magnetic field AL does not let the light beamed on it pass (because of the crossed nicols). It yields a dark field when observed or photographed. The situation does not change in a magnetic field parallel to the optical axis of AL. In a field perpendicu-lar to the optical axis the following picture can be observed: the tray's central part remains dark while interference rings can be seen along the sides. The stronger the field the closer they are to the center. This allows a conclusion that "for every value of field H there is a limiting thickness of layer z_{cr} within which the original normal position of the optical axis is preserved." And vice versa, each value of z can be put in correspondence with value $H = H_{thr}$ so that, when $H > H_{thr}$, the direction of the AL layer's optical axis changes and it lets the light pass. This is explained by the fact that, when $z < z_{cr}$, cohesion of the AL molecules with the surface of the glass is strong enough not to allow the magnetic field to reorient the molecules, that is, to shift the optical axis of AL. The authors published quite impressive photographs of a sample made in polarized light. One can clearly see a dark part (growing narrower as field H increased) and the interference rings in the AL sectors with the shifted optical axes. The liquid behaved similarly in the longitudinal and transverse magnetic fields, when its optical axis was perpendicular to the slide's surface. Finally, Frederiks and Repieva (1927b) demonstrated that this behavior was typical for both types of AL.

Anyone aware of Frederiks's previous research has every ground to insist that the extensive theoretical part of the article was written by him. When analyzing the experimental data on the electrical properties of a flat "liquid plate" of AL by comparing them with experimental data for its crystallized phase, he concludes that the effects and dependencies observed by Kast were not caused by a magnetic moment inherent in AL particles (or their aggregates, "swarms" and "drops," to use modern terms). They were caused by the anisotropy of diamagnetic properties, that is, by the inequality $M_{\parallel} \neq M_{\perp}$ (where $M_{\parallel}$ and $M_{\perp}$ are, respectively, the values of the magnetic permeability orthogonal and parallel to the optical axis of AL). By matching the values of the physical parameters that determine the dependence $\Delta(H)$, Frederiks obtained a good correspondence between the experimental and theoretical data. He believed that in this way he, first, proved the diamagnetic nature of the effect and, second, estimated the characteristic size of "swarms." Further, he analyzed the forces that affected AL particles in a magnetic field. He distinguished three different kinds of forces. The forces of the first kind are associated with the interaction between the AL molecules (or "swarms") and a solid surface on which AL is placed. This force determines the orientation of AL molecules, mainly those that are close to the surface and also those inside the layer. The farther the molecules are from the surface the lesser the influence of this force. The force of the second sort is related to the "internal field" and determined by the dipole (and multipole) moments. Finally, the force of the third kind is associated with the external magnetic field; the experimenter can vary its strength within wide limits. These forces are responsible for the elastic deformations of AL in a magnetic field.

In their later articles Frederiks and his associates (Frederiks and Zolina 1930,[12] 1931) continued to investigate various ALs in magnetic fields. They also established a relation according to which the product Hz for the perpendicular orientation of the optical axis is constant:

$$(Hz_{\text{thr}}) = const. = C_1. \tag{1}$$

H_{thr} corresponds to a given z and a given H to z_{cr}. Thermodynamic considerations (Frederiks and Tsvetkov 1935b) show that coefficient C_1 in this formula is given by

$$C_1 = \pi \sqrt{A/\Delta x},$$

where $\Delta\chi = \mu_{\parallel} - \mu_{\perp}$ is the diamagnetic anisotropy and A is the constant of elasticity. This relation shows the threshold nature of elastic deformations of AL in the magnetic field: the field deforms AL and turns its molecules (or their groups) only in the layers whose thickness exceeds a certain critical parameter z_{cr}. This effect was called "Frederiks's effect" or "the

Frederiks transition." The work (Frederiks and Tsvetkov 1934) established a similar "threshold" correlation for torsional strain.

It was demonstrated further that similar effects can be found in nematic AL placed in electric fields (Frederiks and Tsvetkov 1934). Here too the deformations of AL reveal their threshold nature while the critical thickness is related to a corresponding (threshold) value of the electric field intensity E_{thr} via a hyperbolic correspondence analogous to (1):

$$(Ez)_{thr} = C_2 = \pi\sqrt{A/\Delta k}, \qquad (2)$$

where Δk is the dielectric anisotropy of AL: $\Delta k = \varepsilon_\| - \varepsilon_\perp$. Equation (2) is called the Frederiks–Tsvetkov equation. As in case (1), here, too, the situation is typical for the Frederiks transition: the value of E_{thr} corresponds to a given value of z and, vice versa, a given E corresponds to z_{cr}.

Finally, let us note another significant effect discovered in these works (Frederiks and Tsvetkov 1935a and 1935b), the dynamic diffusion of light in nematics. It turns out that, under a definite critical value of the electric field (that usually exceeds E_{thr} of equation (2) for a given thickness of AL layer), AL begins to move, which can be seen through a microscope. Here is what the authors had to say: "Our observations have shown that, under sufficient field intensity, the matter in the AL phase begins to move and that the intensity of motion grows with the field intensity" (Frederiks and Tsvetkov 1935b, p. 123). Specks of dust make this motion ("bubbling," to borrow a metaphor from the authors) even more perspicuous. Once again, Frederiks and Tsvetkov established a hyperbolic correlation between field E and thickness of the layer z for the effect they discovered, that is, they drew an important conclusion about the effect's threshold nature. It should be noted that the critical value of the electric field potential is about 10^4 V/cm. For sufficiently thin layers, this intensity can be obtained under a very insignificant potential difference. This proved to be very important in applications which the authors (Frederiks and Repieva 1927a and 1927b, Repieva and Frederiks 1927, Frederiks and Zolina 1930, Frederiks and Tsvetkov 1935a and 1935b) did not discuss at that time. Let us note here that the mentioned works of Frederiks and Tsvetkov studied the joint effect of the magnetic and electric fields; the authors demonstrated that in different conditions, the magnetic field could diminish or amplify the effect.

The phenomenon of light diffusion can be used as an effective lecture demonstration. One should take two pieces of glass with a thin isolating film of AL between them. The inner surfaces must be covered with transparent electrodes. (In their works of the 1930s the authors used a mesh with a pitch of about .5 mm and the total area of about 1 cm².) When no field is applied, AL is transparent. As the intensity exceeds the critical value, AL turns milky and the vortices dissipate light.

Frederiks and his colleagues performed some research on AL's electric conductivity and dielectric losses in it. Their aim was to obtain information on the role of dipoles and the "swarms" in AL. These are the fields (domanis) of uniform orientation of molecules in a liquid crystal (Beneshevich, Frederiks, and Mikhailov 1935a and 1935b). (The concept of such domains was first introduced in 1909.) One can even insist that the results they obtained served as a stimulus to abandon the theory of swarms. In the early 1950s it was replaced with a continuum theory. Another series of articles was devoted to the studies of optical (interference) effects that appear when mechanical forces, both constant and variable, influence AL (Frederiks and Zolina 1933,[13] Zolina 1936).

In the seventy-some years that separate us from the pioneering works of Frederiks and his colleagues, the studies of liquid crystals and their non-linear properties gave rise to a new branch of molecular physics bordering the physics of polymers and biology. His name is invariably present in books and surveys in this area; the transitions named after him are discussed in scientific publications. More and more refined experiments make it possible to formulate ever new and more exact theories of what is going on in AL. Let us mention one example of AL's behavior in the field of a light wave—in a laser beam. It turns out that the electric field of this wave causes a reorientation of AL's optical axis. This brings forth some vivid physical effects including the light-induced (threshold) Frederiks's transition that can be observed when AL interacts with a powerful laser beam.[14]

While Frederiks and his colleagues at the Institute of Physics and Technology (Repieva, Zolina, Mikhailov, Beneshevich, and Shulvas-Sorokina) and at the Institute of Physical Research of the Leningrad University (Tsvetkov) were carrying out the research described above, their efforts seemed to be restricted to "pure" physics and totally divorced from any practical applications. In his report of 1936 Frederiks outlined, in more general terms, possible technological applications. By the early 1960s the situation had dramatically changed with a rapid development of computer technology. In looking at a display of an electronic clock or a micro-calculator few of us realize that they are the result of the discoveries made by Frederiks and his colleagues in the 1930s. The figures in clocks and micro-calculators are formed by sending weak electric voltage to electrode stripes the combination of which produces one of the ten digits. In the systems that employ the effect of dynamic light diffusion, the incident light is reflected from a mirror surface covered with a layer of nematic. Light is not allowed to pass through a part that is placed above the stripes that are under super-threshold voltage. The result is dark digits against a lighter background. If an indicator system includes crossed polarizers, an analogous effect appears due to deformation (turn of the optical axis) in the super-threshold electric field: light fails to pass through the AL stripes above the electrodes. In the first case the situation is described by equation (1), in the second by equation (2). The effect described by (1) is not widely used, yet it is popular in experiments. It is very important from the historical point of view as the starting point from which the studies of the Frederiks thresholds began.

In 1935 he was 50. Even if this was not his creative peak, he was very active in science and enjoyed authority among his colleagues, both theoreticians and experimenters. He headed laboratories at the Institute of Physics and Technology and the University, was a professor at the Leningrad Polytechnical and a lecturer at the University. There was a rapid progress in the physics of liquid crystals and Frederiks had a group of talented young pupils and colleagues.

In 1936, Frederiks and Tsvetkov sent another article on this subject to the Soviet journal *Acta Physicochimica*. By the time the article appeared Frederiks had been arrested on a false pretext (September 1936) and sent to prison. His family keeps the letters he wrote from imprisonment. There is not a single word about the appalling conditions he had to survive. These letters provide an evidence of his noble nature and staunch spirit. In the late 1930s the conditions improved somewhat. He wrote that he could even do some research work. In his letters he thanked his friends who sent him latest publications on liquid crystals. We do not know whether he was informed that his friends and relatives were fighting for him. Dmitri Shostakovich, his brother-in-law, took close to heart the misfortune of his sister who was a pianist, and tried to help Frederiks. On 22 April 1941, Frederiks wrote to her from Ukhta (a town in the northern part of European Russia): "As for me, I continue my research and am quite satisfied with it. I feel quite well. To save you from much of unnecessary trouble I'll consult the local doctors and send you all the details. The other day I heard Dmitri's quintet over the radio: all of us, me included, were delighted."

One should not take this letter at face value. Evidently, he tried to protect his family against unnecessary anguish. His research was restricted to manipulations with samples of oil, something that any high-school student could perform, while the local doctors were either his fellow prisoners or the personnel from the concentration camp's medical department. All these details became known from Prof. L. Polak who had served a term in the same camp. He testified that the high moral qualities of Frederiks produced a great impression on all those who met him.

Igor Kurchatov's colleagues from Laboratory No. 2[15] recounted that their patron who held Frederiks and his research work in high esteem spared no effort to get him out of the camp.

In 1943, victory seemed close at hand: according to some people he was discharged from prison, according to others he was just transferred from the northern camp to some other place to do physical research. He was never destined to reach it: on his way there he fell ill with pneumonia and died in a train while passing through Gorky. He was buried in the city where he spent part of his youth. The last pages in his personal file at the Institute of Physics and Technology are an official confirmation of his rehabilitation.

5. Appendices

5.1 A LETTER FROM HILBERT TO THE CURATOR OF GÖTTINGEN UNIVERSITY (THE GEORGE AUGUSTUS UNIVERSITY)[16]

Vitznau (Switzerland)[17]
Thursday, 28 September 1916

Dear Mr. Curator,

In the coming winter semester I am going to read a lecture course on the mathematical foundations of physics (four hours a week).[18] It will place high demands on those who will attend. So I would like, your permission granted, to restrict myself to a carefully selected number of students, mainly from among the members of the mathematical and physical societies.

Dr. von Frederiks, a Russian citizen and a former assistant at the Physical Institute, has already helped me on a private basis.[19] I shall find everything much easier if I have him at my lectures. Therefore, I ask Your Honor to allow me this. I think that your oral consent to Dr. von Frederiks's presence at my lectures, pending your special instructions, will suffice with an understanding that he will never replace me as a lecturer and will be present only as a private assistant [als private Hülfe]. I refuse to think that any of my audience would be displeased with his presence. Still, I can always find out its attitude through my assistants.

With the most profound feelings and greatest respect,

D. Hilbert.[20]

5.2 THREE LETTERS FROM FREDERIKS TO HILBERT [21]

Moscow, 21 July 1918

Esteemed Mr. Counselor,[22]

I am back home at last after a lot of traveling and troubles. What your newspapers write about us has nothing to do with reality. Be that as it may, I live here, and I am very glad to find myself at home. Best regards to you and your wife.

Yours,
V. Frederiks

Moscow, 14 September 1922

Esteemed Mr. Counselor,

Thank you very much for your March letter.[23] I deeply regret that I was unable so far to come to Göttingen; I would like to go there very much but there were circumstances that prevented me from doing so. Let's hope that the next year proves more favorable for my foreign trip. There is no doubt that Göttingen will be my first destination.[24] A friend of mine, Georg Krutkov,[25] professor of theoretical physics at Petersburg University,[26] will come to Germany some time this year. At present,[27] he is one of the best Russian theoreticians. This is probably not much, yet he is a very talented scientist. He is working in statistical mechanics and quantum theory in connection with adiabatic invariants.[28] People abroad know little about him, as the conditions for publication of Russian works there are beginning to improve only now. Unfortunately he could not come to Göttingen in summer. He would have been much interested in the very important seminar you wrote me about.[29] Krutkov will tell everything about me: after all, one cannot tell a lot in one letter.

I found myself in Moscow by pure chance on my way to Nizhni Novgorod to attend the congress of Russian physicists.[30] Normally I live in Petersburg with my closest relatives. I regret to say that I lecture too much to the detriment of my other work. I use this opportunity to thank you once again for these splendid field glasses (*Fernstecher*).[31] I will be most indebted to you if you give them to Mr. Krutkov.

My best and most cordial regards to you and your wife from your sincere friend,

V. Frederiks

Petersburg
Physical Institute,[32] 1 August 1923

Most Esteemed Mr. Counselor,

About a year ago I wrote to you about my friend Mr. Krutkov who wanted to meet you in Göttingen. I got a short letter from him the other day in which he wrote about your cordial reception. When I think about your friendly attitude and kindness you showed to me I always feel profound and sincere gratitude.

One of my friends, mathematician A. Friedmann, is going to Germany this year. He also would like to meet you in Göttingen.[33] Along with other problems he works a lot on the relativity theory and pays much attention to the trend associated with the names of Weyl and Eddington.[34] While introducing a concept of a special geometrical image—a certain "plane"—for an arbitrarily curved space, he invoked one more gauge vector, in addition to that suggested by Weyl (*Eichungsvektor*).[35] This made it possible to interpret the electrodynamic equations without recurring to the Mie theory.[36] We have decided to write a book on the relativity theory.[37] True, much has been written about it, yet clarity of exposition and interpretation leave much to be desired. To my mind, the situation with electrodynamics is especially bad—we are going to write about it, too.

Cordial greetings to you and your wife from your friend

V. Frederiks

Acknowledgments. In conclusion the authors want to express their gratitude to those who made this article possible: Dmitri Frederiks, Dr. Haenel (Göttingen University), R. Rami, associate at the CERN Archive in Geneva, Dr. L. Belloni (Milan University), L. Polak, A. Yushkevich, and B. Bulyubash for their help in obtaining archive materials used in the article and also L. Polak and G. Gorelik for their valuable suggestions and B. Lange (The Institute of the Theory, History, and Organization of Science of the GDR Academy of Sciences in Berlin) who was kind enough to translate the letters of V. Frederiks and D. Hilbert reproduced in Appendices into Russian.

NOTES

[1] This section is based on materials from the archives of the Ioffe Institute of Physics and Technology, Kalinin Polytechnical Institute in Leningrad, and from family archives.

[2] Not infrequently, Charles Guye is taken for a French physicist Louis Guye. It is not surprising, as they both worked in molecular and statistical physics. There was one more Guye among the professors of chemistry at the University of Geneva during Frederiks's time there: Philip August (1862–1922), well-known for his works in stereo-chemistry. Probably, he was Charles's elder brother.

[3] Frederiks's thesis, which he dedicated to his teachers, did not reveal his Russian origins. Since the name index in Khovlson's first volume of *A Course Of Physics* mentioned him among Russian authors one may conclude that the author had known Frederiks before the revolution of 1917.

[4] Prof. E. Shtrauf is far less known than his colleagues. He used the chapters he wrote for the *Course* as the basis for two more books for physics students (Shtrauf 1949, 1950).

[5] It is interesting to note that, while he was a student in Geneva, he had an opportunity to attend a special course on Huygens offered by A. Berneau under the history of science curriculum in 1906–07.

[6] See the Biographical Essay and the Appendices to this article.

[7] In his book Frederiks wrote: "A full and complete presentation of this theory requires a highly complex mathematical apparatus that is way above the heads of virtually all physicists" (Frederiks 1921, p. 162). It goes without saying that when need arose physicists had to master Riemannian multi-dimentional geometry and tensor analysis. Still, the new mathematics was extremely difficult and went far beyond the limits of the habitual. In his survey Frederiks was wrong when he wrote that, in the absence of matter (that is, when the energy-momentum tensor vanishes), the metric is always flat (see Frederiks 1921, p. 176).

[8] "I believe that Einstein's work of 1905 can be regarded as a work of genius: he let us see the profound and at the same time simple physical meaning of Lorentz's hypothesis on the contraction of bodies as the effect of motion and the meaning of the transformations of coordinates introduced by Lorentz and of some other formal calculations. . . . Einstein demonstrated that our previous concepts of length, time, and certain other physical quantities were wrong because they did not take into account the true nature of such quantities and the methods of their measurement" (Frederiks 1935, p. 5).

[9] "It seems that the following two circumstances were responsible for the failures [in creating a unified geometrized field theory]: the geometrical ideas on parallelism, the direction of shifts of line segments, and the change of the length scale are devoid of physical content when applied to real objects. The second explanation is even more significant: there is a fundamental obstacle to creating a unified field theory: it requires a purely macroscopic theory" (Frederiks 1935, p. 16). By the mid-1930s, the quantum nature of electromagnetic radiation had been firmly established and the foundations of quantum electrodynamics set in.

[10] In conclusion Frederiks wrote: "On the whole, unsuccessful attempts at a unified field theory testify, nevertheless, that we are not far from a discovery of new amazing and significant physical laws. I believe that these works are methodologically valuable: they show how closely the success of a physical theory

is related to the circumstances of its creation on a physically solid basis and how different its development is in the absence of the relevant supporting evidence" (1935, p. 17).

[11] The expression "anisotropic liquid" is no less paradoxical.

[12] In August 1930, on Frederiks's instructions, Zolina delivered this paper at the First National Congress of Physicists.

[13] This article was written because Frederiks was invited to take part in a discussion on the physics of AL held by the Faraday Society in April 1933 in Great Britain. A special issue of the Society's *Transactions* published works by the leading specialists in the field, including J. Bernal, W. Ostwald, T. Stewart, and others.

[14] A detailed description of the experiment and the theory can be found in Zel'dovich and Tabiryan 1985, together with an extensive bibliography on this and related problems.

[15] Later to become the Kurchatov Institute of Atomic Energy, the cradle of the Soviet atomic project.

[16] From "Hilbert Personalakte" kept at the Göttingen University Archives. The Editors thank Dr. Ulrich Hunger (Universitätsarchiv Göttingen) for permission to cite this letter in English here.

[17] Vitznau is a small settlement on the shores of Lake Luzern in Switzerland, which Hilbert frequented during World War I, especially when there were food shortages (the worst year was 1916).

[18] There is no doubt that the lecture course was built around general relativity theory and Hilbert's ideas regarding the axiomatic structure of theoretical physics. Between late 1915 and 1918, Hilbert was tirelessly working on these problems. Max Born (1922) later recalled Hilbert's lectures on relativity theory.

[19] According to the documents found in the Göttingen University Archives (Akte IIID, 335 (58)), Frederiks was admitted to the University's Physical Institute (Department of Theoretical Physics) as a *Hilfsassistent* on 1 September 1911. He remained in this post until World War I. In 1913, Department of Theoretical Physics was renamed Department of Mathematical Physics; at that time it was headed by Voigt. All aliens who had failed to leave Germany (including Frederiks) were forbidden to work at the University. This sentence shows that Hilbert officially regarded Frederiks as his private assistant, which he probably was, throughout the war.

[20] In submitting his request, Hilbert displayed a wealth of kindness, courage, and internationalism. On 25 October 1916, however, the University administration turned his request down.

[21] Copies of these letters were kindly provided by the Hilbert Archive kept in the Göttingen State and University Library (*Nachlass Hilbert* 106). The Editors wish to thank Dr. Helmut Rohfling (Niedersächsische Staats- und Universitätsbibliothek Göttingen) for his kind permission to cite these letters in English here.

[22] Hilbert obtained this title (*Geheimrat*) from the German government for his services to German science. For his attitude to this high rank, see Reid 1970.

[23] So far Hilbert's letters to Frederiks have not been found.

[24] As far as we know, Frederiks never traveled abroad upon his return to Russia.

[25] Frederiks refers to Yuri Krutkov (1890–1952), Professor of Petrograd University since 1921. Frederiks calls him by the German equivalent of his name. In 1922–23 he travelled abroad and visited Berlin, Göttingen, and Leiden. In May 1923, we had a discussion with Einstein in Leiden, in which he was defending Friedmann's cosmological results (see Frenkel 1970, 1974, and this volume).

[26] Frederiks continued to call Petrograd by its old name "St. Petersburg."

[27] The German text of the letter contains a spelling error here: *augenglicklich* instead of *augenblicklich*.

[28] Paul Ehrenfest, Krutkov's teacher, laid the foundations of the theory of adiabatic invariants and its application to quantum theory (see, for example, Frenkel 1977). Later Krutkov wrote a detailed survey of the theory (Krutkov 1931). For its present state, see Bakay and Stepanovskiy 1981.

[29] One can guess that Frederiks refers here to the famous Göttingen seminar on the structure of matter. It was started by Hilbert and Debye during World War I. In the early twenties it was headed by Born and J. Franck.

[30] Frederiks refers to the Third Congress of the Russian Association of Physicists held in Nizhni Novgorod from 17 to 21 September 1922.

[31] Hilbert probably purchased binoculars, at Frederiks's request. Frederiks called them *Fernstecher*, not *Fernglas* or *Feldstecher*.

[32] That is, the Physical Institute at Petrograd University, where Frederiks worked, together with Friedmann, in the Atomic Commission.

[33] We know that, in August–September 1923, Friedmann visited Germany and, in particular, Berlin. It remains unknown, however, whether he was in Göttingen and met Hilbert. It is equally unclear whether he met Einstein.

[34] By the time Frederiks wrote this letter, Friedmann had no published works in the field associated with the names of Weyl and Eddington, that is, on unified geometrical field theories. The only work on this subject, which he wrote together with J. Schouten, appeared in 1924 (Friedmann and Schouten 1924). An introduction to the book (Frederiks and Friedmann 1924) makes it clear that they both pinned great hopes on the concept of a unified geometrized field theory.

[35] One can surmise that Frederiks refers to symmetrical and semi-symmetrical connections (according to Schouten's terminology). It should be noted that in the symmetrical case, when $\Gamma^{\alpha}_{\mu\nu} = \Gamma^{\alpha}_{\nu\mu}$, semi-symmetrical connection turns into the Weyl connection. Various geometries of this sort were discussed in the article by Friedmann and Schouten mentioned in the previous note. They had not, however, led to promising prospects for unified field theories. See Vizgin 1994.

[36] In 1915, Hilbert made an attempt to create a unified field theory of gravitation and electromagnetism relying on GRT and on Mie's non-linear electrodynamics (see Vizgin 1994). All unified geometrized field theories starting with Weyl's sought to interpret both the gravitational and electromagnetic fields as

various aspects of a unified geometrical structure. They did not involve the assumptions of the Mie theory.

[37] The "book on the relativity theory," which Frederiks mentions in this letter, was never finished. Only a small part of it was published (Frederiks and Friedmann 1924).

REFERENCES

Afanasiev, A. P., and Frederiks, V. K. (1935). *Kurs obshchey fiziki*. [*A Course of General Physics*] Leningrad: Kubuch.

Bakay, A. S., and Stepanovskiy, Y. P. (1981). *Adiabaticheskie invarianty*. [*Adiabatic Invariants*] Kiev: Naukova dumka.

Beneshevich, D., Frederiks V., and Mikhailov, G. (1935a). "Ob elektroprovodnosti anizotropnoy zhidkosti," ["On the Electric Conductivity of Anisotropic Liquids"] *Doklady AN SSSR* 2: 208–212.

— (1935b). "O dielektricheskikh poteryakh v anizotropnoy zhidkosti." ["On Dielectrical Losses in Anisotropic Liquids"] *Doklady AN SSSR* 2: 469–473.

Born, Max (1922). "Hilbert und die Physik." *Naturwissenschaffen* 10: 588–93.

Dobronravov N. I., Nasledov D. N., Khariton, Y. B., and Shtrauf, E. A. (1941). *Kurs obshchey fiziki* [*A Course of General Physics*], vol. 1. Moscow and Leningrad: Gostekhteorizdat.

Eddington, Arthur S. (1923). *The Mathematical Theory of Relativity*. Cambridge: Cambridge University Press.

Einstein, Albert (1925). "Einheitliche Feldtheorie von Gravitation und Elektrizität." *Preussischen Akademie der Wissenschaften. Physikalisch-Mathematische Klasse. Sitzungsberichte* 22: 414–419.

Fock, Vladimir A. (1926). "Über die invariante Form der Wellen- und der Bewegungsgleichungen für einen geladenen Massenpunkt." *Zeitschrift für Physik* 39: 226–232.

— (1966). "Raboty Fridmana po teorii tyagoteniya Einsteina." ["Friedmann's Works on Einstein's Theory of Gravity"] In A. A. Fridman, *Izbrannye Trudy*. [A. A. Friedmann, *Selected Works*] Moscow: Nauka, pp. 398–402.

Frederiks, Vsevolod K. (1910). *Sur le frottment intérieur des solides aux basses températures*. Thése No. 433. Geneva.

— (1921). "Obshchiy printsip otnositel'nosti Einshteina." ["Einstein's General Principle of Relativity"] *Uspekhi Fizicheskikh Nauk* 2: 162–186.

— (1924). "Predislovie k russkomu izdaniyu." ["Introduction to the Russian Edition"] In E. Freundlich, *Osnovy teorii tyagoteniya Einshteina*. [*The Basics of Einstein's Theory of Gravity*] Moscow and Petrograd, pp. 5–6.

— (1925). "O silakh tsentrobezhnykh i Coriolisa v obshchei teorii otnositel'nosti." ["On the Centrifugal and Coriolis Forces in General Relativity"] *JRPCS. Physics Section* 57: 475–483.

—— (1927a). "Nachala mekhaniki N'yutona i printsip otnositel'nosti." ["Foundations of Newtonian Mechanics and the Principle of Relativity"] *Uspekhi Fizicheskikh Nauk* 7: 75–86.

—— (1927b). "Teoria Schredingera i obshchaya teoriya otnositel'nosti." ["Schrödinger's Theory and General Relativity"] In *Osnovaniya novoy kvantovoy mekhaniki [Foundations of New Quantum Mechanics]*. Moscow and Leningrad, pp. 83–98.

—— (1929). *Elektricheskaya razvedka poleznykh iskopaemykh po metodu izmereniya peremennykh magnitnykh poley. [Electrical Search of Earth Metals Based on Measurements of Variable Magnetic Fields]* Leningrad.

—— (1933). "Predislovie k russkomu izdaniyu." ["Introduction to the Russian Edition"] In A. Kopff, *Osnovy teorii otnositel'nosti Einshteina. [The Basics of Einstein's Theory of Relativity]* Leningrad and Moscow: Gostechteorizdat, p. 3.

—— (1934). *Elekrodinamika i vvedenie v teoriyu sveta. [Electrodynamics and Introduction to the Theory of Light]* Leningrad.

—— (1935). "Sovremennoe sostoyanie voprosa o teorii otnositel'nosti (K 30-letiyu teorii otnositel'nosti)." ["The Present State of the Theory of Relativity (On the Occasion of its Thirtieth Anniversary)"] *Sorena* 8: 3–17.

Frederiks, Vsevolod K., and Bronstein, Matvei P. (1931). "Otnositel'nosti, teoria.' ["The Theory of Relativity"] In *Tekhnicheskaya entsiklopediya [Encyclopedi of Technology]*, vol. 15. Moscow and Leningrad, pp. 352–359, 362–367.

Frederiks, Vsevolod K., and Friedmann, Aleksandr A. (1924). *Osnovy teor otnositel'nosti [The Basics of Relativity]*, part 1. Leningrad.

Frederiks, Vsevolod K., and Izakson, A. A. (1926a). "Über die einheitliche Fel theorie Einsteins." *Zeitschrift für Physik* 38: 48–60.

—— (1926b). "Zur Frage des räumlich ausgedehnten Elektrons in der allgemein Relativitästheorie." *Zeitschrift für Physik* 38: 788–802.

Frederiks, Vsevolod K., and Ivanenko, Dmitri D., eds. (1935). *Printsip otnc tel'nosti: Sbornik rabot klassikov relativizma. [The Principle of Relativity Collection of Works by the Founders of Relativity]* Leningrad: ONTI.

Frederiks, Vsevolod K., and Mandel, G. A. (1926). "Tvyordyi elektron Born pole tyagoteniya." ["Born's Rigid Electron in a Gravitational Field"] *JRP Physics Section* 58: 387–393.

Frederiks, Vsevolod K., and Repieva, A. (1927a). "K teorii anizotropnykh zhic stey i nekotorye novye nablyudeniya nad nimi." ["On the Theory of An tropic Liquids and Some New Observations of Them"] In *V S"yezd russ fizikov. Moscow, 15–20 December 1926. [Proceedings of the Fifth Cong of Russian Physicists]* Moscow and Leningrad, pp. 16–17.

—— (1927b). "Theoretisches und Experimentelles Frage nach der Natur der ar tropen Flüssigkeiten." *Zeitschrift für Physik* 42: 532–546.

Frederiks, Vsevolod K., and Shekhter, Anna B. (1928). "Vychislenie astron cheskoy aberratsii i parallaksov v mirakh Einsteina, de Sittera i Fridma tochki zreniya obshchey teorii otnositel'nosti." ["Calculation of Astronc

Aberration and Parallaxes in the World Models of Einstein, de Sitter, and Friedmann from the Point of View of General Relativity"] *JRPCS. Physics Section* 60: 469–484.

Frederiks, Vsevolod K., and Tsvetkov, V. (1934). "Über die Orientierung der anisotropen Flüssigkeiten in dünnen Schichten and die Messung einiger die elastischen Eigenschaften charakterisierenden Konstanten." *Physikalische Zeitschrift der Sowjetunion* 6: 490–504.

—— (1935a). "Über die Einwirkung des elektrischen Feldes auf anistrope Flüssigkeiten." *Acta Physicochimica URSS* 3: 879–912.

—— (1935b). "O dvizhenii voznikayushchem v anisotropnoy zhidkosti pod deystviem elektricheskogo polya." ["On the Motion Arising in an Anisotropic Liquid in the Electric Field"] *Doklady AN SSSR* 4: 123–125.

Frederiks, Vsevolod K., and Zolina, V. (1930). "O primenenii magnitnogo polya k izmereniyu sil, orientiruyushchikh anizotropnye zhidkosti v tonkikh odnorodnykh solyakh." ["On the Use of Magnetic Fields to Measure Forces Orienting Anisotropic Liquids in Thin Uniform Salts"] *JRPCS. Physics Section* 62: 457–464.

—— (1931). "Über die Doppelbrechung dünner anistropflüssiger Schichten im Magnetfelde und die diese Schicht orientierenden Kräfte." *Zeitschrift für Kristallographie* 79: 255–267.

—— (1933). "Forces Causing the Orientation of an Anisotropic Liquid." *The Faraday Society. Transactions* 29: 919–928.

Frenkel, Viktor Ya. (1970). "Yuri Aleksandrovich Krutkov." *Soviet Physics. Uspekhi* 13: 816–825.

—— (1974). "Novye materialy o diskussii Einshteina i Fridmana po relyativistskoy kosmologii." ["New Materials on the Einstein–Friedmann Debate on Relativistic Cosmology"] In *Einshteinovskiy Sbornik, 1973*. V. L. Ginzburg and U. I. Frankfurt, eds. Moscow: Nauka, pp. 5–18. The English version is reprinted in this volume.

—— (1977). *Paul Ehrenfest.* 2nd ed. Moscow: Atomizdat.

Friedmann, Aleksandr A., and Schouten, Jan Arnoldus (1924). "Über die Geometrie der halbsymmetrischen Übertragungen." *Mathematische Zeitschrift* 21: 211–223.

Friedmann, D. Z., and Nieuwenhuizen, P. van (1985). "The Hidden Dimensions of Spacetime." *Scientific American* 252: 74–81.

Fresnel, O. J. (1928). *O svete.* [*On Light*] V. K. Frederiks, ed. Moscow and Leningrad: ONTI.

Ginzburg, Vitaly L. (1969). "Ob izluchenii i sile radiatsionnogo treniya pri ravnomerno uskorennom dvizhenii zaryada." ["On the Radiation and Radiative Damping of a Uniformly Accelerated Charge"] *Uspekhi Fizicheskikh Nauk* 98· 569–585.

—— (1975). *Teoreticheskaya fizika i astrofizika. Dopolnitel'nye glavy.* [*Theor Physics and Astrophysics. Addenda*] Moscow: Nauka.

— (1979). "Geliotsentricheskaya sistema i obshchaya teoriya otnositel'nosti (Ot Kopernika do Einshteina)." ["Geliocentric System and General Relativity (From Copernicus to Einstein)"] In *O teorii otnositel'nosti*. [*On the Theory of Relativity*] V. L. Ginzburg, ed. Moscow: Nauka, pp. 7–62.

Hilbert, David (1915). "Die Grundlagen der Physik: erste Mitteilung." *Köngliche Gesellschaft der Wissenschaften zu Göttingen. Mathematische-physikalische Klasse. Nachrichten*: 395–407.

— (1917). "Die Grundlagen der Physik: zweite Mitteilung." *Köngliche Gesellschaft der Wissenschaften zu Göttingen. Mathematische-physikalische Klasse. Nachrichten*: 55–76.

Huygens, C. (1935). *Traktat o svete*. [*A Treatise on Light*] V. K. Frederiks, ed. Moscow and Leningrad: ONTI.

Ioffe, Abram (1936). "Raboty po issledovaniyu zhidko-kristallicheskogo veshchestva." ("Works on Liquid Crystals") In *Problemly sovremennoy fiziki v rabotakh Fiziko-Tekhnicheskogo Instituta akademika A. F. Ioffe*. Moscow and Leningrad: USSR Academy of Sciences Press, pp. 37–40.

Ivanenko, D. D., Pronin, P. I., and Sardanashvili, G. A. (1985). *Kalibrovochnaya teoriya gravitatsii*. [*Gauge Theory of Gravitation*] Moscow: Moscow State University Press.

Kaluza, Theodor (1921). "Zum Unitätsproblem der Physik." *Preussische Akademie der Wissenschaften* (Berlin). *Sitzungsberichte*: 966–972.

Khvolson, Orest D. (1914). *Kurs Fiziki* [*A Course of Physics*], vol. 1. St. Petersburg: Braunschvieg F. Vieneg.

Klein, O. (1926). "Quantentheorie und fünfdimensionale Relativitätstheorie." *Zeitschrift für Physik* 37: 895–906.

Kopff, August (1921). *Grundzüge der Einsteinschen Relativitätstheorie*. Leipzig: S. Hirzel.

— (1933). *Osnovy teorii otnositel'nosti Einshteina*. [*The Basics of Einstein's Theory of Relativity*] Leningrad and Moscow: Gostekhteorizdat.

Krutkov, Yuri A. (1931). "Adiabaticheskie invarianty i ikh primeneniya v teoreticheskoy fizike." ["Adiabatic Invariants and their Applications in Theoretical Physics"] *JRPCS. Physics Section* 53: 3–171.

Lorentz, Hendrik Antoon (1916). "On Einstein's Theory of Gravitation" *Proceedings of the Section of Sciences, Koninklijke Akademie van Wetenschappen te Amsterdam* 19 (1916–1917): 1341–1354; 1354–1369; 20 (1917–1918): 2–19; 20–34 (communication of February 26 and April 1916).

Mandel, H. (1926). "Zur Herleitung der Feldgleichungen in der allgemeinen Relativitätstheorie." *Zeitschrift für Physik* 39: 136–145.

Reid, Constance (1970). *Hilbert*. Berlin and New York: Springer-Verlag.

Repieva, A., and Frederiks, Vsevolod K., (1927). "K voprosu o prirode anizotropno-zhidkogo sostoyaniya veshchestva." ["On the Nature of Anisotropic Liquids"] *JRPCS. Physics Section* 59: 183–200.

Schrödinger, E. (1963). *Space-Time Structure*. Cambridge: Cambridge University Press.

Shtrauf, E. A. (1949). *Molekulyarnaya fizika*. [*Molecular Physics*] Leningrad and Moscow: Gostekhteorizdat.

—— (1950). *Elektrichestvo i magnetizm*. [*Electricity and Magnetism*] Leningrad and Moscow: Gostekhteorizdat.

Sonin, A. S. (1983). *Vvedenie v fiziku zhidkikh kristallov*. [*Introduction to the Physics of Liquid Crystals*] Moscow: Nauka.

—— (1980). *Kentavry prirody*. [*Centaurs of Nature*] Moscow: Atomizdat.

Thirring, H. (1918). "Über die Wirkung rotierender ferner Massen in der Einstein-schen Gravitationstheorie." *Physikalische Zeitschrift* 19: 33–39.

Vizgin, Vladimir P. (1981). *Relativistskaya teoriya tyagoteniya: Istoki i formirova-nie: 1900–1915*. [*Relativistic Theory of Gravitation: Origins and Formation: 1900–1915*] Moscow: Nauka.

—— (1994). *Unified Field Theories in the First Third of the 20th Century*. Julian B. Barbour, trans. Basel and Boston: Birkhäuser.

Weyl, Hermann (1918). *Raum. Zeit. Materie*. Berlin: Springer-Verlag.

Zel'dovich, B. Y., and Tabiryan, N. V. (1985). "Orientatsionnaya opticheskaya nelineynost' zhidkikh kristallov" ["Orientational Optical Non-Linearity of Liquid Crystals"] *Uspekhi Fizicheskikh Nauk* 147: 633–647.

Zolina, V. (1936). "Uprugie kolebaniya anizotropnoy zhidkosti." ["Elastic Oscil-lations of Anisotropic Liquids"] *The Lomonosov Institute Transactions* 8: 11–13.

Einstein's Fluctuation Formula and the Wave-Particle Duality

Alexei Kojevnikov

1. Introduction

1.1 THE PROBLEM OF ENERGY FLUCTUATIONS

Apart from the history of physical concepts, one can also write the history of persistent problems to which various approaches were applied and, therefore, tested. In the theory of radiation during the first thirty years of this century, one of the most important problems of this kind was the problem of energy fluctuations. The theory of radiation was then a troublesome area of physics. The apparently well-established classical wave theory of light unexpectedly found itself in a crisis, which lasted several decades until a new fundamental theoretical scheme, quantum electrodynamics, became gradually established. In the meantime, many alternative approaches were suggested, yet none of them, despite partial successes, developed into a complete, consistent and universally accepted theory.[1]

The debate centered around several key problems of the classical theory. These anomalies were pushing the theory ahead, suggesting some interpretations, while confirming or refuting others. The list of these problems, in the chronological order of their appearance, included: the derivation and interpretation of the Planck formula, fluctuations of the energy and momentum of radiation, frequencies and intensities (transition probabilities) of atomic spectral lines, the Compton effect, and the dispersion of light.

Contemporary textbooks usually discuss Einstein's fluctuation formula in connection with the wave-particle duality, which the formula is supposed to illustrate. The standard story runs approximately as follows: Einstein believed in the duality of waves and particles from relatively early on. At

Yuri Balashov and Vladimir Vizgin, eds., *Einstein Studies in Russia.*
Einstein Studies, Vol. 10. Boston, Basel, Birkhäuser, pp. 181–228.

least in the case of light, he already thought about duality in 1909, when he derived the formula for fluctuations of electromagnetic energy with its two equally important terms, one wave and the other corpuscular (Einstein 1909a, Einstein 1909b). But Einstein had not discussed his dualistic beliefs publicly until 1924, when he became acquainted with similar ideas by Louis de Broglie (de Broglie 1924). In his second paper on the gas of Bose-particles, Einstein sided with the dualistic model proposed by the young French physicists (Einstein 1925). Thanks to this support, the previously unknown de Broglie's thesis began to be widely read, his idea of matter waves, in particular, was developed by Erwin Schrödinger into wave mechanics, and the principle of the wave-particle duality became generally accepted as one of the basic foundations of quantum physics.

Some important parts of the above story lack sufficient documentary support, especially regarding ideas in which Einstein believed but which he did not express publicly. Moreover, Einstein did not use the word "duality" either before or after 1925, nor did he make any clear assertion of the principle of the wave-particle duality. Rather than reflecting his personal views on the matter, the story seems to be telling more about perceptions of Einstein's position that were widespread among the physics community. Historians disagreed in their interpretations of the events. John Hendry assumes that Einstein was advocating the wave-particle duality already in 1909 (Hendry 1980, pp. 59, 66–67, 74) and also attributes the same view to Martin Klein (1963). Klein, however, is more careful in his statements and, without disagreeing with the standard interpretation explicitly, stops short of asserting it either, and so do major histories of quantum mechanics (Jammer 1966, Mehra and Rechenberg 1982). Among historically minded physicists, Friedrich Hund (1967, p. 44) writes: "With his paper on the statistical fluctuations in radiation Einstein introduced the duality of waves and particles for light," only to add immediately that "he certainly could not yet see the full consequences of this revolutionary idea."[2] Leon Rosenfeld, on the contrary, asserts that Einstein's 1909 paper was generally understood as, and indeed was, an argument in favor of the corpuscular model of light (Rosenfeld 1973, p. 252), while according to Abraham Pais, Einstein in 1909 "was prepared for a fusion theory (of wave and corpuscles). . . . This fusion now goes by the name of complementarity" (Pais 1982, p. 404).

This spectrum of opinions reflects the fact that the wave-particle duality was and remains a rather vague concept that has neither been well defined nor used with sufficient consistency. Leaving aside the question of whether a precise definition is possible, the present paper looks more carefully into the development of Einstein's and others' views regarding the energy fluctuations in radiation, which could and have been interpreted in a variety of different ways. In fact, the history of one single formula can lead us through practically all major rival approaches in the theory of radiation, duality being only one of them.

1.2 FORMULAE

The formula for the energy fluctuations in radiation appeared first in Einstein 1909a. Consider a thermodynamical system divided into two parts: one, relatively small, which from now on will be called the "system," and the remaining larger one, the "reservoir." Their total energy is conserved, but the energy of the smaller part fluctuates around certain equilibrium value E due to exchanges with the reservoir. Regardless of the nature of the system, general laws of thermodynamics give the following expression for the mean square of these fluctuations:

$$\overline{\varepsilon^2} = kT^2 \frac{\partial E}{\partial T}, \tag{1}$$

where k is the Boltzmann constant and T the temperature in equilibrium.

This generally valid formula can be applied to the specific case of a system consisting of a blackbody radiation with temperature T in the frequency interval $(v, v + dv)$ placed in a cavity of the total volume V. In this case, E is determined by the Planck law of the spectral distribution of radiation energy

$$E = \frac{8\pi h v^3}{c^3} \frac{1}{e^{hv/kT} - 1} V dv. \tag{2}$$

Inserting (2) into (1) leads directly to Einstein's main fluctuation formula

$$\overline{\varepsilon^2} = hvE + \frac{c^3}{8\pi v^2 V dv} E^2. \tag{3}$$

It can also be rewritten in the following convenient way:

$$\overline{\left(\frac{\varepsilon}{E}\right)^2} = \frac{hv}{E} + \frac{c^3}{8\pi v^2 V dv} = \frac{1}{N} + \frac{1}{Z}, \tag{4}$$

where N is the number of energy quanta $h\nu$ in E, and Z the number of modes of stationary waves within volume V and frequency interval $d\nu$.

The above derivation can be called "thermodynamical." One can also proceed from the "statistical" point of view which considers microscopic states of the system and their probabilities. To derive (3) this way, one has to postulate certain microscopic models of the structure of radiation, and this is where difficulties arise. Each of the two terms in either (3) or (4) can be obtained rather easily, but only separately and from two apparently incompatible microscopic models: the first one from the model of radiation as an ideal gas of particles with energy $h\nu$; the second one from the classical model of radiation as waves with frequency ν. In the first case, fluctuations of energy arise because the number of particles within the volume V fluctuates. In the second case, the energy of the system fluctuates due to the interference between waves with accidental amplitudes and phases. The subsequent history dealt primarily with the problem of how to obtain the complete result with both of its terms from one unambiguous microscopic model of radiation.

Two other fluctuation formulas, also coming from Einstein, are relevant to the story. The first one is quite similar to (3) and describes the fluctuations of the momentum transmitted by radiation. A mirror with surface area f placed in a cavity will have its momentum fluctuate due to the fluctuations of radiation pressure. From the Planck law, one can obtain the following expression for the mean square of these fluctuations during the time interval T:

$$\overline{\Delta^2} = \frac{f\tau}{cV}\left(h\nu E + \frac{c^3}{8\pi\nu^2 V d\nu} E^2\right), \qquad (5)$$

which has a form similar to (3) and poses a similar difficulty with regard to its microscopic derivation and interpretation.

The last of the fluctuation formulas applies only in the limiting case of short waves (the so-called "corpuscular" region). The Planck law in this limit can be approximated by the Wien formula

$$E = \frac{8\pi h\nu^3}{c^3}(e^{-h\nu/kT}) V d\nu, \qquad (6)$$

from which one obtains only the first, quantum or corpuscular, term of energy fluctuations (3). In this limit, one can also calculate the probability of another, quite different and extremely rare fluctuation: an event in which the entire energy of radiation concentrates in a fraction U of the total volume V of the cavity:

$$W = (U/V)^{E/h\nu}. \tag{7}$$

Formula (7) has the same form as the expression for the probability for all atoms of an ideal gas of N particles to be found within the same fraction of the total volume V:

$$W = (U/V)^{N}. \tag{8}$$

This striking similarity constitutes a challenge for the classical wave theory of radiation.

2. Fluctuations before Duality

2.1 THE ATOMICITY OF LIGHT (1905–1909)

The very way Equations (3–5) are written explains why it is so tempting for textbook authors looking for pedagogical simplicity to connect the fluctuation formula with the wave-particle duality and to attribute the same interpretation to Einstein's paper of 1909. However, Einstein's early papers and letters reveal, rather surprisingly, little concern over the opposition of waves and particles. Later he would recognize the dilemma and would become troubled in a most serious way by the impossibility of combining the two concepts, but during the first several years he formulated the main problem regarding light quanta and the structure of radiation in somewhat different terms.

His most persistent idea was that of an analogy between radiation and matter and he sought a uniform treatment for both. His very first paper introducing the hypothesis of light quanta opens with an observation on the existing dissimilarity between theoretical descriptions of ponderable matter and of electromagnetic phenomena (Einstein 1905, p. 132), which he then attempts to overcome. "Light carries mass with it," was his immediate conclusion after the discovery of the equivalence of mass and energy in relativity theory (Einstein to Conrad Habicht, June–September 1905, Einstein 1993b, Doc. 28, p. 33). "The theory of relativity," Einstein added further in his 1909 Salzburg presentation, "has thus changed our view of the nature of light insofar as it does not conceive of light as a sequence of states of a hypothetical medium, but rather *as something having an independent existence just like matter*" (Einstein 1909b, p. 490, italics added).

In his first paper (Einstein 1905) on the analogy between radiation and an ideal gas, Einstein described radiation with the help of the Wien formula, which perfectly suited the analogy, and derived there the fluctuation formula (7). Planck's more general law of the energy spectrum of radiation complicated the analogy and, therefore, Einstein regarded it as somewhat contradicting his views.[3] Already in 1905, he possessed all the necessary tools for the derivation of the first term of the fluctuation formula (3),[4] but he did not make this step until 1909. In the meantime (Einstein 1906), he accepted the Planck law and thus obtained formulas (3) and (5) with both their terms (Einstein 1909a). Yet even in 1909 Einstein still regarded primarily the Wien part of the energy spectrum as experimentally verified, while apparently retaining some reservations about the exact validity of the Planck formula and mentioning in passing a possibility that it might later be found incorrect.[5]

In a letter to Hendrik Antoon Lorentz, Einstein referred to formulas (3) and (5) as "das geringfügige Ergebnis von jahrelangem Nachdenken" ("the trifling result of years of reflection"), the only real progress he had achieved in the problem of radiation since presenting his initial 1905 hypothesis on the structure of light (Einstein to Lorentz, 30 March 1909, Einstein 1993b, Doc. 146, p. 166). The low tone of his remark reflected an acknowledgment that he had not yet succeeded in developing a comprehensive new theory of radiation and an attempt to enlist Lorentz into effort. Despite obtaining both terms of (3), Einstein devoted most of his attention, rhetoric, and actual thoughts to the first one, which, "if present alone, would yield a fluctuation of the radiation energy equal to that produced if the radiation consisted of point quanta of energy $h\nu$ moving independently of each other."[6] Yet he constantly avoided—in the letter to Lorentz as well as in other papers and private letters of the time—referring to light quanta as either "particles" or "corpuscles." Rather than invoking the visual image of a mechanical particle, Einstein repeatedly compared his idea of light quantum to the (non-mechanical) electron: "Still for the time being the most natural interpretation seems to me that the occurrence of electromagnetic fields of light is associated with singular points just like the occurrence of electrostatic fields according to the electron theory" (Einstein 1909b, p. 499); "one should keep in mind the possibility of conceiving of light quanta and electrons as mathematically identically defined singularities" (Einstein to Lorentz, 23 May 1909, Einstein 1993b, Doc. 163, p. 195)

Lorentz's reply was, simultaneously, also the first draft of a paper published the following year (Lorentz 1910). Lorentz pointed out that the latest experimental results had confirmed the exact validity of the phenomenon of interference of light over very large distances in space, which could not be brought into agreement with the idea of spatially localized light quanta. Lorentz's rejection of Einstein's hypothesis was based on careful consideration rather than a prejudice or an a priori conviction, and he even expressed regret: "It is a real pity that the light quantum hypothesis encounters such serious difficulties, because otherwise the hypothesis is very pretty." In the published paper, however, these emotions were omitted and replaced by a categorical statement: "What is said is enough to demonstrate that there can be no discussion about the light quanta which are permanently indivisible and localized within a small space during the motion."[7]

Einstein was already aware of the substance of Lorentz's critique when he discussed the light quantum hypothesis at the annual meeting of the *Gesellschaft Deutscher Naturforscher* in Salzburg in September 1909. This did not make him change the content or tone of his argument in the main part of his paper, but only add two soothing statements at the beginning and at the end of his talk: "It is therefore my opinion that the next stage in the development of theoretical physics will bring us a theory of light that can be understood as a kind of fusion of the wave and emission theories of light"; and "All I wanted is briefly to indicate that the two structural properties (the undulatory structure and the quantum structure) simultaneously displayed by radiation according to the Planck formula should not be considered as mutually incompatible."[8]

These two remarks are often cited as an indication of Einstein's early dualistic views, yet a closer look at Einstein's own description of the kind of "fusion" he was looking for (in the last paragraph of his talk and in more detail in his reply to Lorentz's letter) reveals that his visions were rooted in classical electrodynamics and in Lorentz's concept of the electron. Einstein wrote to Lorentz:

> I believe that the light groups around singular points [a note appended at this point at the foot of the page reads: "not necessarily singular in the mathematical sense"] in a way similar to what we are accustomed to assume about the electrostatic field. Thus I think of a single light quantum as a point surrounded by a greatly extended vector field that somehow decreases with distance. The point is a singularity without which the vector field cannot exist. I wouldn't know to say whether one has to envision a simple superposition of the vector fields when many light quanta with mutually overlapping fields are present. In

any case, in order to determine the processes one would also have to have equations of motion for the singular points in addition to the differential equations for the vector field, if mathematical singularities are introduced. The energy of the electromagnetic field—at least in the case of sufficiently diffuse radiation—should be related, in some way or another, to the number of these singular points. Absorption would take place only in association with the disappearance of such a singular point or degeneration of the radiation field belonging to this point [a note appended at this point at the foot of the page reads: "Better: 'radiation based upon this point'"]. By specifying the motions of all singularities one would completely determine the vector field, so that the number of variables necessary for the characterization of radiation would be *finite.* . . .

By the way, the essential thing seems to me to be not the assumption of singular points but the assumption of field equations of a kind that permit solutions in which finite quantities of energy propagate with velocity c in a specific direction without dispersion.

One would think that this goal could be achieved through a slight modification of Maxwell's theory.[9]

In fact, two different models are discussed in this paragraph: one of light quanta as sources of the field, like electrons in electrodynamics, and the other of particle-like solutions of field equations, similar to many turn-of-the-century attempts to construct field models of the electron and to late Einstein's struggles with the field theory of quanta. Field models of the electron eventually disappeared from classical electrodynamics largely because relativity theory made them unnecessary, yet Einstein's attitude towards them was more positive than one could have derived from this fact. Einstein regarded the relationship between relativity theory and the electron models not as an opposition or rivalry but as being similar to the relation-ship between the thermodynamics of Thomson and Clausius and the underlying microscopic molecular dynamics of Boltzmann. Like the former, relativity theory and Maxwell's electrodynamics were deductive theories based on universal principles and formally complete. The Maxwell theory would remain, according to Einstein, valid as the description of average properties of macroscopic systems, yet this did not eliminate the need for an underlying microscopic theory of electrons and light quanta.[10]

Rather than either "dualistic" or "corpuscular," Einstein's early views on radiation should better be characterized as "atomistic," and his light quantum as modeled after the (non-mechanical, non-corpuscular) Lorentz electron. The electron—the singularity, the atom of electricity—was initially as foreign to continuous Maxwell's electrodynamics as Einstein's atom of light was to the wave theory of light, yet Lorentz succeeded in this case to marry the two seemingly incompatible concepts into a general theoretical scheme. This feat was exemplar for Einstein, and Lorentz was his hero. Therefore, when his persistent attempts at a comprehensive theory failed, Einstein turned to Lorentz in the hope that the latter would take the job. The success would have meant adding the quantum of radiation to the already existing quantum of electric charge and thus completing the atomistic reform of electrodynamics started by Lorentz at the end of the nineteenth century. As for the precise combination of continuous and discontinuous, field and quantum, aspects in the future theory of radiation, Einstein was ready to try different possibilities which were usually similar to the variety of field and corpuscular combinations discussed in contemporary theories of the electron[11] and hoped that either one or another would eventually work. Only gradually did he become aware of the difficulty, or futility, of the task.

2.2 ACKNOWLEDGING THE ANOMALY

Unlike Einstein's other arguments in favor of light quanta, fluctuation formulas did not offer any realistic hope for experimental test and thus remained purely theoretical considerations. Yet they were, in some sense, Einstein's strongest arguments: all his other examples could be downplayed as dealing with discontinuities in the process of the interaction between light and matter. Fluctuations, on the contrary, occurred in pure radiation and apparently were pointing directly towards its microscopic structure.

As is well known, most of Einstein's colleagues initially did not accept his arguments. His first papers on light quanta (Einstein 1905 and Einstein 1906) appear to have remained practically unnoticed. By the time of his report at the Salzburg meeting, Einstein was already a recognized scholar whose opinions had to be heard, and he did get a number of responses, but mostly critical ones (Discussion 1909, Lorentz 1910, Planck 1910, Ehrenfest 1911). Spokesmen of the theoretical community—Lorentz, Wilhelm Wien, Max Planck—criticized Einstein's suggestion in a strong tone that indicated the existence of a certain consensus. The only theoretician who expressed a more cautious attitude was Paul Ehrenfest, who was as young as Einstein and did not have the authority of the opponents. The list of main objections derived from a variety of publications, including Lorentz 1916, Einstein 1911, Discussion 1911, Wien 1913, Ehrenfest and Kamerlingh Onnes 1915, can be summarized as follows:

1. The division of fluctuation into two terms is artificial (Planck 1910, Discussion 1911). With the help of dimensional considerations, Einstein estimated that the classical theory could only account for the second term

in (3), but he failed to provide a direct proof. It might still be possible to derive the complete formula (3) from the wave theory of light (Planck 1910).

2. Radiation in a cavity with perfectly reflecting walls without absorption or emission is non-ergodic. Wave modes with different frequencies cannot exchange energy among themselves and hence thermal equilibrium cannot be established. Therefore, in general, statistical methods are not applicable to a system consisting of pure radiation without taking into account the presence of emitting and absorbing matter.

3. The first term in (3) refers to the interaction between matter and radiation rather than to the structure of radiation itself. Planck thought that "first of all one should attempt to transfer the whole problem of the quantum theory to the area of *interaction* between matter and radiation energy; the processes in pure vacuum could then temporarily be explained with the aid of the Maxwell equations" (Discussion 1909, pp. 585–586). This opinion was a rather widespread one, shared by most participants of the debate.

4. In order to derive formula (1), one had to assume the additivity of entropies of the system and the reservoir. It was not clear whether this was, indeed, the case for the blackbody radiation. For example, one could not make such an assumption in the case of coherent rays (Discussion 1911).

5. Statistically independent light quanta (independent in the sense of Boltzmann's statistics) do not correspond to the Planck law. Einstein's hypothesis is consistent only with Wien's approximate formula (Ehrenfest 1911, Ehrenfest and Kamerlingh Onnes 1915).

Most of these objections are very serious, reflecting deep understanding of the problems involved, and they should not be downplayed as a failure to understand Einstein's genius. By and large, they reflected the prevailing mood of justified conservatism. There were still too many blank spots in quantum theory that could have possibly caused the difficulty. Moreover, light quanta failed to provide any better alternative to the classical theory because, as had been already pointed out by Lorentz, they could not account for solidly verified phenomena of interference and diffraction. It looked as if it was still too premature to declare the end of the wave theory of light.

The fates of these objections were different, as will be discussed in subsequent sections. Lorentz refuted the first one by a direct and explicit calculation. The fifth one proved to be entirely correct and was resolved, eventually, in the adoption of Bose statistics instead of the classical Boltzmann statistics for light quanta. Lorentz's attempt to formulate the third idea in mathematical terms succeeded only partially. The issue remained unresolved and doubts about the need to quantize pure radiation would still be expressed at least until the late 1920s – early 1930s. They disappeared only gradually with the gradual establishment of quantum electrodynamics. Debates pro and contra the second and the forth objections remained on the level of verbal rather than mathematical discourse and did not disappear until many years later. They would still be remembered as a possible source of problem on a number of occasions, for example by Ehrenfest (1925a, 1925b) and by Heisenberg (1931), whenever the emerging quantum electrodynamics encountered new difficulties.

In view of the existing conflict of opinions, the fluctuation problem remained without an accepted solution for a number of years. Taking one side or the other was a matter of taste or fashion. The fashion, however, was changing. In the early period, it was established by the first Solvay meeting in 1911, which achieved a certain compromise concerning the theory of radiation. On the one hand, fluctuations were recognized as an "anomaly," but on the other hand, the line of "justified conservatism" prevailed and the small community of concerned theoretical physicists preferred to stick for the time being to the wave theory of light while placing the causes of all encountered difficulties in the yet unknown process of the interaction between light and matter.

Einstein's position also shifted towards a compromise, although he probably did not agree with any of the aforementioned critiques. In his report to the Solvay meeting (Einstein 1911), he modified the derivation of the fluctuation formula in order to circumvent the second objection, but the final result remained unchanged. He also became much more cautious in interpretation. Whereas previously Einstein had emphasized the importance of the first term in equation (3), by 1911 he had become aware that the second term contradicted the hypothesis of light quanta just about as strongly as the first one contradicted the wave theory of light. He thus called the problem an "unsolved puzzle," and the hypothesis of light quanta "a provisional attempt, . . . auxiliary idea, which does not seem compatible with the experimentally verified conclusions of the wave theory."[12] No longer insisting on his initial proposal, Einstein had to acknowledge his inability to solve the problem.

Most of the opponents of light quanta conceded, on their part, the general correctness of formula (3) as well as the impossibility of obtaining its first term from the wave theory of light. They hoped the solution would be found later and were inclined to see the first term as originating from the obscure process of the interaction between radiation and matter. This opinion, which was very close to consensus, was presented in the most careful and clear way in Lorentz's Paris lectures of 1912, which were published with a significant delay caused by the war in Europe (Lorentz 1916).

Although Lorentz's lectures covered a textbook topic, statistical mechanics, some parts, and in particular the treatment of fluctuations, constituted original new contributions to the field. In Lorentz's presentation, fluctuations were crucial for the understanding of the entire theory of blackbody radiation. Equation (3) was for him as fundamental as the Planck law: both were strictly equivalent and could be derived from each other, both were challenging the established theories in most serious ways (in the case of the Planck formula, the problem was in its derivation and in the interpretation of the constant h), and both, in Lorentz's expectation, were the most important keys towards further progress in theoretical physics. Subsequent developments confirmed his judgement: various approaches to the derivation of the Planck formula had their parallel attempts at resolving the difficulty with fluctuations, similar models were applied to both problems and usually with a similar degree of success. The following sections discuss these attempts to find the solution. There was quite a variety of them, as is expected at the time of crisis.

2.3 RESOURCES OF THE WAVE THEORY

Was the wave theory of light, in fact, incapable of explaining both terms in formula (3)? Einstein's assertion was not indisputable. Despite his firm conviction that classical theory only accounts for the second term, he was unable to provide a direct calculation and, at least for the time being, there remained some room for doubt. The question was clarified only gradually.

The first step was made by Einstein and Ludwig Hopf in 1910 (Einstein and Hopf 1910). In order to circumvent Planck's objection, they did not apply statistical methods to wave modes of the electromagnetic field, but only to material oscillators. Having diligently calculated the radiant pressure of light and the fluctuations of the momentum transmitted from the field to an oscillator according to the classical theory, they obtained the result that corresponded to the second, wave term in the fluctuation formula and to the Rayleigh–Jeans, instead of Planck's, law of energy distribution. Because of the then almost universal rejection of the hypothesis of light quanta, Einstein and Hopf did not dare to mention them explicitly in the paper, concluding it with a modest statement that intrinsic contradictions of the theory of radiation could not be resolved by forbidding the application of statistics to radiation.

Planck's former student Max von Laue came out in defense of the wave theory (Laue 1915b). Although he did not directly discuss fluctuations in his paper, his objections were obviously rooted in disagreement over the interpretation of the fluctuation formula in the preceding paper (Laue 1915a). Laue questioned one of the basic assumptions made by Einstein and Hopf, namely, that the Fourier coefficients of the wave modes are statistically independent variables. As a possible physical cause of additional fluctuations, Laue suggested, without elaborating, that the system's effective number of degrees of freedom decreases with the transition to quantum domain due to some yet unknown mechanism.

The debate continued in a couple of further papers (Einstein 1915, Laue 1915c) without any noticeable progress. Subsequent polemic died off either by itself or because of the publication of Lorentz's lectures. Lorentz furnished Einstein's claim that the classical theory is responsible only for the second term in energy fluctuations with a long, elaborate, and straight-forward mathematical proof (Lorentz 1916, Appendix IX). He considered radiation in a cavity as a superposition of Fourier modes with statistically independent coefficients. Their interference with each other made the energy within a portion of the cavity fluctuate in time. Lorentz managed to calculate these fluctuations directly, and his result coincided with Einstein's earlier estimates leading exactly to the second term in formula (3).

Lorentz's mastery of mathematical calculations and his authority convinced almost everybody, as is obvious from references in later publications. Planck alone stuck to his earlier ideas and repeated once again Laue's argument as late as 1924. But even he had to acknowledge in the introduction to his paper that, because of Lorentz's calculations, "an opinion about the incompatibility between the classical theory of propagation of light and the requirements of the quantum theory has found much stronger support and is presently widespread" (Planck 1924, p. 273). By that time it was much more than merely an "opinion," but a generally accepted point of view.

2.4 QUANTIZED INTERACTION WITH MATTER

In the same course of Paris lectures, Lorentz also managed to translate into mathematical equations the idea that the first term of fluctuations originates from discontinuities in the processes of the emission and absorption of light by matter rather than by the structure of light itself. Lorentz did not require any precise model of such interaction: already the assumption that matter and radiation exchange energy in finite quantitites $h\nu$ sufficed for him to derive the first, quantum term in equation (3) (Lorentz 1916, pp. 73–76). With the second term already calculated from the interference of radiation waves, Lorentz thus explained the complete formula (3) as resulting from the combination of two independent causes.

In his own judgment, the explanation was only partially satisfactory. The idea worked well in the case of the system represented by the radiation inside the cavity of volume V and the reservoir represented by matter, or the walls of the cavity. But Einstein's formula was also supposed to be equally valid in the case of the system consisting only of a portion U of the whole volume V of the cavity and the reservoir represented by the walls and by the remaining volume $V - U$. Thermodynamical calculation would give for this case

$$\overline{\varepsilon^2} = h\nu e + \frac{c^3}{8\pi\nu^2 U d\nu} e^2, \qquad (9)$$

where $e = EU/V$ is the average energy in the volume U. Leonard Salomon Ornstein and Frits Zernike (Ornstein and Zernike 1919) applied Lorentz's approach to this case but obtained for the first fluctuation term only a portion of the required result

$$\frac{U}{V} h\nu e = h\nu \frac{e^2}{E}. \qquad (10)$$

Despite its failure to deliver the full formula (3) in this specific case, Lorentz's calculation appears to have inspired Einstein to another major breakthrough in the field. Lorentz's basic approach consisted in considering the emission and absorption of radiation as statistically random acts and the resulting radiation in the cavity as statistical equilibrium. He assumed that the exchange of energy between matter and radiation takes place only in quanta $h\nu$ and that the absorption of radiation is proportional to the radiation density $\rho = E/V$. These assumptions are practically identical with the assumptions in Einstein's famous 1916 derivation of the Planck law in which he first introduced the probability coefficients $A''_m \downarrow$ for spontaneous, $B_n^m \uparrow$ and $B''_m \downarrow$ for induced radiation transitions. Einstein's paper appeared later the same year in which Lorentz's lectures were published, after several years of impasse in Einstein's quantum studies during which he failed to achieve any substantial progress in solving the riddle of radiation and was mainly preoccupied with the completion of his theory of general relativity (arriving at its basic equations in November 1915 and studying further consequences, such as gravitational waves and conservation laws, throughout 1916). The published version of the paper (Einstein 1916) contains no references and practically no clues to what could be the incentive and the starting point for this work. It appears very likely that Einstein was influenced by reading Lorentz's derivation of the fluctuation formula.

There is a strong internal relationship between the two calculations. In his discussion of the fluctuation formula, Lorentz characterized it as being essentially equivalent to the Planck law and thus having the same fundamental importance. A possible solution of one of these problems would also offer the way to solve the other one. While Lorentz in his lectures applied the model of statistical equilibrium in discontinuous energy exchanges between matter and radiation to the derivation of the fluctuation formula, Einstein in his paper used a very similar approach in order to produce a new derivation of the Planck law. Starting from Lorentz's basic model, Einstein added to it two important innovations. First, he used a more detailed mechanism of energy exchanges that implicitly relied on Bohr's idea of quantum states of an atom and radiation transitions between them. In Einstein's notations, Lorentz's statistical equation would be written as

$$A\binom{n}{m}\!\downarrow \; = \; \rho B\binom{m}{n}\!\uparrow \, e^{h\nu(n,m)/kT}, \tag{11}$$

where $\rho = E/V$ is the radiation density, while $\nu(n,m)$ is its frequency corresponding to the transition $n \rightarrow m$. Since Lorentz was concerned in this part of his lectures only with the explanation of the first term of the fluctuations, this equation corresponded to the Wien law for ρ. Its mathematical form, however, suggests a one-step modification required in order to obtain the Planck law, namely the subtraction of 1 from the right-side exponent. This would be tantamount to adding a new term, the stimulated emission of radiation, which is the second and the most important novelty of Einstein's paper.

Einstein's correspondence with Lorentz provides some further support for the above historical reconstruction. It appears from it that in the course of 1915 and the first half of 1916, Einstein was working full time on general relativity. In a postscript to the letter of 17 January 1916 he thanked Lorentz "for sending your lectures on statistical mechanics," which had just been published,[13] and returned to the discussion of the book later, in the letter of 17 June, in a paragraph that establishes the transition from Einstein's preoccupation with gravitational theory back to his earlier concerns about quanta:

I am not quite finished yet with the theory of emission in material systems.[14] But this much is clear to me: that the quanta difficulties affect the new theory of gravitation just as much as Maxwell's theory. I was delighted that in your Parisian lectures you gave the fluctuation properties of radiation deservedly thorough discussion; there the theories' inaccuracies become most clearly evident.[15]

Exactly one month later, on 17 July 1916, Einstein's paper "Strahlungs-Emission und -Absorption nach der Quantentheorie" was received by the editors of the *Verhandlungen der Deutschen Physikalischen Gesellschaft.*

2.5 QUANTIZATION OF WAVES

Classical waves and light quanta were not the only models of radiation applied to the explanation of the fluctuation formula. Another, more complex one, relied on the idea of quantized waves, or stationary wave modes of the cavity with discrete energy levels separated by the interval hv. The model was first proposed by Ehrenfest as a possible interpretation of the Planck law (Ehrenfest 1906), and later also mentioned by Lorentz (Lorentz 1910) and Debye (Debye 1910), although in the latter case in a combination with light quanta. At that initial stage, physicists' attention was devoted mostly to the problem of discontinuities in energy values, but once that feature of the new theories had become firmly established, discussion shifted to the problem of statistics.

Two different ways of introducing statistics of microscopic states appeared already in Planck's *Vorlesungen* (Planck 1906), where they were both applied to material oscillators.[16] According to the first of them, the energy quantum hv is chosen as a microscopic object that can be distributed over different oscillators (whether material oscillators of atoms, or radiation modes, or vibrations of a solid body). To introduce statistics, one assigns equal probabilities to different microscopic distributions of quanta over the oscillators. In order to obtain the Planck formula, one had to deal with these quanta as indistinguishable entities, assuming, either explicitly or implicitly, that microscopic distributions are defined only by the number of quanta in each oscillator state,[17] a procedure essentially equivalent to the later Bose–Einstein statistics for light quanta.

The other way to introduce statistics was to choose oscillators as microscopic objects and distribute them over the discrete set of energy values 0, hv, $2hv$, $3hv$, . . . as possible states. One can derive the Planck formula this way without postulating microscopic indistinguishability: probabilities for oscillators correspond to the usual Boltzmann statistics of distinguishable objects. Both methods lead to the same result at the end, the Planck law (2). I will call the first of them "the statistics of quanta," and the second "the statistics of oscillators."[18] The basic combinatorial formula is also the same in both methods, only its interpretation differs: in the first case, energy quanta are chosen as "distributed objects" while oscillators represent "boxes over which the objects are distributed"; in the second case, it is the other way round.

The statistics of oscillators had already been used by Planck in application to material oscillators (Planck 1906). Later the same year, Ehrenfest suggested that it be applied to electromagnetic oscillations in a cavity, thus introducing a new model, that of quantized radiation waves (Ehrenfest 1906, Ehrenfest 1911), which offered a new way of deriving the Planck law (2). The clearest and most laconic presentation of this method came from Laue in (Laue 1915a), where he used the model of quantized waves to derive fluctuation formulas (3) and (4) for blackbody radiation as well as for two other kinds of systems.[19]

The formalism of quantized waves allowed Laue to obtain, for the first time, both fluctuation terms from a single microscopic model. Despite this apparent achievement, the model of quantized waves did not attract much attention in those years. Although it was known to most of concerned experts in quantum theory, the method did not become as famous and widely discussed as the model of light quanta. The advocates of both rival approaches—the wave and the corpuscular ones—apparently regarded the quantization of waves as a merely formal trick. Einstein never discussed it in a published form in connection with the problem of radiation. Although Laue and Planck used this model at some points, they both abandoned it soon thereafter in order to return to the model of classical waves (Laue 1915b, Planck 1924). Bohr discussed it once as a plausible one, still preferring classical waves and definitely rejecting light quanta (Bohr 1921).

Among the main experts, Ehrenfest appears to be the only one who took the approach seriously and returned to it quite regularly over an extended period of time. He applied it once again in 1925 to the remaining difficulty in the derivation of the fluctuation formula (3). After von Laue's 1915 paper, only one problem remained: the case when the system was represented by a partial volume U of the whole cavity V. Ehrenfest's detailed calculations on quantized waves led to the same unsatisfactory result as the earlier attempt by Ornstein and Zernike with the model of quantized energy exchange between radiation and matter: he obtained the full second term but only a portion of the first term (Ehrenfest 1925a and 1925b).

2.6 MOLECULES OF LIGHT

If quantized waves can be regarded as a quantum mutation of classical waves, one could similarly try to modify somewhat the model of light quanta in order to bring it in correspondence with the Planck law. The existing contradiction, which amounted to differences in statistics, was clarified largely thanks to the efforts by Ehrenfest. He explained that statistically independent energy quanta led directly to the Wien law, while in order to obtain the Planck law, one had to assume that quanta were not independent (in the classical sense of the term, which was then the only available one) but indistinguishable objects (Ehrenfest 1911).

This peculiarity of Planck's combinatorics was also understood around the same time by Ladislas Natanson (Natanson 1911) and a few years later explained with ultimate clarity by Ehrenfest and Kamerlingh Onnes (1915), who formulated the statistics of indistinguishable objects in comparison with the statistics of independent, or distinguishable objects in exactly the same way in which contemporary textbooks explain the difference between the Bose–Einstein and Boltzmann statistics.[20] Their understanding, however, did not immediately become part of the common knowledge in the field, which led, in particular, to further polemics in 1914, between Mieczyslaw Wolfke and Yuri Aleksandrovich Krutkov (Wolfke 1914a and 1914b, Krutkov 1914a and 1914b). Wolfke had published a derivation of the Planck law from realistically interpreted light quanta and came under critique by Krutkov, Ehrenfest's student from St. Petersburg who was visiting Leiden at the time,[21] for the unawareness of the exact kind of statistics required for the derivation. Krutkov demonstrated explicitly that the application of Boltzmann statistics to light quanta immediately yields the Wien law. Krutkov mentioned an additional possibility in his paper: one could derive the Planck law from the model of independent particles if, in addition to the ordinary light quanta $h\nu$, the quanta of multiple energies $2h\nu$, $3h\nu$, . . . were also present.

The idea originated from a mathematical identity

$$\frac{1}{e^{\alpha}-1} = \sum_{i=1}^{\infty} e^{-i\alpha}, \tag{12}$$

which, if used to reformulate the Planck law (2), transforms it into an infinite series with the first term given by the Wien law (6) and with similar further terms only involving quanta $2h\nu$, $3h\nu$, ... Krutkov mentioned this idea only as an illustration of his (or rather Ehrenfest's) thesis that the original hypothesis of light quantum could not fully account for the blackbody radiation, but without actually arguing for the existence of multiple light quanta. According to Krutkov, the idea was suggested by his St. Petersburg colleague Abram Ioffe. Indeed, Ioffe was apparently the first to mention multiple quanta in print (Ioffe 1911, p. 550), if only in passing and as something already known. It is likely that the possibility of interpreting the Planck law this way had been discussed around 1910 within the St. Petersburg circle of physicists—in which Ioffe and Ehrenfest were major participants—but considered of no real importance.

Wolfke, on the contrary, having learned the idea from Krutkov, started to advocate the real existence of the "molecules of light." This became possible due to a general change of attitude in physics following the end of World War I. At the time of the first Solvay Conference of 1911 and for several years thereafter, light quanta were considered too radical, viewed rather negatively and, as a result, practically disappeared from physical journals. Einstein himself either changed his mind or yielded to the general mood. This attitude, however, changed dramatically after the end of the war: by 1920 light quanta grew out of oblivion into an extremely popular concept and began to be widely understood as particles or corpuscles. Traditional historiography saw the explanation of this change in the discovery of the Compton effect in 1923, but the development had already been in place for several years before that and was crowned by, rather then caused by, Compton's landmark achievement. Before the new experimental evidence became available, the post-war wave of publications on light quanta usually justified them by reference to Einstein's theory of transition coefficients (Einstein 1916, Einstein 1917).[22]

Although that paper by Einstein neither provided any new argument in favor of light quanta nor even dared to mention them explicitly, it could still be used as a kind of psychological argument since its results could very easily be visualized and interpreted with the help of light quanta. Rather than being caused by new experimental or theoretical developments, the revival of light quanta appears more like a shift in the prevailing fashion among the physicists. Most of the authors who started using this concept soon after the end of the war actually belonged to a younger generation who also favored different approaches to physical problems. If the first generation of quantum theorists (including Planck, Einstein, and Ehrenfest) were strongly influenced by statistical mechanics and thermodynamics,[23] the younger generation often lacked a thorough background in statistical methods, preferred dynamics to statistics, and tended to discuss physical processes in terms of individual microscopic events. The comeback of dynamical theories occurred in a number of different ways, the revival of light quanta being only one of them.

The spectrum of problems that occupied physicists also shifted from the discussion of blackbody radiation towards atomic spectra and the interaction between radiation and atoms. However, some of the older unsolved problems, including the derivation of the Planck law and Einstein's fluctuations, were not completely forgotten. The general rise in the popularity of light quanta as realistically interpreted corpuscles paved the way to considering such modifications of them as the combination of quanta into "molecules of light."

Wolfke recollects the idea in (Wolfke 1921). For each Wienian term in the series, he calculated the probability of spatial concentration of the entire energy of radiation in a smaller fraction of volume V and obtained the series of the fluctuation formulas (7) for each type of multiple quanta, concluding from this that light molecules really existed in radiation. Apparently, Louis de Broglie independently arrived at the same model of multiple quanta and used it in de Broglie 1922a and 1922b to derive the Planck law and the fluctuation formula (3). Once again, the same possibility was discovered by Walther Bothe (Bothe 1923 and 1924), although Bothe preferred the term "multiples" rather than "molecules" and did not consider multiple light quanta as bound to one another, but only as coherent and moving closely in the same direction. He thought that such "multiples" could actually be created in induced radiation transitions.

Although the "molecules of light" approach resembles the later Bose–Einstein statistics in this particular respect that light quanta within such a molecule would lose their individuality, one should not confuse molecules of light with the mathematical equivalent of Bose cells with several indistinguishable quanta.[24] The number of Bose cells with i quanta, as derived by Satyendra Nath Bose (1924),

$$p_i = A(1 - e^{-h\nu/kT})e^{-ih\nu/kT}, \tag{13}$$

where $A = 8\pi V\nu^2 d\nu/c^3$, differs from the number of light molecules $ih\nu$ required for the derivation of the Planck formula

$$n_i = A\frac{e^{-ih\nu/kT}}{i}. \tag{14}$$

It is difficult to assign any reasonable statistics to the model of light molecules. For example, if one assumes that quanta form a molecule when they are put in one cell, so that each molecule and each single quantum occupy a separate cell, then the total number of available cells A will not be sufficient to contain all the quanta

$$N = \sum_{i=1}^{\infty} i n_i \qquad (15)$$

for the region of short waves ($h\nu \ll kT$). This explains, perhaps, why no attempt was made to formulate any kind of statistics in the papers on light molecules. The proposal completely disappeared from journals after the introduction of Bose–Einstein statistics.

3. Duality and Dualism

3.1 BOSE STATISTICS

Ehrenfest was unlucky in that he understood some problems, in particular, the discrepancy between the statistics of independent quanta and the statistics of indistinguishable energy quanta, too well, perhaps better than anybody else. Bose, on the contrary, was luckily unaware of the difficulty and of the boldness of his calculations when he authored a paper contradicting the classical statistics (Bose 1924): "I had no idea that what I had done was really novel. . . . I was not a statistician to the extent of really knowing that I was doing something totally different from what Boltzmann would have done, from Boltzmann statistics."[25]

Like many other younger enthusiasts of light quanta, Bose was not as well trained in statistical mechanics as the generation of Einstein and Ehrenfest. Thus, it can be said that an "erroneous or opportunistic transposition of [combinatorical] formulas resulted in what we now call the Bose–Einstein statistics" (Darrigol 1991, p. 239).

But what was obscure to Bose should have been well known to Einstein, at least from earlier publications by Ehrenfest, his close friend and colleague. Einstein, however, had demonstrated an astonishing indifference to the problem: neither his publications nor, to the best of my knowledge, correspondence, contain a response to Ehrenfest's criticism or any other remark on the issue of independence or distinguishability. Even if he did not agree with Ehrenfest, he did not express the disagreement either. Darrigol suggests that Einstein and Ehrenfest stood behind the polemics between Wolfke and Krutkov in 1914 (Darrigol 1991, p. 256), but while Krutkov was obviously representing Ehrenfest's view that light quanta, when taken as independent particles, cannot explain the Planck law, it is much less clear to what degree the somewhat obscure ideas of Wolfke— that quanta must not be necessarily independent, or strictly independent, or independent in one sense or another—represented Einstein's views, despite Wolfke's allusion to his discussion with Einstein (Wolfke 1921).

Even in his first paper on the ideal gas with Bose statistics, Einstein did not mention the peculiarity of statistics at all (Einstein 1924b). As the main result of his paper, Bose regarded the purely corpuscular method of deriving the factor $8\pi v^2 V dv/c^3$ in the Planck law (2). Rather than representing the number of wave modes of radiation, the factor corresponded, according to Bose, to the number of cells in the phase space of light quanta. Einstein, apparently, also saw, in this aspect, the main novelty of the approach:[26]

> Bohr, Kramers and one more have abolished the "loose" quanta. These, however, do not allow us to get along without them. The Indian Bose has given a beautiful derivation of Planck's law *together with the constant* on the basis of loose quanta. Derivation is elegant, but its nature remains unclear. I have applied his theory to the ideal gas. Strict theory of "degeneracy." No zero-point energy and no energy defect. God knows whether this is so.[27]

Einstein first commented on the issue of statistics only three months later, in his second publication on the Bose gas, responding to a letter from Ehrenfest:

> Ehrenfest and other colleagues blame Bose's theory of radiation and my theory of ideal gas because in these theories quanta and, correspondingly, molecules are not treated as statistically independent objects, and what's more, this fact was not mentioned in our papers. That is true. (Einstein 1925, p. 5)

Einstein thus acknowledged for the first time the existence of the problem, but could justify the new procedure only by its successful applications. Long after it had become generally accepted, Bose statistics continued to worry a few concerned physicists, but the interpretation that eventually became predominant consisted in turning the assumption and the peculiarity into a postulate. The quantum statistics was declared fundamentally, a priori, different from the classical one and requiring no further explanation than the statement that "particles are indistinguishable." This possibility was suggested almost immediately, for example, in Landè 1925, and was later introduced and became standard in the new quantum mechanics (Dirac 1926), together with the second kind of indistinguishable statistics, the Fermi statistics.

The Bose method of deriving the Planck law can also be applied to fluctuations and deliver the full formula (3) with both its terms interpreted as fluctuations in the number of particles N within the given volume V.[28] Einstein demonstrated this immediately in his paper (Einstein 1925), obtaining

$$\frac{\overline{(\Delta N)^2}}{N^2} = \frac{1}{N} + \frac{1}{Z}. \tag{16}$$

No additional difficulty emerged even for the case of a system represented by the part U of the entire volume V of radiation. If one accepts the fundamentality of the statistics of indistinguishable particles, the problem can be regarded as solved without any reference to either waves or the wave-particle duality. This was not, however, the option chosen by Einstein. For him, acknowledging that Bose quanta were not independent did not mean that the classical notion of statistical independence had to be abandoned and indistinguishability accepted instead as a fundamental postulate. Like Ehrenfest and several other theoreticians, mostly of the statistically conscious older generation, Einstein was trying to find a deeper explanation behind the appearance of indistinguishable statistics.

The other available option consisted in switching from the statistics of (indistinguishable) quanta to the statistics of (distinguishable) oscillators. This possibility of interpreting the results of Bose was most probably preferred by Ehrenfest and quickly proposed in a published form by Schrödinger and independently by Pascual Jordan (Schrödinger 1926; Born, Heisenberg, and Jordan 1926). This would also mean choosing quantized waves rather than light quanta as the fundamental object of the new quantum theory, thus shifting the wave-particle balance towards the wave side. Einstein, who had never used Ehrenfest's model of quantized waves, apparently continued to have reservations about this option in 1925 and did not choose it, either. Instead, he suggested that the failure of independence of light quanta should be explained by some yet unknown kind of interaction between them, the interaction that had properties of a wave field. At this point Einstein referred to the dualistic model of de Broglie (de Broglie 1924), who had proposed to associate every particle with a scalar field.[29]

3.2 TURNING TO DE BROGLIE. EINSTEIN ON FIELD AND QUANTA

After his response to Lorentz's objections in the 1909 Salzburg talk, Einstein's vocabulary changed. He was becoming increasingly aware of the difficulty in combining the concepts of waves and quanta (he still avoided using the word "particle" or "corpuscle"), which occupied much of his thought. At first, he did not view the problem as insoluble and actively tried to develop a field-like theory of light quanta and thus to understand the meaning of the Planck constant "in an intuitive (visual) way" ["in anschaulicher Weise"] (Einstein to Sommerfeld, 14 January 1908, Einstein 1993b, Doc. 73, p. 87). Brief remarks in Einstein's letters give the chronology of his multiple attempts. Early in 1909 he looked for a nonlinear and nonhomogenous differential equation that would have explained both the electron and the light quantum, by reducing the number of fundamental constants from three to two (Einstein 1909a). He then tried linear and homogenous equations with singularities, as explained in his letter to Lorentz. After a year of unsuccessful attempts, Einstein attempted a radically different approach:

> At the moment I am very hopeful that I will solve the radiation problem, and that I will do so *without light quanta*. I am awfully curious how the thing will turn out. One would have to give up the energy principle in its present form. . . . I no longer believe (at the present) in spatial light quanta.[30]

Even sacrificing the strict conservation of energy did not help, and in 1911 Einstein stopped his active search for models and turned to more formal applications of quantization:

> I no longer ask whether these quanta really exist. Nor do I try to construct them any longer, for I now know that my brain cannot get through in this way. But I rummage through the consequences as carefully as possible so as to learn about the range of applicability of this conception. (Einstein to Michele Besso, 13 May 1911, Einstein 1993b, Doc. 267, p. 295)

Over years, after many unsuccessful attempts to solve the riddle of the theory of radiation, Einstein's views shifted from earlier revolutionary emphasis on the atomicity of light towards gradual acknowledgment of the symmetrical nature of the difficulty. He became increasingly concerned about the inability of light quanta to account for wave-like properties such as interference and diffraction, which, for him, rose to the same level of difficulty as the inability of the classical theory to account for quantum effects in radiation. Already in 1917, he made a remark that his theory of spontaneous and induced radiative transitions made "the establishment of a quantum-like theory of radiation appear as almost unavoidable," followed by "[T]he weakness of the theory is . . . that it does not bring us closer to a link-up with the undulation theory."[31] Even with corpuscular light quanta becoming immensely popular among the younger generation of post-war physicists, Einstein remained less enthusiastic, expressing views that the ultimate solution would have to combine features of both classical and quantum theories of light.

Einstein repeated these cautious attitudes in the midst of the triumph of light quanta after the discovery of the Compton effect in 1923:

> Newton's corpuscular theory is reanimated again, although it proved to be completely unsound in the domain of geometrical properties of light. We have now, therefore, two theories of light; both are necessary, and both, we have to admit this, exist without any logical connection, despite the twenty years of tremendous efforts of theoretical physicists. (Einstein 1924a)

Establishing this logical connection between two equally indispensable theories became Einstein's foremost concern. Solving the wave-quantum problem, in his view, was the biggest challenge for quantum theory and the criterion of its success. In 1921 he hoped to design an experiment that could distinguish between the quantum and classical mechanisms of light emission (Einstein 1921). In connection with it, he discussed, in correspondence with Lorentz and Ehrenfest, the idea of *Gespensterfeld* ("ghost field"), according to which both waves and quanta were emitted in a process of radiative transition. Waves were responsible for interference but carried no energy (perhaps because their amplitudes were small). Waves, however, directed the motion of quanta, which transported energy (see Lorentz to Einstein, 13 November 1921 and Einstein to Ehrenfest, 11 January 1922, EA 10-003). Einstein did not publish his theory and abandoned it soon after his proposed experiment failed. He mentioned once again the idea of the statistical conservation of energy, but rejected it again and even more strongly than in 1911.[32] Another idea, which he considered repeatedly, was that quantum restrictions arise due to overdetermination in the system of differential equations, if the number of equations is larger than the number of independent variables. Although unable to develop this idea very far, Einstein mentioned it in a published paper (Einstein 1923), as well as in his later attempts at a unified field theory.

Altogether, combining field and quantum aspects in a single model of radiation was the most important goal for Einstein. He was open to more than one possibility, yet always looked towards realistic, *anschauliche*, combinations. Although de Broglie did not achieve such a definite solution, Einstein regarded his attempt as a very promising hint pointing in the right direction: "He lifted the edge of a large veil" ["Er hat einen Zipfel des grossen Schleiers gelüftet"] (Einstein to Langevin,16 December 1924, EA 15-377), and in this sense referred to de Broglie's work when trying to answer Ehrenfest's critique regarding the difficulty with the new statistics (Einstein 1925).

Einstein's choice, however important, was not a logical necessity but reflected his reluctance to accept any of the other two available options: the fundamentality of indistinguishable particles or quantized waves. His supportive reference to de Broglie was interpreted by most readers as an assertion of the wave-particle duality. What was the meaning of duality in popular interpretation and how true was Einstein's reputation as a dualist? The answer depends on some boring terminological distinctions.

3.3 EINSTEIN'S REPUTATION AS A DUALIST

Before it entered physics, the word *Dualismus* was used in theology and philosophy. Articles in encyclopedic dictionaries, borrowing from one another in a sequence, usually attribute the first use of the term to Thomas Hyde's 1700 history of the Persian religion. The original source of this attribution, Rudolf Eisler, referred to Hyde's book more cautiously as merely an example of the early meaning of the term which was related to the fundamental theological problem of Evil (Eisler 1909). From the point of view of Christian theology, Persian Zoroastrianism, as well as a number of Christian heresies, were dualistic because they regarded the Good and the Evil as the two equally fundamental elements of the world. Christian orthodoxy, on the contrary, accepted St. Augustine's monistic solution, according to which only the Good was real, while the Evil had only a negative existence, as the relative lack of Good.

In philosophy, Christian Wolff applied the term *dualistae* in 1732 to those who, like himself, accepted the existence of two fundamental substances, material and immaterial. The philosophical meaning of the word as the opposition to both idealism and realism became standard, and philosophically educated German physicists of the early twentieth century were certainly aware of it. It is not easy to establish who first brought the term into physics. Neither Einstein nor de Broglie, to my knowledge, had used it, but by 1927, *Dualismus* had already been present in a number of German-language physical papers and typically attributed to Einstein's and de Broglie's views (sometimes with disapproval). From Bohr's Copenhagen perspective, at least, dualism was not consistent as a philosophy of quantum physics and had to be replaced by complementarity. Professional philosophers concentrated on debates about complementarity and remained largely aloof to the issue of the wave-particle duality, which figures more prominently in popular and textbook accounts of quantum physics. As a result, duality remained a rather vaguely defined, or rather undefined, concept.

A minimal pragmatic definition would probably be one that would distinguish between the typical reasoning about waves and about particles in, correspondingly, classical and quantum physics. Classical physics at the turn of the century knew both waves and particles, as well as a variety of their combinations, yet the tension between the two concepts was largely derived from a more fundamental opposition between action-at-a-distance and field theories, or continuous and discontinuous descriptions. Whether certain phenomena and objects were particle-like or field-like in nature was often unclear. In classical electrodynamics of the late 19th and early 20th century there were many attempts to get rid of the particles by treating the electron as an artefact of the field, while on the other hand, electrodynamics could also be reformulated as an action-at-a-distance theory without any field at all. A variety of more complex combinations of fields and particles was also discussed, the assumption being, however, that each element would be of either one or the other distinctive type. Most of Einstein's earlier and later proposals to solve the basic problem of radiation belonged to this category, many of his models had direct roots in the theories of the electron at the turn of the century. If one wants to call them "dualistic," the same term should probably also be applied to the early twentieth-century electrodynamics.

The failure of his and others' many attempts resulted in the opposition of waves and particles attaining more fundamental importance in quantum physics compared to what it had in the classical period and, gradually, to the widespread acknowledgment of the impossibility of making the ultimate *either-or* distinction. Complementarity tends to say *neither-nor* instead, regarding microscopic quantum objects as "*unanschaulich*" in principle. Werner Heisenberg addressed the problem directly in his classical Chicago lectures (Heisenberg 1930). According to him, both waves and particles are classical notions that prove equally inadequate when applied to quantum phenomena, yet both are also equally indispensable as the only available means for physicists to express their intuitions and interpretations. The ultimate quantum description is, therefore, abstract, while visual models of waves and particles can only be used with certain reservations and should not be taken too literally.

De Broglie, Schrödinger, as well as many popular and textbook expositions of quantum physics, on the contrary, tended to say *"both together"*: quantum objects have the properties of both wave and particles, which could not be separated. In the actual practice of physics, one encounters unrestricted opportunism. Two mathematical formalisms—one involving a particle obeying quantum laws of motion and the other involving quantized waves—exist, none of which uses waves and particles on an equal basis. Both formalisms are used rather interchangeably, depending on the context and on convenience, with the assumption that results would always be equivalent (and, at least in many cases, this can be proven). Physicists learned which of the existing visual intuitive languages to apply to which problem for the discussion and interpretation of results, and how to do this without getting into arguments or contradictions. It might be convenient to use two somewhat different words, 'dualism' and 'duality', in order to distinguish between ontological statements by some physicists about the centaur-like nature of quantum objects, which is a kind of philosophical interpretation, and the opportunistic freedom of the actual practice of modern physics in using different languages while avoiding the polemics.

What would Einstein's place be in this classification? He realized that both wave- and particle-like descriptions were indispensable but did not fit well together, but was reluctant to abandon the classical ideal of visualization and to accept as an ultimate solution either any of the available quantum mutations of classical models—indistinguishable particles or quantized waves—or even their interchangeable, opportunistic use. Einstein recognized duality in a negative rather than positive sense, as a crucially important problem rather than as a basic principle, a feature of quantum theory with which he struggled for most of his life without much success.

Einstein's 1925 paper on quantum gas in which he acknowledged duality as the main problem which had to be solved in, or rather eliminated from, the quantum theory was his last great contribution to the field's mainstream development. With the nascence of the new quantum mechanics, he was becoming more of an outsider looking for deviant strategies. Throughout 1926, while most quantum theorists were occupied with Heisenberg's and Schrödinger's new schemes, Einstein was struggling with the fatal problem of distinguishing between waves and quanta. He designed another experiment aimed to clarify whether the process of the emission of radiation by an atom occurs instantaneously or takes a certain amount of time, and spent much time and effort discussing the details and the preparation of this experiment with Emil Rupp. Rupp's experimental data and Einstein's more refined theoretical considerations forced him to abandon the initial hope of finding new disagreements between experimental results and the predictions of the classical theory, and no further clarification of the respective roles of waves and quanta ensued (Einstein 1926a, Einstein 1926b, Rupp 1926).

Many representatives of the new quantum mechanics—not only Schrödinger, but also such important authors of matrix mechanics as Max Born and Jordan—were inspired by reading Einstein's 1925 paper, followed his advice to take de Broglie's ideas seriously, and accepted dualism as a positive principle and one of the basic foundations of quantum theory. Due to their works and frequent references to dualism as Einstein's (and de Broglie's) idea, Einstein's reputation as a dualist became solidly established. For many of them, quantum mechanics (or at least wave mechanics) was, in some way, a realization of the Einstein–de Broglie program. Einstein, however, viewed the situation differently. From early on, already at the end of 1925, he expressed some skepticism about matrix mechanics, and in early 1926 he did not become ultimately satisfied with wave mechanics either.

Einstein's criticism had developed already before quantum mechanics took the acausal turn and thus must have had different reasons than Einstein's later disagreement with the mature theory expressed in philosophical notions of causality and completeness. At the early stage, his cautious welcome of the emerging theory seems to be related to the fact that neither of its versions—even wave mechanics, Schrödinger's own claims notwithstanding—was able to solve satisfactorily, by Einstein's strict criteria, the crucial problem of waves and particles. A newspaper report of his lecture on 23 February 1927 on recent developments in quantum theory makes this point rather clearly:

> The principal issue arising before us in the field of light phenomena is as follows: either to show that the corpuscular theory captures the true essence of light, or that the wave theory is right and the quantum properties are only illusory, or, finally, that both conceptions correspond to the true nature of light and that light has both wave and quantum characteristics. There were attempts to find a synthesis of both these features, but they have not succeeded mathematically so far. The latest great progress in the theory of light has been achieved through returning back again from the corpuscular conception and making a step in a direction opposite to the one which had led us from the wave to the corpuscular theory. Einstein refers here to the works of de Broglie and Schrödinger. . . . Nature requires from us a synthesis of both properties, but so far, this has been way beyond the intellectual abilities of physicists.[33]

Just at the time when the wave-particle duality was about to be trium-phantly proclaimed as one of the basic (some thought the most basic) and verified pillars of quantum mechanics, Einstein was still regarding it as the most troublesome mystery, which both the old and new quantum theories ultimately failed to solve.

4. Fluctuations in Quantum Electrodynamics

4.1 QUANTIZATION OF ELECTROMAGNETIC FIELD

Once the problem of the derivation of Planck's law could be considered solved by Bose statistics, the parallel problem of fluctuations lost its prominence as well. After 1925, formula (3) appeared more frequently in textbooks than in original research papers. The story could have ended at this point, if it were not for the role the fluctuation formulas played in the emergence of the new discipline of quantum electrodynamics.

Jordan, who would become one of the main creators of the new theory, had completed his doctorate with Max Born in Göttingen (Jordan 1924) and soon thereafter, in summer 1925, happened to attend a visiting lecture by Ehrenfest on the then still unpublished paper dealing with fluctuations and quantized waves (Ehrenfest 1925a, Ehrenfest 1925b). Jordan's thesis had been on the conflict between light quanta and the Bohr–Kramers–Slater theory of 1924, the last major attempt to rescue the classical theory of light at the expense of sacrificing the strict validity of energy conservation. Jordan thus had some background in the problem of radiation in the old quantum theory, though as a recent student he was not familiar with many earlier details and ramifications. From Ehrenfest, he learned about the remaining problem in calculating fluctuations: the case of radiation in the partial volume U of the entire cavity V.

Trying to solve the problem by his method of quantized waves, Ehrenfest obtained the same partial, and therefore unsatisfactory, result as Ornstein and Zernike with the Lorentz model of quantized interaction between radiation and matter. Both papers tried to explain the discrepancy verbally, referring to the uncertainty about whether it was justified to assume the additivity of entropies of the two parts of the volume. Ehrenfest also discussed with Göttingen physicists Einstein's recent theory of quantum gas, whereupon Born wrote to Einstein:

> Your brain, heavens know, looks much neater; its products are clear, simple and to the point. With luck, we may come to understand them in a few years'

time. This is what happened in the case of your and Bose's gas degeneracy statistics. Fortunately, Ehrenfest turned up here and cast some light on it. Then I read Louis de Broglie's paper, and gradually saw what they were up to. I now believe that the wave theory of matter could be of very great importance.[34]

Together with Walter Elsasser, another student of Born's, Jordan, too, became an explicit advocate of the wave-particle dualism and of the symmetry between radiation and matter. These ideas influenced to a great extent his subsequent contributions to new quantum mechanics and electrodynamics.

Towards the end of August 1925, Jordan was already collaborating with Born on the development of Heisenberg's proposal of the new theory into its first relatively completed form, matrix mechanics. Both his paper with Born and the subsequent one, the famous "Dreimännerarbeit," which they wrote together with Heisenberg, included sections on electrodynamics authored by Jordan and dealing with the problem of the quantization of electromagnetic field and its energy fluctuations (Born and Jordan 1925; Born, Heisenberg, and Jordan 1926).[35]

Jordan, and almost simultaneously and independently Schrödinger, observed that there was an exact equivalence and a one-to-one correspondence between the two methods of introducing statistics, called above the "statistics of quanta" and the "statistics of oscillators," or the statistics of indistinguishable particles and (distinguishable) quantized oscillators (Born, Heisenberg, and Jordan 1926, p. 609; Schrödinger 1926).[36] Calculations with Bose particles and with quantized waves produced equivalent results, which, according to Jordan, constituted the mathematical formulation of the wave-particle dualism. Only in the case recently treated by Ehrenfest there did seem to be a remaining discrepancy between the two methods. Jordan hoped that quantum mechanics would prove more successful in dealing with the problem than the old quantum theory and took the method of quantized waves over into the new theory. Jordan considered energy fluctuations in a simple one-dimensional case of an oscillating string and, following closely Ehrenfest's procedures, calculated fluctuations in a small segment a of the whole string of length L. In the classical case, he obtained, like Lorentz and Ehrenfest before (Lorentz 1916, Ehrenfest 1925a), the result corresponding to the second term in (9)

$$\overline{\varepsilon^2} = \frac{e^2}{2a}. \tag{17}$$

Quantization of an oscillator according to matrix mechanics added an extra zero-point energy $h\nu/2$ to all its energy values. Jordan thus substituted total energy e in (17) with the sum $e_t + h\nu/2$ of the "real thermal energy" e_t and zero-point energy for each available mode of oscillation. Only the thermal energy was supposed to figure in Einstein's formulas (1), (3), and (9). For its fluctuations, Jordan's simple substitution delivered the required result with both expected terms. The origin of the quantum term in fluctuations thus seemed to be connected with the zero-point energy of quantized waves.

At that early stage, the new quantum theory could still claim only very few successes. Enthusiastic about the apparent advantage of his calculation over the methods of the old quantum theory, Jordan wrote a postcard to Einstein, but Einstein was not very impressed:

> I have been occupied myself much with Heisenberg–Born. I tend more and more to consider it as inappropriate despite all the admiration for the idea. The zero-point energy of the black body radiation cannot exist. The corresponding argument by Heisenberg, Born and Jordan (fluctuations) I consider to be untenable, already because the probability of large fluctuation (for instance, for the total energy in the part v of volume V) surely cannot be derived in this way.[37]

Einstein made the same two objections in his reply to Jordan: the sum of zero-point energies $h\nu/2$ for all oscillating modes would give an infinitely large contribution to the total energy, and, besides, the new procedure did not offer a way of deriving the other fluctuation formula (7). Einstein's skepticism did not discourage Jordan, who was hoping to develop "a systematic matrix theory of electromagnetic field."[38] He considered the infinitely large energy of zero-point modes as being of no real physical importance. The methods of matrix mechanics, indeed, did not allow one to handle formula (7), but Jordan returned to the problem two years later and solved it with the help of the method of second quantization and the probability interpretation of the wave function (Jordan 1927b).

The quantization of waves eventually became the standard method in quantum field theory. In the view of at least some historians, Jordan's 1925 quantization of electromagnetic waves marks the beginning of quantum electrodynamics, its first successful application being the solution of the problem of energy fluctuations (Pais 1986). Yet it remains somewhat uncertain to what extent that success was unambiguous. Initially, not only Einstein, but also Born, Heisenberg, and Pauli, Jordan's collaborators on matrix mechanics, were not entirely confident in the result and in the usefulness of the entire procedure of wave quantization. The quantization of electromagnetic waves was generally accepted only after Dirac's impressive success with it in calculating Einstein's radiation transition coefficients A and B (Dirac 1927). As for the calculation of fluctuations, doubts still remained and were raised on various occasions.

Unlike triumphant quantum mechanics, quantum electrodynamics had a rather bumpy start. Its obvious successes mixed with equally obvious failures, old and newly found problems, most of them connected with divergences. Around 1930, the theory passed through a major crisis when its entire foundation was called in question and criticized once again. In this situation, Heisenberg reconsidered Jordan's calculation and found divergent integrals in it, which made fluctuations infinite and which were previously overlooked. The divergences appeared in the ultraviolet region, were connected with the first term of fluctuations and did not depend on the existence of the zero-point energy, which was also infinite. Heisenberg realized that one could get rid of them mathematically by considering the volume with "smeared-out" rather than precise boundaries and interpreted this result along the lines of the uncertainty principle: an experimental attempt to precisely determine the volume in space leads to great uncontrolled disturbances in the physical system (Heisenberg 1931).

Several years later, both Jordan's and Heisenberg's treatment came under a strong critique by Born and Klaus Fuchs, who claimed to have shown that all divergences were caused by zero-point oscillations, yet were disastrous, leading to observable deviations from Einstein's formula (Born and Fuchs 1939a). Both these conclusions rested upon a mistake in calculation, which was found by Markus Fierz (Born and Fuchs 1939b). In 1933 Niels Bohr and Leon Rosenfeld proposed a way of ignoring the problem by arguing that the energy of the electromagnetic field within a particular volume cannot be measured. What could be measured was the intensity of the field whose fluctuations are relatively small in the physically interesting domain (Bohr and Rosenfeld 1933). The latter claim, however, was refuted by later calculations (Corinaldesi 1953).

The present status of the problem of energy fluctuations in quantum electrodynamics does not seem to have improved much since Heisenberg. The divergences which he had found remained in renormalized quantum electrodynamics (Dyson 1951). Some authors could ignore this by insisting that volume boundaries have to be "smooth," while others could use this as a way of challenging conventional quantum field theory. At least one later calculation dealt with the question directly and led to a confusing result that "the fluctuation of the spectral density is always finite, but it is not in general given by Einstein's formula" (González and Wergeland 1973, p. 1). Overall, it seems that the problem of fluctuations of electromagnetic energy, once very acute and most actively debated, has not been resolved in modern quantum electrodynamics, after all, but rather has been marginalized and has disappeared from the center of attention of field theorists.

4.2 FLUCTUATIONS IN FERMI GAS AND SECOND QUANTIZATION

In early 1926, Enrico Fermi understood that electrons obeyed a different kind of statistics from light quanta. Later that year Dirac rediscovered this feature and included it into the general formalism of emerging quantum mechanics (Dirac 1926). Particles obeying Fermi statistics cannot occupy the same cell in phase space, each cell therefore contains either 0 or 1 particle. Once the difference between Bose and Fermi statistics was understood, it was not so difficult to derive the formula for fluctuations in the number of electrons within a particular volume. It differed from the analogous formula for Bose particles (16) only in the sign (minus instead of plus):

$$\frac{\overline{(\Delta N)^2}}{N^2} = \frac{1}{N} - \frac{1}{Z}. \tag{18}$$

Formula (18) was first published in (Pauli 1927) among other important results in the theory of the electron gas. As in the Bose case, one could use the statistics of either quanta or oscillators. Pauli chose the latter and distributed oscillators over two possible energy states (0 or 1 particle). Following Einstein's example, he interpreted the second term as the interference of de Broglie waves.

The model of quantized waves did not seem to offer any straightforward way of deriving the formula (18), but in the fall of 1927 Jordan realized how the method should be altered in order to account for Fermi statistics. This constituted his other great contribution to quantum electrodynamics, the method of second quantization for fermions. Instead of the standard quantization with the help of the commutation relations

$$a_r a_r^* - a_r^* a_r = 1, \tag{19}$$

one had to use a modified anticommutator set

$$b_r b_r^* + b_r^* b_r = 1. \tag{20}$$

The difference again could be reduced to one sign, but Jordan's path towards this discovery was a very difficult one, since it required a change in the standard form of canonical commutator, which was regarded as the very essence of quantum mechanics. Jordan arrived at the anticommutator via a rather complicated route, working at first with quaternions, finding in (Jordan 1927a) a concrete matrix representation of what is currently known as the operators of creation and annihilation

$$ b_r = \begin{pmatrix} 0 & 1 \\ 0 & 0 \end{pmatrix}, \quad b_r^* = \begin{pmatrix} 0 & 0 \\ 1 & 0 \end{pmatrix} \tag{21} $$

and only later, in (Jordan and Wigner 1928), realizing that they correspond to the commutation relations (20).[39]

In Jordan's understanding, he was quantizing the electron's de Broglie wave (or the Schrödinger function in its initial wave-mechanical rather than probabilistic interpretation). He was guided in this search by his own version of the "dualism between corpuscles and quantized waves in quantum mechanics" ["quantenmechanishe Dualismus von Korpuskeln und gequantelten Wellen"] (Jordan 1927b, p. 766). As he had already demonstrated in late 1925, quantized electromagnetic oscillations delivered results equivalent to those obtained with Bose quanta. Once Schrödinger's wave mechanics of the electron appeared in early 1926, Jordan developed hopes to prove that the same equivalence can be demonstrated for particles of matter and quantized matter waves, though he did not publish the idea, partly because of Heisenberg's and Pauli's criticism and partly because he did not see the mathematical solution then.[40] He hoped that the quantization of wave would provide a mathematical method of describing an ensemble of many particles, as it did in the case of radiation, but it took him almost two years to find out how to make this ensemble obey Fermi statistics instead of Bose statistics.

In the meantime, the many-body problem in quantum mechanics was developing along a different route, by considering symmetrical and antisymmetrical ψ-functions in $3N$-dimensional space (N is the number of particles) (Dirac 1926). This trend undermined Schrödinger's initial *anschaulich* interpretation of the ψ-function as a matter wave, since it was no longer possible to visualize the multi-dimensional mathematical function in the ordinary three-dimensional space. The probabilistic interpretation of the ψ completed the demise of realistically interpreted waves. Although Jordan was among the first proponents of the probabilistic interpretation and readily sacrificed the multidimensional ψ-function for this purpose, he simultaneously maintained a hope of restoring the initial visualization of the three-dimensional wave. He achieved this, at least partly, in his method of second quantization for fermions, since the description of many-body ensemble of electrons was achieved there by the quantization of a three-dimensional wave for one electron.

As he did two years earlier, Jordan tested his new method on the problem of energy fluctuations. Repeating his previous calculations from (Born, Heisenberg, and Jordan 1926) but with both kinds of commutators, he derived expressions for the fluctuations in the number of particles within a given volume,

$$a_r^* a_r^2 a_r^* = N_r (1 + N_r) \qquad (22)$$

in the Bose case, and

$$b_r^* b_r^2 b_r^* = N_r (1 - N_r) \qquad (23)$$

in the case of Fermi statistics, which are equivalent to Equations (16) and (18) (Jordan 1927a). Like his earlier calculation of fluctuations, the result did not look very impressive to others, but Jordan felt encouraged by it enough to formulate, in the last section of his paper, a bold vision of the future relativistic quantum theory based on the fundamental concept of the quantized wave (electromagnetic waves for radiation and matter waves for electrons). Although expressed a little too prematurely and publicly rebuffed by Pauli, Jordan's proposal eventually became realized in modern theory.

Although fluctuations once again played an important role as the pretext for the very first programmatic proposal of quantum field theory, Jordan's calculation later encountered the same kind of difficulties as his derivation of the Einstein formula. Heisenberg, again, pointed out that the actual integrals diverge (in the case of electrons' fluctuations, this is related to the processes of pair creation in the vacuum state) and suggested the same mathematical trick of making results finite by "smearing out" the boundaries of the volume (Heisenberg 1934). Bohr and Rosenfeld could not declare the number of electrons (or electric charge) unmeasurable in their discussion of the possibilities of measurement in quantum electrodynamics (Bohr and Rosenfeld 1950). With regards to the difficulty with infinite fluctuations, they simply repeated Heisenberg's argument about volume boundaries. Their paper reminded the post-war generation of physicists of the existence of the problem. Some discussion arose, which led to more precise calculations of the divergences, but not to their elimination.[41]

5. Conclusions

Einstein initially used fluctuations as an argument for an atomistic theory of light that would complete Lorentz's atomistic reform of classical electrodynamics of the late nineteenth century. The closest analog of Einstein's light quantum was the non-mechanical electron of contemporary electrodynamics. In exactly what way field and quanta combined was not known, but combinations actually tried by Einstein in his multiple attempts to design a model for the light quantum—although someone might want to call them dualistic—were actually rooted in models of the electron discussed in classical electrodynamics around 1900.

Responding to criticism, first of all by Lorentz, Einstein changed his views towards a gradual recognition of the grave contradiction between the wave and quantum aspects of radiation. Many of his critics came to a similar conclusion from the opposite direction, after many unsuccessful attempts to defend the wave theory of light. A compromise achieved at the first Solvay meeting in 1911 saw the origin of quantum effects in discontinuities in the interaction between light and matter rather than in the structure of radiation itself. Lorentz developed a corresponding explanation for the formula of energy fluctuations, which influenced Einstein's 1916 theory of spontaneous and stimulated radiation transitions.

Despite the enormous rise in popularity of corpuscular light quanta after the end of World War I, due to the generational change and the change of prevailing mood within the physics community, Einstein remained acutely aware of their insufficiency and of the necessity to resolve the contradiction with the wave theory. His 1924 reference to de Broglie's dualistic hypothesis reflected both the inability to find a solution that would satisfy him and his reluctance to accept either of the two available alternatives: the fundamentality of Bose statistics that could explain both terms in fluctuations without any reference to waves and Ehrenfest's model of quantized waves. Einstein continued to hope for a solution in some kind of an *anschaulich* combination of waves and quanta, seeing in de Broglie's work a promising step in that direction. It can be said that Einstein accepted the wave-particle duality only in a negative sense, as a fundamental difficulty of quantum theory that had to be resolved rather than turned into a postulate. By his strict criteria, neither matrix nor wave mechanics succeeded in this crucial task, which was an important source of his critical attitude towards the emerging quantum mechanics.

Many of the authors of quantum mechanics, however, became aware of the idea of the wave-particle duality via the discussion of fluctuations in (Einstein 1925) and accepted it in a positive sense, as one of the most basic principles of the new theory. This historical contingency established a permanent link in popular perception between fluctuations and duality, which was later transformed into textbook statements about the necessary logical relationship between them. The difficulties with the interpretation of the fluctuation formula, however, did not come to an end with the establishment of quantum mechanics. Fluctuations played an important role in Jordan's early proposals of relativistic quantum theory, including such basic methods as field quantization and second quantization, yet even in renormalized quantum electrodynamics, the problem of energy fluctuations does not seem to have found an entirely satisfactory solution. The problem that once was crucial and moved the theory ahead became largely forgotten rather than ultimately resolved.

Acknowledgments. The paper was originally published in Russian (Kojevnikov 1990). The current revised version is based on two preprints written in 1992 during my stay in München as a Humboldt fellow at the Max-Planck-Institut für Physik. I am grateful to my doctoral adviser, Igor Yur'evich Kobzarev, for suggesting this topic for research and invaluable support, to Olivier Darrigol, Helmut Rechenberg, and participants at the Einstein Symposium in Ulm (March 1992) for discussions, to the Alexander von Humboldt Stiftung for financial support, and to Einstein Archive (Hebrew University), the Collected Papers of Albert Einstein (Boston University), Niels Bohr Archive (Copenhagen), and Niels Bohr Library (American Institute of Physics) for the permission to use and quote archival documents.

NOTES

[1] The Kuhnian word "crisis" was also the term widely used by participants in the events themselves, and the entire situation, indeed, is reminiscent of a crisis in science as described by Kuhn. For some time, I thought that the history of quantum theory of radiation and of quantum electrodynamics offers one of the very few—if not the only—example in the history of physics, when historical actors, in their explicit words and deeds, behaved as if they were consciously playing according to the Kuhn scenario of crisis. Now [1999] I rather suspect that the very process of the historical development of quantum electrodynamics and the eventual resolution of its crisis, which Kuhn witnessed directly while doing his graduate studies in physics, could have served as an important hidden source for his later philosophical doctrine.

[2] "Mit seiner Abhandlung über die statistischen Schwankungen in der Strahlung hat Einstein den Dualismus Welle-Teilchen beim Licht eingeführt. Die volle

Tragweite dieses umwälzenden Gedankens konne er natürlich noch nicht sehen" (Hund 1967, p. 44).

[3] "Damals schien es mir, als ob die Plancksche Theorie der Strahlung in gewisser Beziehung ein Gegenstück bildete zu meiner Arbeit" (Einstein 1906, p. 199).

[4] Indeed, one year earlier Einstein derived, independently of Gibbs, the general formula (1) and all he needed was to apply it to the Wien law.

[5] "Es ist hervorzuheben, daß die angegebenen Überlegungen im wesentlichen keineswegs ihren Wert verlieren würden, falls die Plancksche Formel noch als ungültig erweisen sollte; gerade der von der Erfahrung genügend bestätigte Teil der Planckschen Formel (das für große v/T in der Grenze gültige Wiensche Strahlungsgesetz) ist es, welcher zur Lichtquantentheorie führt" (Einstein 1909a, p. 191).

[6] "Das erste Glied, wenn es allein vorhanden wäre, (würde) eine solche Schwankung der Strahlungsenergie liefern, wie wenn die Strahlung aus voneinander unabhängig beweglichen, punktförmigen Quanten von der Energie hv bestünde" (Einstein 1909a, p. 189).

[7] "Deze bezwaren jammer want theorie lichtquanta wel mooi" (Lorentz to Einstein, 6 May 1909, draft in Dutch, EA 16-417). For the actual letter, in German, and its English translation, see Einstein 1993b, p. 176, and the accompanying volume of translations. "Das Gesagte dürfte genügen, um zu zeigen, daß von Lichtquanten, die bei der Fortbewegung in kleinen Räumen konzentriert und stets ungeteilt bleiben, keine Rede sein kann" (Lorentz 1910, p. 354).

[8] "Deshalb ist es meine Meinung, daß die nächste Phase der Entwickelung der theoretischen Physik uns eine Theorie des Lichtes bringen wird, welche sich als eine Art Verschmelzung von Undulations- und Emissionstheorie des Lichtes auffassen läßt"; and "Ich wollte durch dasselbe nur kurz veranschaulichen, daß die beiden Struktureigenschaften (Undulationsstruktur und Quantenstruktur), welche gemäß der Planckschen Formel beide der Strahlung zukommen sollen, nicht als miteinander unvereibar anzusehen sind" (Einstein 1909b, pp. 482–483, 500).

[9] "Ich glaube vielmehr, daß sich das Licht in ähnlicher Wiesse um singuläre Punkte (es brauchen nicht notwendig mathematisch singuläre Punkte zu sein) herum gruppiert, wie wir das vom elektrostatischen Felde anzunehmen gewohnt sind. Ich denke mir also ein einzelnes Lichtquant als einen Punkt, der von einem sehr ausgedehnten Vektorenfeld umgeben ist, das mit der Entfernung irgendwie abnimmt. Der Punkt ist eine Singularität, ohne welche das Vektorenfeld nicht existieren kann. Ob man sich beim Vorhandensein vieler Lichtquanten mit einander überdeckenden Feldern eine einfache Superposition der Vektorenfelder vorzustellen hat, das kann ich nicht sagen. Jedenfalls müsste man zur Bestimmung der Vorgänge ausser den Differentialgleichungen für das Vektorfeld auch noch Bewegungsgleichungen für die singulären Punkte haben. Die Energie des elektromagnetischen Feldes müsste—wenigstens bei genügend verdünnter Strahlung—mit der Anzahl dieser singulären Punkte in gewisser Weise zusammenhängen. Absorption fände nur statt beim Verschwinden eines derartigen singulären Punktes bezw. bei Degenerieren des zu diesem Punkt gehörigen Strahlungsfeldes (besser "von auf

diesen Punkt stützender Strahlung"). Durch die Angabe der Bewegungen aller Singularitäten wäre das Vektorenfeld vollkommen bestimmt, sodass die Anzahl der zur Charakterisierung einer Strahlung nötigen Variabeln eine *endliche* wäre. . . . Das Wesentliche scheint mir übrigens gar nicht in der Annahme singulärer Punkte zu liegen, sondern in der Annahme solcher Feldgleichungen, welche Lösungen zulassen, bei welchen sich endliche Engergiemengen ohne Zerstreuung in einer bestimmten Richtung mit der Geschwindigkeit c fortpflanzen. Man sollte meinen, dass das Ziel mit einer geringen Modifikation der Maxwell'schen Theorie zu erreichen sei" (Einstein to Lorentz, 23 May 1909, Einstein 1993b, Doc. 163, pp. 193–194).

[10] See Einstein to Arnold Sommerfeld, 14 January 1908, Einstein 1993b, Doc. 73, pp. 86–89. The same year, Einstein also turned to a spectroscopist August Hagenbach asking whether experiment could hint towards any possible decrease in the ability of two rays to interfere with the decrease in their intensity. Despite Hagenbach's disappointing reply, Einstein did not become convinced and retained a hope that interference would come out as a result of the interaction of many quanta, while vanishing with the decrease in their number. (See Einstein to and from Hagenbach, 6, 9, and 14 July 1908, Einstein 1993b, Docs. 109–111, pp. 128–130.)

[11] For a further discussion of this point, see Kojevnikov 1994.

[12] "ungelösten Rätsel" (Einstein 1911, p. 347) and "Ein provisorischer Versuch, . . . Hilfsvorstellung, die sich mit den experimentell gesicherten Folgerungen der Undulationstheorie nicht vereinigen zu lassen scheint" (Discussion 1911, p. 359).

[13] Einstein to Lorentz, 17 January 1916, Einstein 1998, Doc. 184, p. 247; cf. also Lorentz to Ehrenfest, 18 January 1916 (Archive for the History of Quantum Physics).

[14] This is a reference to the theory of emission of gravitational waves from a massive body in general relativity.

[15] "Mit der Theorie der Ausstrahlung materialler Systeme bin ich noch nicht ganz fertig. Aber soviel ist mir klar, dass die Quanten—Schwierigkeiten auch die neue Gravitationstheorie treffen, ebenso gut wie die Maxwell'sche Theorie. Es hat mich sehr gefreut, dass Sie in Ihren Pariser Vorträgen die Schwankungs—Eigenschaften der Strahlung einer eingehenden Behandlung gewürdigt haben; hier treten die Unrichtigkeiten der Theorieen am reinsten zutage" (Einstein to Lorentz, 17 June 1916, Einstein 1998, Doc. 226, p. 300). Pais has already suggested that the study of the emission of gravitational waves could have spurred Einstein's return to quantum electrodynamics in 1916 (Pais 1982, p. 280), but he did not pay attention to the reference to Lorentz's book in the same paragraph of Einstein's letter.

[16] For a discussion of Planck's combinatorics, see Bergia 1987 and Darrigol 1991.

[17] Early quantum theorists would say that quanta in the oscillator lose their individuality or that they are not independent from one another. I am using all three

words, "individuality," "independency," and "distinguishability," as synonyms, as they often were in those days.

[18] In Darrigol 1991, they are referred to by symbols W_2 and W_4 correspondingly.

[19] In 1923 Planck developed this method a bit further: Laue had given the expression for the probability $w(N)$ of one oscillator to have the energy value Nhv, Planck then calculated the probability $W_m(N)$ for a system of m oscillators to have the total energy Nhv (Planck 1923b).

[20] For a detailed discussion of these papers and the history of the concept of indistinguishability, see Darrigol 1991.

[21] Ehrenfest lived in St. Petersburg in 1907–1912 and initiated the tradition of theoretical physics in Russia.

[22] See the citation analysis in Small 1986. On the history of experimental aspects of the wave-particle dilemma, see Wheaton 1983.

[23] On the role of thermodynamics in Einstein's thought, see Klein 1967.

[24] Bergia (1987) seems to have overlooked this difference.

[25] An undocumented later interview, reported by J. Mehra. Quoted from Bergia 1987, p. 226.

[26] Bose to Einstein, 4 June 1924, EA 6-127. As Mehra and Rechenberg (1982, vol. 1, p. 565) have already noted, de Broglie had derived this coefficient earlier in (de Broglie 1922a) also from the entirely corpuscular perspective. De Broglie also divided phase space into quantum cells, but he used this to derive the Wien law only, proceeding then to the hypothesis of light molecules in order to obtain the Planck law. Another part of Bose's calculations, the expression for the number of cells with r particles (Equation (13)) is exactly the same as the expression for the number of oscillators with r quanta in Schrödinger 1924. Schrödinger was aware that such quanta were indistinguishable.

[27] "Bohr, Kramers und noch einer haben die "losen" Quanten abgeschaft. Werden sich aber nicht entbehren lassen. Der Inder Bose hat eine schöne Ableitung des Planckschen Gesetzes samt Konstante auf Grund der losen Lichtquanten gegeben. Ableitung elegant, aber Wesen bleibt dunkel. Ich habe auf Grund seine Theorie auf ideales Gas angewendet. Strenge Theorie der "Entartung." Keine Nullpunktsenergie und oben kein Energiedefekt. Gott weiss ob es so ist" (Einstein to Ehrenfest, 12 July 1924, EA 10-090).

[28] Bose attempted to calculate these fluctuations in his manuscript "Fluctuations in density," undated, but evidently written in 1924–1925 (EA 6-133).

[29] On de Broglie's work leading to his dualistic model and the introduction of matter waves, see Kubli 1971 and Darrigol 1993.

[30] Einstein to Jakob Laub, 4 November 1910, Einstein 1993b, Doc. 231, pp. 260–262. Einstein once mentioned the idea of energy non-conservation in print (1911, p. 348).

[31] "lassen die Aufstellung einer eigentlich quantenhaften Theorie der Strahlung fast unvermeidlich erscheinen. . . . Die Schwache der Theorie liegt einerseits darin,

daß sie uns dem Anschluß an die Undulationstheorie nicht näher bringt" (Einstein 1917, pp. 127–128).

[32] He also rejected the proposal that energy is not conserved, when it appeared in print in the 1924 theory of Bohr, Kramers, and Slater: "This idea is an old acquaintance of mine, whom I don't consider as a real guy." ["Diese Idee ist ein alter Bekannter von mir, den ich aber fur keinen reelen Kerl halte."] (Einstein to Ehrenfest, 31 May 1924, EA 10-088)

[33] "die Fragestellung prinzipieller Natur, die wir nun auf dem Gebiete der Lichterscheinungen haben, gipfelt darin, entweder, zu zeigen, daß die Korpuskular-theorie das wahre Wesen des Lichtes erfaßt, oder, daß die Undulationstheorie richtig ist und das Quantenhafte nur scheinbar ist, oder endlich, daß beide Auffassungen dem wahren Wesen des Lichts entsprechen und das Licht sowohl Quanteneigenschaften als undulatorische Eigenschaften hat. Man suchte nun eine Synthese dieser beiden Eigenschaften zu finden, was bisher mathematisch noch nicht gelungen ist. Der letzte große Fortschritt, der in der Physik des Lichts gemacht wurde, ist dadurch erreicht worden, daß man sich wieder von der Korpuskularauffassung entfernt hat und wieder einen Schritt gemacht hat, der umgekehrt ist demjenigen, der von der Undulationstheorie zur Korpuskulartheorie geführt hat. Einstein verweist hier auf die Arbeiten von De Broglie und Schrödin-ger. . . . die Natur fordert von uns eine Synthese beider Auffassungen, die bis jetzt allerdings noch über die Denkkräfte der Physiker hinausgegangen ist" (Einstein 1927, p. 546). On Einstein's reluctance to accept duality in the sense of quantum mechanics, see also Bach 1989, p. 174.

[34] "Dein Gehirn sieht, weiß der Himmel, reinlicher aus. Seine Produkte sind klar, einfach und treffen die Sache. Wir kapieren es dann zur Not ein paar Jahre später. So ist es uns auch mit Deiner Gasentartung und der Boseschen Statistik gegangen. Glücklicherweise erschien Ehrenfest hier und hat uns ein Licht aufgesteckt. Darauf habe ich die Arbeit von Louis de Broglie gelesen und bin allmählich auch hinter Deine Schliche gekommen. Jetzt glaube ich, daß die 'Wellentheorie der Materie' eine sehr gewichtige Sache werden kann" (Born and Einstein 1969, p. 120). [English translation from Born and Einstein 1971, p. 83]

[35] The following discussion is restricted to the problem of fluctuations only. For a detailed analysis of these papers in the context of matrix mechanics, see Mehra and Rechenberg 1982, vol. 3, and van der Waerden 1967, and in the context of the emerging quantum electrodynamics, Darrigol 1986 and Pais 1986.

[36] Jordan also criticized in passing the idea of light molecules, calling it "Debye statistics." The mistaken attribution of the idea to Debye is due to a wrongly interpreted obscure footnote in Schrödinger 1924.

[37] "Mit Heisenberg–Born habe ich mich noch viel beschäftigt. Ich neige mehr und mehr dazu, bei aller Bewunderung des Gedankens diesen für unzutreffend zu halten. Nullpunkts-Energie der Hohlraumstrahlung kann es nich geben. Das diesbezügliche Argument von Heisenberg, Born und Jordan (Schwankungen) halte ich für hinfällig, schon deshalb weil die Wahrscheinlichkeit für grosse Schwank-ungen (zum Beispiel Antreffen der *ganzen* Energie in Teilvolumen U von V) so

gewiss nicht richtig herauskommt." (Einstein to Ehrenfest, 12 February 1926, EA 10-131)

[38] "Ich will, sobald ich Zeit habe, eine systematische Matrizentheorie des elektromagnetischen Feldes zu überlegen versuchen" (Jordan to Einstein, 15 December 1925, EA 13-474).

[39] He also made a mathematical mistake on the way, multiplying the operators by $-i$ and i correspondingly, but this did not affect the final result.

[40] ". . . zumal Pauli und Heisenberg nicht davon wissen wollten, während Born zwar anfanglich sehr zustimmte, aber später auch nichts mehr davon hielt" (Jordan to Schrödinger, undated, probably autumn 1927, Archive for History of Quantum Physics).

[41] See Corinaldesi 1953 for further references. There was also an attempt to substitute the dualistic interpretation of the fluctuation formula with another one based on classical statistics in Bach 1989 and literature cited there.

REFERENCES

Bach, Alexander (1989). "Eine Fehlinterpretation mit Folgen: Albert Einstein und der Welle-Teilchen Dualismus?" *Archive for History of Exact Sciences* 40: 173–206.

Bergia, Silvio (1987). "Who Discovered the Bose–Einstein Statistics?" In *Symmetries in Physics (1600–1980)*. Manuel G. Doncel, Armin Hermann, Louis Michel, and Abraham Pais, eds. Barcelona: Bellaterra, pp. 221–250.

Bohr, Niels (1921). "Zur Frage der Polarisation der Strahlung in der Quantentheorie." *Zeitschrift für Physik* 6: 1–9.

Bohr, Niels, Kramers, Hendrik, and Slater, John (1924). "The Quantum Theory of Radiation." *Philosophical Magazine* 47: 785–802.

Bohr, Niels, and Rosenfeld, Leon (1933). "Zur Frage der Messbarkeit der Elektromagnetischen Feldgrössen." *Matematisk-fysiske Meddelelser af Det Kongelige Danske Videnskabernes Selskab*. Københann. Vol. 12, no. 8.

—— (1950). "Field and Charge Measurements in Quantum Electrodynamics." *Physical Review* 78: 794–798.

Born, Max, and Fuchs, Klaus (1939a). "On Fluctuations in Electromagnetic Radiation." *Royal Society of London. Proceedings* A170: 252–265.

—— (1939b). "On Fluctuations in Electromagnetic Radiation." *Royal Society of London. Proceedings* A172: 465–466.

Born, Max and Hedwig, and Einstein, Albert (1969). *Briefwechsel. 1916–1955*. München.

Born, Max and Hedwig, and Einstein, Albert (1971). *The Born–Einstein Letters*. New York: Walker and Company.

Born, Max, Heisenberg, Werner, and Jordan, Pascual (1926). "Zur Quantenmechanik II." *Zeitschrift für Physik* 35: 557–615.

Born, Max, and Jordan, Pascual (1925). "Zur Quantenmechanik." *Zeitschrift für Physik* 34: 858–888.

Bose, Satyendra Nath (1924). "Plancks Gesetz und Lichtquantenhypothese." *Zeitschrift für Physik* 26: 178–181.

Bothe, Walther (1923). "Die räumliche Energieverteilung in der Hohlraumstrahlung." *Zeitschrift für Physik* 20: 145–152.

— (1924). "Über die Wechselwirkung zwischen Strahlung und freien Elektronen." *Zeitschrift für Physik* 23: 214–224.

Broglie, Louis de (1922a). "Rayonnement noir et quanta de lumière." *Journal de Physique* 3: 422–428.

— (1922b). "Sur les interférences et la théorie des quanta de lumière." *Comptes Rendus* 175: 811–813.

— (1924). *Recherche sur la théorie des quanta.* Thèses. Paris: Masson et Cie.

Corinaldesi, E. (1953). "Some Aspects of the Problem of Measurability in Quantum Electrodynamics." *Nuovo Cimento* 10, Suppl. 2: 83–100.

Darrigol, Olivier (1986). "The Origin of Quantized Matter Waves." *Historical Studies in the Physical and Biological Sciences* 16, Part 2: 197–253.

— (1991). "Statistics and Combinatorics in Early Quantum Theory, II: Early Symptoms of Indistinguishability and Holism." *Historical Studies in the Physical and Biological Sciences* 21, Part 2: 237–298.

— (1993). "Strangeness and Soundness in Louis de Broglie's Early Works." *Physis. Rivista Internazionale di Storia della Scienza* 30: 303–372.

Debye, Peter (1910). "Der Wahrscheinlichkeitsbegriff in der Theorie der Strahlung." *Annalen der Physik* 33: 1427–1434.

Dirac, Paul A. (1926). "On the Theory of Quantum Mechanics." *Royal Society of London. Proceedings* A112: 661–677.

— (1927). "The Quantum Theory of the Emission and Absorption of Radiation." *Royal Society of London. Proceedings* A114: 243–265.

Discussion (1909). "Diskussion" following the lecture (Einstein 1909b). *Physikalische Zeitschrift* 10: 825–826. [Reprinted in Einstein 1989, vol. 2, Doc. 61, pp. 585–587]

Discussion (1911). "Diskussion" following the lecture (Einstein 1911). In *Die Theorie der Strahlung und der Quanten.* Arnold Eucken, ed. Halle a. S.: Knapp, 1914, pp. 353–364. [Reprinted in Einstein 1993a, Doc. 27, pp. 549–562]

Dyson, Freeman (1951). "Heisenberg Operators in Quantum Electrodynamics." *Physical Review* 82: 428–439.

Ehrenfest, Paul (1906). "Zur Planckschen Strahlungstheorie." *Physikalische Zeitschrift* 7: 528–532.

— (1911). "Welche Züge der Lichtquantenhypothese spielen in der Theorie der Wärmestrahlung eine wesentliche Rolle?" *Annalen der Physik* 36: 91–118.

— (1925a). "Energieschwankungen im Strahlungsfeld oder Kristallgitter bei Superposition quantisierter Eigenschwingungen." *Zeitschrift für Physik* 34: 362–373.

— (1925b). "Bemerkungen betreffs zweier Publikationen über Energie-schwankungen." *Zeitschrift für Physik* 35: 316.

Ehrenfest, Paul and Kamerlingh Onnes, Heike (1915). "Simplified Deduction of the Formula from the Theory of Combinations which Planck Uses as the Basis of his Radiation Theory." *Proceedings, Koninklijke Akademie van Weten-schappen te Amsterdam* 17: 1184–1190.

Einstein, Albert (1905). "Über einen die Erzeugung und Verwandlung des Lichtes betreffenden heuristischen Gesichtspunkt." *Annalen der Physik* 17: 132–148.

— (1906). "Zur Theorie der Lichterzeugung und Lichtabsorption." *Annalen der Physik* 20: 199–206.

— (1909a). "Zum gegenwärtigen Stand des Strahlungsproblems." *Physikalische Zeitschrift* 10: 185–193.

— (1909b). "Über die Entwickelung unserer Anschauungen über das Wesen und die Konstitution der Strahlung." *Physikalische Zeitschrift* 10: 817–825.

— (1911). "Zum gegenwärtigen Stande des Problems der Spezischen Wärme." In *Die Theorie der Strahlung und der Quanten.* Arnold Eucken, ed. Halle a. S.: Knapp, 1914, pp. 330–364.

— (1915). "Antwort auf eine Abhandlung M. v. Laues 'Ein Satz der Wahrschein-lichkeitsrechnung und seine Anwendung auf die Strahlungstheorie'." *Annalen der Physik* 47: 879–885.

— (1916). "Strahlungs- Emission und -Absorption nach der Quantentheorie." *Verhandlungen, Deutsche Physikalische Gesellschaft* 18: 318–323.

— (1917). "Zur Quantentheorie der Strahlung." *Physikalische Zeitschrift* 18: 121–128.

— (1921). "Über ein den Elementarprozess der Lichtemission betreffendes Experiment." *Sitzungsberichte der Preussischen Akademie der Wissen-schaften. Physikalisch-mathematische Klasse*: 882–883.

— (1923). "Bietet die Feldtheorie Möglichkeiten für die Lösung des Quanten-problems?" *Sitzungsberichte der Preussischen Akademie der Wissenschaften. Physikalisch-mathematische Klasse*: 137–140.

— (1924a). "Das Comptonsche Experiment." *Berliner Tageblatt* (20 April 1924). 1. Beiblatt.

— (1924b). "Quantentheorie des einatomigen idealen Gases." *Sitzungsberichte der Preussischen Akademie der Wissenschaften. Physikalisch-mathematische Klasse*: 261–267.

— (1925). "Quantentheorie des einatomigen idealen Gases. Zweite Abhandlung." *Sitzungsberichte der Preussischen Akademie der Wissenschaften. Physi-kalisch-mathematische Klasse*: 3–14.

— (1926a). "Vorschlag zu einem die Natur des elementaren Strahlungsemissions-prozesses betreffenden Experiment." *Naturwissenschaften* 14: 300–301.

— (1926b) "Über die Interferenzeigenschaften des durch Kanalstrahlen emittierten Lichtes." *Sitzungsberichte der Preussischen Akademie der Wissenschaften. Physikalisch-mathematische Klasse*: 334–340.

— (1927). "Mathematisch-Physikalische Arbeitsgemeinschaft an der Universität Berlin." *Zeitschrift für angewandte Chemie* 40: 546.

— (1989). *The Collected Papers of Albert Einstein*. Vol. 2, *The Swiss Years: Writings, 1900–1909*. John Stachel, et al., eds. Princeton, New Jersey: Princeton University Press.

— (1993a). *The Collected Papers of Albert Einstein*. Vol. 3, *The Swiss Years: Writings, 1909–1911*. Martin J. Klein et al., eds. Princeton, New Jersey: Princeton University Press.

— (1993b). *The Collected Papers of Albert Einstein*. Vol. 5, *The Swiss Years: Correspondence, 1902–1914*. Martin J. Klein et al., eds. Princeton, New Jersey: Princeton University Press.

— (1996). *The Collected Papers of Albert Einstein*. Vol. 6, *The Berlin Years: Writings, 1914–1917*. Martin J. Klein et al., eds. Princeton, New Jersey: Princeton University Press.

— (1998). *The Collected Papers of Albert Einstein*. Vol. 8, *The Berlin Years: Correspondence, 1914–1918*. Robert Schulman et al., eds. Princeton, New Jersey: Princeton University Press.

Einstein, Albert, and Hopf, Ludwig (1910). "Statistische Untersuchung der Bewegung eines Resonators in einem Strahlungsfeld." *Annalen der Physik* 33: 1105–1115.

Eisler, Rudolf (1909). *Wörterbuch der philosophischen Begriffe*. Berlin: E. S. Mittler.

González, J. J., and Wergeland, H. (1973). "Einstein–Lorentz Formula for the Fluctuations of Electromagnetic Energy." *K. Norske Vidensk. Selsk. Skf.* 4: 1–4.

Heisenberg, Werner (1930). *The Physical Principles of the Quantum Theory*. Chicago: The University of Chicago Press.

— (1931). "Über Energieschwankungen in einem Strahlungsfeld." *Berichte über der Verhandlungen der Sächsischen Akademie der Wissenschaften zu Leipzig* 83: 3–9.

— (1934). "Über die mit der Entstehung von Materie aus Strahlung Verknupfen Ladungsschwankungen." *Berichte über der Verhandlungen der Sächsischen Akademie der Wissenschaften zu Leipzig* 86: 317–322.

Hendry, John (1980). "The Development of Attitudes to the Wave-Particle Duality of Light and Quantum Theory, 1900–1920." *Annals of Science* 37: 59–79.

Hund, Friedrich (1967). *Geschichte der Quantentheorie*. Mannheim: Bibliographisches Institut.

Ioffe, Abram (1911). "Zur Theorie der Strahlungserscheinungen." *Annalen der Physik* 36: 534–552.

Jammer, Max (1966). *The Conceptual Development of Quantum Mechanics*. New York: McGraw-Hill.

Jordan, Pascual (1924). "Zur Theorie der Quantenstrahlung." *Zeitschrift für Physik* 30: 297–319.

— (1927a). "Zur Quantenmechanik der Gasentartung." *Zeitschrift für Physik* 44, 473–480.

— (1927b). "Über Wellen und Korpuskeln in der Quantenmechanik." *Zeitschrift für Physik* 45, 766–775.

Jordan, Pascual, and Wigner, Eugene (1928). "Über das Paulische Äquivalenz-verbot." *Zeitschrift für Physik* 47: 631–651.

Klein, Martin (1963). "Einstein and the Wave-Particle Duality." *The Natural Philosopher* 3: 182–202.

—— (1967). "Thermodynamics in Einstein's Thought." *Science* 157: 509–516.

Kojevnikov, Alexei (1990). "Einshteinovskaya formula fluktuatsii i korpus-kulyarno-volnovoy dualizm." In *Einshteinovskiy Sbornik, 1986–1990.* I. Yu. Kobzarev and G. E. Gorelik, eds. Moscow: Nauka, pp. 102–124.

—— (1994). "Light Quanta and Waves: Einstein's Reputation as a Dualist" In *La découverte des ondes de matière. Colloque organisé à l'occasion du centenaire de la naissance de Louis de Broglie.* Paris: Lavoisier, pp. 11–23.

Krutkov, Yuri A. (1914a). "Aus der Annahme unabhangiger Lichtquanten folgt die Wiensche Strahlungsformel." *Physikalische Zeitschrift* 15: 133–136.

—— (1914b). "Bemerkung zu Herrn Wolfkes Note: 'Welche Strahlungsformel folgt aus der Annahme ser Lichtatome?'" *Physikalische Zeitschrift* 15: 363–364.

Kubli, Fritz (1971). "Louis de Broglie und die Entdeckung der Materiewellen." *Archive for History of Exact Sciences* 7: 26–68.

Landè, Alfred (1925). "Lichtquanten und Kohärenz." *Zeitschrift für Physik* 33: 571–578.

Laue, Max von (1915a). "Die Einsteinschen Energieschwankungen." *Deutsche Physikalische Gesellschaft. Verhandlungen* 17: 198–202.

—— (1915b). "Ein Satz der Wahrscheinlichkeitsrechnung und seine Anwendung auf die Strahlungstheorie." *Annalen der Physik* 47: 853–878.

—— (1915c). "Zur Statistik der Fourierkoeffizienten der natürlichen Strahlung." *Annalen der Physik* 48: 668–680.

Lorentz, Hendrik Antoon (1910). "Die Hypothese der Lichtquanten." *Physikalische Zeitschrift* 11: 349–354.

—— (1916). *Les théories statistiques en thermodynamique. Conférences faites au Collège de France en novembre 1912.* Leipzig and Berlin: B. G. Teubner.

Mehra, Jagdish, and Rechenberg, Helmut (1982). *The Historical Development of Quantum Theory.* Vols. 1–3. New York: Springer-Verlag.

Natanson, Ladislas (1911). "Über die statistische Theorie der Strahlung." *Physikalische Zeitschrift* 12: 659–666.

Ornstein, Leonard Salomon, and Zernike, Frits (1919). "Energiewisselingen der zwarte straling en licht-atomen." *Verslagen van de gewone vergaderingen der Afdeeling natuurkunde, Koniklijke Nederlandse Akademie van wetenschappen te Amsterdam* 28: 280–292.

Pais, Abraham (1982). *"Subtle is the Lord. . ." The Science and Life of Albert Einstein.* Oxford and New York: Oxford University Press.

—— (1986). *Inward Bound of Matter and Forces in the Physical World.* London and New York: Oxford University Press.

Pauli, Wolfgang (1927). "Über Gasentartung und Paramagnetismus." *Zeitschrift für Physik* 41: 81–102.

Planck, Max (1906). *Vorlesungen über die Theorie der Wärmestrahlung.* Leipzig: J. A. Barth.

— (1910). "Zur Theorie der Wärmestrahlung." *Annalen der Physik* 31: 758–768.

— (1923a). "Die Energieschwankungen bei der Superposition periodischer Schwingungen." *Sitzungsberichte der Preussischen Akademie der Wissenschaften. Physikalisch-mathematische Klasse*: 350–354.

— (1923b). "Bemerkung zur Quantenstatistik der Energieschwankungen." *Sitzungsberichte der Preussischen Akademie der Wissenschaften. Physikalisch-mathematische Klasse*: 355–358.

— (1924). "Über die Natur der Wärmestrahlung." *Annalen der Physik* 73: 272–288.

Rosenfeld, Leon (1973). "The Wave-Particle Dilemma." In *The Physical Conception of Nature.* Jagdish Mehra, ed. Dordrecht: D. Reidel, pp. 251–263.

Rupp, Emil (1926). "Über die Interferenzeigenschaften des Kanallichtes." *Sitzungsberichte der Preussischen Akademie der Wissenschaften. Physikalisch-mathematische Klasse*: 341–351.

Schrödinger, Erwin (1924). "Über das thermische Gleichgewicht zwischen Licht- und Schallstrahlen." *Physikalische Zeitschrift* 25: 89–94.

— (1926). "Zur Einsteinschen Gastheorie." *Physikalische Zeitschrift* 27: 95–101.

Small, Henry (1986). "Recapturing Physics in the 1920's through Citation Analysis." *Czechoslovak Journal of Physics* 36: 142–147.

van der Waerden, Bartel Leendert, ed. (1967). *Sources of Quantum Mechanics.* New York: Dover.

Wien, Wilhelm (1913). *Vorlesungen über neuere Probleme der theoretischen Physik, gehalten an der Columbia-Universität in New York in April 1913.* Leipzig and Berlin: B.G. Teubner.

Wheaton, Bruce R. (1983). *The Tiger and the Shark: Empirical Roots of Wave–particle Dualism.* Cambridge: Cambridge University Press.

Wolfke, Mieczyslaw (1914a). "Welche Strahlungsformel folgt aus der Annahme der Lichtatome?" *Physikalische Zeitschrift* 15: 308–310.

— (1914b). "Antwort auf die Bemerkung Herrn Krutkows zu meiner Note: 'Welche Strahlungsformel folgt aus der Annahme der Lichtatome?'" *Physikalische Zeitschrift* 15: 463–464.

— (1921). "Einsteinsche Lichtquanten und raumliche Struktur der Strahlung." *Physikalische Zeitschrift* 22: 375–379.

Dirac's Quantum Electrodynamics

Alexei Kojevnikov

Dirac's relationship with quantum electrodynamics was not an easy one. On the one hand, the theory owes to him much more than to anybody else, especially if one considers the years crucial for its emergence, the late 1920s and early 1930s, when practically all its main concepts, except for that of renormalization, were developed. After this period Dirac also wrote a number of important papers, specifically, on indefinite metrics and quantum dynamics with constraints. On the other hand, since the early 1930s he was an active critic of the theory and tried to develop alternative schemes. He did not become satisfied with the later method of renormalization and regarded it as a mathematical trick rather than a fundamental solution, and died unreconciled with what, to a large extent, was his own brainchild.[1]

In the present paper I examine Dirac's contribution to quantum electrodynamics during the years 1926 to 1933, paying attention to the importance and the specificity of his approach and also tracing the roots of his dissatisfaction with the theory, which goes back to the same time and which, as I see it, in many ways influenced his attitude to its subsequent development. Some of Dirac's crucial accomplishments of that period, in particular his theory of the relativistic electron, have already been studied by historians in much detail. I will describe them more briefly, placing them in the context of Dirac's other works and of the general situation in quantum theory and leaving more room for other, less studied works, such as the 1932 Dirac–Fock–Podolsky theory.

1. 1926: The Compton Effect

This was the year when quantum mechanics was still being created and, at the same time, attempts were already being made to develop a relativistic quantum theory and to study problems related to radiation. It was not yet

Yuri Balashov and Vladimir Vizgin, eds., *Einstein Studies in Russia.*
Einstein Studies, Vol. 10. Boston, Basel, Birkhäuser, pp. 229–259.
© 2002 The Center for Einstein Studies.

clear whether all this would lead to one unified quantum theory or, as it eventually happened, would split into two separate disciplines. Dirac was at the center of all these developments.[2]

On the subjects that would later become part of quantum electrodynamics, he wrote two papers containing the theory of the Compton effect. The discovery of that effect in 1923 led to the triumph of Einstein's light quanta and provoked a crisis of the old quantum theory. But that was just the beginning rather than the end of the history of the problem.[3] In the experimental studies prior to 1926, the formula for frequency change and the validity of conservation laws were reliably established, but the formula for the intensity of scattered radiation (i.e., the scattering cross section) and its dependence on the angle and frequency was not yet fully clarified. Theoreticians were proposing competing models of the effect, which continued to be one of the most actively debated problems in physics.

In April 1926, Dirac completed a paper that became, in the framework of new quantum mechanics, the first attempt at a relativistic generalization of the theory. Dirac was not yet employing Schrödinger's formalism there but followed his own, the so-called algebra of q numbers, which was a generalization of the Göttingen matrix mechanics.[4] According to Dirac's analytical approach, the transition to a relativistic theory had to be achieved by getting rid of the special treatment of the time variable t and by making it appear in the formalism in a way similar to space variables x, y, z. He described the electron according to quantum mechanics with relativistic corrections, while radiation was treated classically as an electromagnetic wave. Dirac considered a system consisting of an electron and an incident wave, and calculated the emitted radiation with the help of the conventional formula of matrix mechanics, obtaining through this procedure expressions for the frequency and the intensity of scattered light (Dirac 1926a).

His formula for the intensity,

$$I = \frac{I_0}{R^2} \left(\frac{e^2}{mc^2} \right)^2 \left(\frac{v'}{v} \right)^3 \sin^2\theta, \tag{1}$$

where v, I_0 are the frequency and the intensity of incident and v', I of scattered radiation, R the radius and θ the angle of scattering, was only approximately correct. The exact formula would be derived by Oskar Klein and Yoshio Nishina by the end of 1928 on the basis of the 1928 Dirac relativistic wave equation for the electron with spin. All previous attempts (apart from Dirac, the problem was also attacked by Klein and by Walter

Gordon) provided only approximate results, though quite close to the final one (Gordon 1926, Klein 1927, Klein and Nishina 1929).

Dirac's paper was repeatedly used and referred to (see Small 1983), but his approach was not in itself popular. Practically all physicists who dealt, in 1926 and early 1927, with the relativistic problem in quantum theory treated it by the methods of Schrödinger's wave mechanics rather than matrix mechanics. Solutions for the electrodynamical problems were looked for on the basis of the Klein–Gordon equation combined with the interpretation of the wave function as the density of the electromagnetic charge and with the semi-classical description of radiation. Dirac, on his part, viewed this approach with skepticism. He accepted wave mechanics as a mathematical method only, without its physical interpretation, and that is how he used it in his second paper on the Compton effect in November 1926. It practically reproduced the approach and results of the first paper, only that a number of calculations were drastically simplified thanks to the use of Schrödinger's formalism (Dirac 1927a).

2. 1927: The Quantum Theory of Radiation

The first completed formalism of quantum mechanics—with its basic equations, methods of their solution, and rules of how to compare results of calculations with experiment—was developed very quickly after Max Born's proposal of the statistical interpretation of the wave function. Dirac, and simultaneously and independently Pascual Jordan, accomplished this in November 1926 (Dirac 1927b, Jordan 1927a,b) and soon declared quantum mechanics done.[5] "The new quantum theory . . . has by now been developed sufficiently to form a fairly complete theory of dynamics," Dirac wrote in February 1927, adding: "On the other hand, hardly anything has been done up to the present on quantum electrodynamics" (Dirac 1927c, p. 243).

The quotation reflects both Dirac's refusal to accept the electrodynamics of Klein and Gordon and the conclusion that quantum mechanics had developed as an essentially non-relativistic theory. Dirac's formulation of its basic principles was grounded in the Hamiltonian method, in which time had a special role as a parameter and could not be treated symmetrically with space variables. Therefore, according to Dirac, there was a need for a new separate theory of quantum electrodynamics that would account for relativistic problems as well as for a consistent description of the electromagnetic field.

In the same paper, Dirac made a partial step towards this future theory, proposing a quantum theory of electromagnetic radiation and its interaction with matter. In later accounts, this paper is often referred to as "the beginning" of quantum electrodynamics, although, of course, one cannot say that quantum mechanics had never up to that point considered radiation. A number of phenomena could be accounted for by describing radiation according to the classical wave theory and considering its interaction with a quantum-mechanical particle. These methods were applied to calculate the effects of the scattering of light by an electron (dispersion and the Compton effect) and of atomic quantum transitions caused by incident radiation. In addressing the latter problem, several authors, including Dirac (Born 1926, Dirac 1926b, Slater 1927), had been able to calculate the value of the Einstein coefficient for stimulated emission and absorption of radiation (the so-called B coefficient), but not the other, A coefficient for the probability of spontaneous emission.

To obtain the A coefficients, which determined the intensities of spectral lines, quantum mechanics used an additional rule, called Heisenberg's hypothesis, which was, in fact, the very first postulate from which quantum mechanics had started in 1925. It asserted that the probability of radiative transition is proportional to the square of the corresponding matrix element of the coordinate of the electron. Dirac eliminated the need for an additional postulate, having derived both coefficients together based on a single approach, in which waves of radiation themselves were described quantum-mechanically.[6] Each harmonical component of the radiation was quantized as an oscillator according to the rules of quantum mechanics. This treatment had already been proposed by Jordan in late 1925, in one of the first papers on matrix mechanics (Born, Heisenberg, and Jordan 1926), but only gained wide recognition after Dirac's effective demonstration of its power in 1927. Dirac's approach was instantly welcomed as the first consistent quantum theory of radiation and accepted as the paradigm in a whole series of subsequent studies.

Although very successful in this practical sense and capable of describing an ever increasing scope of phenomena, the theory, by Dirac's own criteria, still lacked a lot to become a consistent quantum electrodynamics. Since radiation was quantized as a Hamiltonian system, the relativistic invariance of the theory was not apparent. Secondly, the quantization dealt only with the radiation part of the electromagnetic field without the Coulomb part for the interaction of charges, and thirdly, the particles themselves were described as non-relativistic. Dirac saw the first of these handicaps as an especially serious one and, because of it, did not consider his accomplishment as the starting point of a future consistent theory, the present appraisal notwithstanding. The latter, he believed, had to be relativistically invariant right from the start and explicitly (Dirac 1927c, p. 243–4).

Things actually developed differently, by way of gradual corrections of the existing shortcomings. In 1928 Jordan and Wolfgang Pauli proved the relativistic invariance of the quantization of radiation, while Dirac found the relativistic wave equation of the electron. The following year, Enrico Fermi and also Werner Heisenberg together with Pauli extended the method of quantization to the full electromagnetic field, together with its Coulomb part (Jordan and Pauli 1928, Dirac 1928a, Fermi 1929, 1930, Heisenberg and Pauli 1929, 1930).[7]

3. 1927: Second Quantization

Apart from the above method of wave quantization, Dirac's 1927 theory contained two other ways of describing radiation. Radiation is treated, in both of them, from a corpuscular perspective as an ensemble of photons, quantum-mechanical Bose particles in the relativistic limit of zero mass and the velocity of light. Dirac derived the wave equation for an ensemble of bosons by two different methods: by imposing the requirement that the wave-functions of many-body systems must be symmetrical and by quantizing the wave function of a single particle, or, in modern terminology, through the second quantization.

This somewhat bizarre term stands for an idea that was also considered bizarre by many: to take the Ψ function of an already quantized system and make an operator out of it, that is, actually, to quantize the system for the second time. With the help of textual analysis, in particular, by paying attention to the evolution of Dirac's system of notation, one can reconstruct the history of the 1927 theory of radiation as originating from the attempt to quantize the wave function.[8] Dirac recalled later that he had not foreseen what would result from it and was sincerely surprised to find out that the procedure transformed the wave equation for one particle into an equation for a system of many Bose particles (Dirac 1983, p. 48). To confirm this unexpected result, he derived it once again by the conventional method of symmetrizing wave functions, meanwhile also obtaining, for the first time in the quantum theory of many-body systems, the wave equation for a Bose ensemble in the external field.

This result suggested to him applying the theory to photons and to their interaction with the atom. Pursuing the theory of quantum-mechanical Bose particles further, and considering its relativistic limit, Dirac managed to calculate the ratio between the two coefficients, A and B, but in order to obtain their absolute values, he had to supplement the photon theory with the method of quantized electromagnetic waves explained above. Both approaches provided results that were in good agreement, but the model of photons did not appear as mathematically powerful as the formalism of quantized waves.

Yet it was, of all the three Dirac methods of describing radiation, the one that satisfied his physical worldview the best, and he invested some more effort into it. In the first edition of *The Principles of Quantum Mechanics,* Dirac presented his quantum theory of radiation as the theory of photons, quantum-mechanical particles described by symmetrical Ψ functions and taken in the limiting case of relativistic velocities (Dirac 1930b). By that time, Dirac managed to develop the mathematics of the model enough to obtain both coefficients, A and B, and the dispersion formula without recourse to the formalism of quantized electromagnetic waves. He also abandoned the method of second quantization immediately after it had served its initial heuristic role. Dirac did not use it in his subsequent papers on quantum electrodynamics and on the many-body theory, leaving it to others to develop the approach further and recognizing it again only in the second edition of *The Principles* (Dirac 1935).

The method of second quantization was picked up by Jordan and then by Vladimir Fock (Darrigol 1986, Kojevnikov 1988b). Jordan's major achievement was to find out how to generalize the quantization of the wave function so as to obtain a system of Fermi particles (Jordan 1927c, Jordan and Wigner 1928). This gave him an opportunity to formulate the program of quantum electrodynamics on the basis of the fundamental concept of quantized waves applied to describe both the electromagnetic field and material particles, the program that has been realized in modern quantum field theory. Fock proposed a special representation (the Fock space) that allowed the translation of the formalism of second quantization at any stage into the language of conventional quantum mechanics, which removed the appearance of strangeness and became important for the method's general acceptance (Fock 1932).

Many modern presentations interpret second quantization wider, as the quantization of waves of any sort, including the electromagnetic waves. From this point of view, the first application of the method should be attributed to Jordan (Born, Heisenberg, and Jordan 1926), while Dirac is credited with its independent invention in the case of material particles. The mathematical procedure, indeed, is very similar in both cases, but it should be noted that this view reflects the modern perspective on second quantization. During the 1920s and 1930s, the quantization of the wave function, or of the matter wave, was viewed as a separate idea distinct from, and much more controversial than, the quantization of radiation waves.

4. 1928: The Dirac Equation

The many difficulties that preceded the development of quantum mechanics were later found to be caused by two separate problems: the electron's wave properties and its spin. Their combination only increased the confusion. Thus Erwin Schrödinger, for example, initially derived his famous equation in a relativistic form, which he immediately rejected because it gave the wrong spectrum for hydrogen. The non-relativistic wave equation proved to be a better approximation, while the relativistic wave was later found to provide good results only if spin was simultaneously taken into account.

In the crucial winter of 1925–26, both issues were clarified independently: George Uhlenbeck and Samuel Goudsmith proposed a visual model of the spinning electron, while Schrödinger published his wave equation. Its relativistic generalization, the Klein–Gordon equation, was quickly suggested by a number of authors (see Kragh 1984), and it was realized that the next step would have to combine both new results together into a quantum mechanical wave equation for the spinning electron. This logical path was taken by many: in 1927 Charles G. Darwin and Pauli solved the problem for the non-relativistic electron (Darwin 1927, Pauli 1927), and soon afterwards Hendrik Kramers, Jordan, Eugene Wigner, Yakov Frenkel, Dmitri Ivanenko, and Lev Landau examined the relativistic case (Kragh 1981b, pp. 61–62). Before any of these attempts succeeded, Dirac arrived at the result in a different way: he was preoccupied with creating a consistent relativistic quantum mechanics rather than with describing spin.[9] Spin came as an extra gift out of his equation, hence Dirac managed not only to unify spin with the wave mechanics, but, in a certain sense, to explain it (Dirac 1928a,b).

Since drafts or archive materials did not survive, historical reconstructions of this landmark achievement are based mainly on Dirac's published papers and his later reminiscences. Opinions differ as to whether he had no intention whatsoever of describing spin (as he himself claimed) or bore it somewhat in mind when he chose to play with the Pauli spin matrices. One way or the other, his primary motivation was his dissatisfaction with the Klein–Gordon relativistic equation

$$\left(\frac{1}{c^2}\frac{\partial^2}{\partial t^2} - \frac{\partial^2}{\partial x^2} - \frac{\partial^2}{\partial y^2} - \frac{\partial^2}{\partial z^2} + \frac{m^2 c^2}{\hbar^2}\right)\Psi = 0, \tag{2}$$

which he considered inconsistent with the basic core of quantum mechanics, the transformation theory.

His objections stemmed from requirements that, as later developments showed, could not be met. This led Dirac to admit, years later, that "the development of the relativistic theory of the electron can now be regarded as an example of how erroneous arguments sometimes lead to a valuable result" (Dirac 1959, p. 32). He presented these arguments in most detail in a little known paper in the summer of 1928 (Dirac 1928c). His objective was to find a wave equation with the always positive probability density. This could be met only for the density defined as $\rho = \Psi^*\Psi$, from which Dirac derived that the desired wave equation should be a differential equation of the first order in time. The equation (2) is of the second order, and it has an entirely different expression for the probability density.[10]

Looking for an equation which would be both linear in time and relativistically invariant, Dirac found it in the form

$$\left(\frac{1}{c}\frac{\partial}{\partial t} + \alpha_1\frac{\partial}{\partial x} + \alpha_2\frac{\partial}{\partial y} + \alpha_3\frac{\partial}{\partial z} + \frac{imc}{\hbar}\beta\right)\Psi = 0, \tag{3}$$

where α_1, α_2, α_3, and β are 4×4 matrices of a special kind, and Ψ is, correspondingly, a 4-component wave function. Dirac further showed that, besides giving the positive probability density, equation (3) describes a particle with spin ½ and, in the first approximation, yields the correct formula for the fine structure of the hydrogen spectrum.

The publication of the Dirac equation had an immediate and tumultuous response. For more than a year, it attracted the attention and absorbed practically all efforts of physicists and mathematicians who dealt with quantum electrodynamics and relativistic quantum theory. Among its subsequent most important applications was the derivation of the final formula for the Compton scattering, the Klein–Nishina formula (Klein and Nishina 1929). The initial euphoria, however, soon gave way to a more sober attitude, especially since the fundamental difficulty of Dirac's theory was mentioned by him explicitly in the very first publication.

Dirac formulated it at first as an additional argument against the Klein–Gordon equation (Dirac 1928a), but he immediately saw that it remained unresolved also in his own theory. The relativistic equation had twice as many degrees of freedom as would have been sufficient for a particle with spin (four components of Ψ in the Dirac equation instead of two) which led to additional solutions corresponding to the states with negative values of energy. Similar solutions also appeared in the classical relativistic theory of a particle, but there they could be put aside and declared non-physical. The quantum theory could not get rid of them as easily, since the formalism did not forbid a transition of the particle into a negative energy state. The situation was further aggravated with the discovery of the "Klein paradox": electrons with positive energy could transform into negative energy electrons when passing through a potential barrier (Klein 1929). It was not clear how to interpret these theoretical electrons. The Dirac equation had great advantages and an enormous potential for application but also such serious problems that it was possible neither to discard it nor to put up with it.

5. 1929: Expectations and Disillusionment

While Dirac chose the dynamics of one particle as his way towards the relativistic generalization of quantum theory, his German colleagues made an attempt to create quantum electrodynamics as a many-body theory from the start. They treated both electromagnetic field and material particles as quantized fields described by Maxwell's equations in the former case and by the Schrödinger–de Broglie matter waves in the latter case. Jordan had formulated the basic principles of this approach in autumn of 1927 (Jordan 1927c), and he, together with Klein, Wigner, and Pauli, took the first important steps towards its realization (Jordan and Klein 1927, Jordan and Wigner 1928, Jordan and Pauli 1928).[11]

Early in 1928 Heisenberg and Pauli attempted to bring this program to completion.[12] At that time they still described particles with the help of the Klein–Gordon wave equation, which they wanted to quantize either according to Dirac (Bose particles) or according to Jordan (Fermi particles). But the general method of field quantization failed during the attempt to extend it from the electromagnetic radiation to the full electromagnetic field with the Coulomb part. Their collaborative work came to a temporary halt and resumed one year later, after Heisenberg had come up with a special mathematical trick to circumvent this difficulty.[13] An unusually long paper, "On the Quantum Dynamics of Wave Fields," was completed in March 1929 (Heisenberg and Pauli 1929).

The Heisenberg–Pauli electrodynamics brought all the earlier achievements—Dirac's radiation theory, second quantization, the Dirac equation—together in a comprehensive scheme and proved its relativistic invariance. But it did not live up to the high level of expectations. In the course of previous years, theoretical physicists had gotten used to quick and extraordinary success in sorting out the difficulties of the old quantum theory. In the case of quantum electrodynamics, however, problems of partial theories did not disappear with their unification into a general scheme; on the contrary, they became even more aggravated. All the difficulties of Dirac's electron theory remained unresolved, Heisenberg's "trick" with the so called ε-term was too artificial for a fundamental method of field quantization. Furthermore, hopes, initially raised by the partial Jordan–Klein theory, to subtract the infinite energy of the point-like electron by switching the order of quantum operators, were not realized in the more general treatment (Jordan and Klein 1927). This was only the first of many other infinities in relativistic quantum theory which would start attracting the attention of theoretical physicists in the subsequent years.

Disillusionment was not long to come. The authors themselves, Jordan first, followed by Pauli, Heisenberg, and by their close collaborators Ivar Waller, Robert Oppenheimer, and Leon Rosenfeld, were quick to announce that something was fundamentally wrong with the basic approach and that new radical changes were required. Such a pattern of behavior was not unprecedented: it also happened during the so-called crisis of the old quantum theory, which preceded the creation of the new quantum mechanics, though it was then confined to a narrower circle of participants and to a largely informal discussion. A similarly critical attitude to quantum electrodynamics became more widespread and more openly pronounced. Expressed readiness to give up some of its basic principles was characteristic of the crisis that lasted from early 1930 to early 1933. Exactly which radical changes to demand was not clear; Pauli, for example, expected the new theory to explain the value of the dimensionless constant $e^2/\hbar c$ (Pauli 1933).[14]

At first, Dirac did not participate in these developments. He wrote nothing on quantum electrodynamics in 1929 besides a section in his textbook *The Principles of Quantum Mechanics* where he presented his previous results. He did not include the Heisenberg–Pauli theory there, only mentioned it once in passing and in rather neutral terms (Dirac 1930b). It was not very likely that he was impressed by it, not only because of its problems, but also because of the underlying program which he did not share. Dirac still preferred the corpuscular version to the quantized waves approach in the treatment of radiation and matter. He also still tried to avoid using second quantization, in particular, in the paper of the same year dealing with the many-body problem in non-relativistic quantum mechanics

(Dirac 1929). But until 1932, he did not express his criticism of the Heisenberg–Pauli theory either.

6. 1930: The Hole Theory

While continental colleagues continued to struggle with quantum field theory, Dirac in Britain kept working on his own, somewhat narrower, topic, the relativistic mechanics of the electron. At the very end of 1929 he took a new important step in it. If one could not get rid of the troubling negative energy solutions of the Dirac equation entirely, it was still possible to try to ban the electron's transition to these states. This would happen, according to quantum statistics, if those states were already occupied with electrons. Dirac suggested that the negative energy solutions had physical meaning, but that the normal state of physical vacuum was the un-observable "sea" consisting of an infinite number of electrons occupying all possible states with negative energies.

   Individual electrons with positive energies moved upon the face of the waters and were prohibited from falling into it by the Fermi statistics, because all positions below were occupied. If, however, a state of negative energy was free, there was a "hole" in the sea, which behaved as if it were a particle with normal, positive energy, but with the opposite sign of the electric charge. Dirac was tempted to identify this particle with the proton. His theory, then, would become a unified theory of matter covering both kinds of fundamental particles known by that time. Of course, the proton mass is almost 2000 times larger than the electron mass, but since the hole moves in the medium of negative energy electrons, Dirac maintained a hope to be able to derive the additional mass from the interaction between the hole and the sea. Dirac made this set of ideas public in a letter to Niels Bohr of 26 November and in lectures at the Institut Henri Poincaré in Paris in December 1929. He published them in a paper in January 1930 (Dirac 1930a).[15]

   An electron of positive energy may fall into a hole, which would be observed as a mutual elimination, annihilation of both particles; and vice versa, if an unobservable negative energy electron gets a quantum of energy enough to jump to a positive state, a pair of an observable electron and a hole is created. Dirac's theory thus described the possibility of mutual creation and annihilation of material particles. The idea itself was not entirely new, since it had been discussed in astrophysics for several years by James Jeans and Arthur Eddington, and had also been mentioned by Jordan, but Dirac's model made it possible to calculate the probability of the process (Bromberg 1976, Jordan 1928). The opposition to Dirac's proposal was chiefly concerned not with this idea, but with the metaphysical, as many physicists thought, concept of the unobserved sea of negative energy electrons with its infinite density of charge and mass.

The response to Dirac's hole theory was as cold as the earlier reception of his electron equation had been enthusiastic. Pauli and Bohr authoritatively disapproved of it. Only a handful of quantum theoreticians supported Dirac: Igor Tamm welcomed the proposal without reservations, Oppenheimer accepted the idea of "sea," but not the identification of the hole with the proton. In his opinion, all holes had to be filled, since their annihilation with electrons occurs very rapidly, and one had to introduce another "sea" for protons. Tamm and Dirac separately calculated the probability of annihilation and realized that it, indeed, allowed the hydrogen atom to live only about 10^{-3} sec (Oppenheimer 1930, Tamm 1930, Dirac 1930c). Hermann Weyl studied mathematical transformations of the theory and became convinced that the hole mass had to be exactly equal to the electron mass; he even mentioned a positively charged electron, but only to say that it was not observed in nature and to return in the physical interpretation of the theory to Dirac's proton hypothesis.[16]

The difficulties of the hole theory or, rather, of the identification of the hole with the proton, were increasing: the mass difference could not be explained: on the contrary, there were hefty arguments in favor of mass equality; the hydrogen atom was stable and did not want to self-annihilate within a split second. All this caused Dirac to reconsider his proposal the following year.

7. 1931: Monopole and Other Particles

An interesting shift occurred in fundamental theoretical physics around 1931. Up to that point, the list of basic ontological entities of the world was likely to be seen as very short, usually consisting of three objects: gravitation, electromagnetic field, and the electron, and occasionally having the fourth one, the proton. And though it had already become apparent that some new forces were acting in the nucleus, and though physicists close to the experiment were occasionally discussing a hypothetical neutral particle, the high theory did not pay any serious attention to that. Much has been written about the great increase in the number of known elementary particles and interactions starting with the "miraculous year" of 1932, which saw the discovery of the neutron and positron. What is more interesting, however, is that the change of prevailing attitudes among theoretical physicists had begun even before those experimental discoveries.

In December 1930 Pauli, in a letter to a conference, put forward an idea that there might be a neutral particle in the nucleus. He needed it in order to solve two difficulties: with the statistics of nuclei and with the continuous spectrum of the β-decay. As would be understood later, these difficulties were to be ascribed to two different particles, a heavy one, the neutron, and an extremely light or massless one, the neutrino. Pauli envisioned the neutral particle to have spin ½, the magnetic moment, and a small mass comparable to that of the electron. He proposed a wave equation for it which was similar to Dirac's equation for the electron.[17]

In May 1931 Dirac submitted a paper that contained a theory of another hypothetical particle, the magnetic monopole (Dirac 1931a). He demonstrated that the idea of magnetic charge, which makes the Maxwell equations fully symmetrical, does not contradict quantum mechanics if the values of charges are connected by the relationship $eg = \frac{1}{2} n\hbar c$, where e is the electric charge, g the magnetic charge, and n an integer number. The monopole, if existing, would thus explain the fact of the quantization of electric charges. Although monopoles did not occur in experiment, Dirac sounded optimistic: "This new development requires no change whatever in the formalism. . . . Under these circumstances one would be surprised if Nature had made no use of it" (Dirac 1931a, p. 71).

Unlike Pauli, who was motivated by experimental difficulties, Dirac apparently came to his idea on the basis of purely theoretical speculation, hoping to explain the quantization of electric charge.[18] His predictions, however, did not stop with the monopole. In the same paper Dirac discussed two more unknown particles. Referring to the difficulties that arose in his hole theory with the proton mass and with annihilation rate, he abandoned the idea that holes were protons, suggesting instead that the theory calls for the existence of a light positively charged particle, the "anti-electron." Likewise, the proton would then require an anti-particle for itself (Kobzarev 1990, Kragh 1990). A real festival of new particles took place on 1 October 1931 in Princeton where both Pauli and Dirac presented reports on their recent proposals.[19]

Dirac was not absolutely sure in his prediction of the anti-electron but rather formulated the dilemma: either the new particle existed, or his electron theory had to be rejected, a possibility which he did not rule out totally.[20] The fact that the antielectron was not observed in experiment jeopardized his entire approach.

8. 1932: A Failed Revolution or How Dirac Nearly Became a New Heisenberg

The wave of critical attitudes towards quantum electrodynamics that was spreading from Germany reached Dirac in 1932, when he joined the work of critical reassessment of existing methods. At this point, his attention shifted from the theory of the electron to the quantum description of the electromagnetic field. The solution of the problem in Dirac's radiation theory (Dirac 1927c) and its further generalizations by Heisenberg, Pauli, and Fermi (Heisenberg and Pauli 1929, Fermi 1929) is considered to be correct now, but in the situation of crisis in the early 1930s, even those results aroused doubts.

In February 1931, Heisenberg proposed a new approach to electro-dynamical problems (Heisenberg 1931). Unlike Dirac's radiation theory, it did not make use of the Hamiltonian function but relied directly on the equations of motion for the field and particles. A quantized electromagnetic wave excited the atomic system; the resulting charge and current densities were calculated using the rules of wave mechanics and determined, according to the classical formulas, the radiation emitted by the system. The theory thus obtained was close to the electrodynamical theory of Klein put forward in the earlier days of quantum mechanics (Klein 1927), with the main difference that incident electromagnetic waves were quantized rather than treated classically. Heisenberg offered his proposal as an alternative to Dirac's 1927 theory of radiation, arguing that a stricter reliance on the correspondence principle would pave the way out of the existing difficulties in quantum electrodynamics.

In a series of papers in 1931, Rosenfeld started developing this approach further and opposing it much more openly to Dirac's radiation theory (Rosenfeld 1931a,b,c). The latter, in his view, was responsible for the infinite values of the results of various calculations in quantum electrodynamics. Within the Heisenberg approach, Rosenfeld was able to reproduce all the basic achievements of the Dirac radiation theory and even to advance it somewhat further by deriving Christian Møller's formula for the scattering of two electrons with the relativistic retardation of the interaction (Møller 1931). It was probably Rosenfeld's critique that drew Dirac's attention to Møller's paper once he returned to Britain from a trip to the U.S. in early 1932.

By the standards of quantum electrodynamics, Møller did not derive his formula rigorously but guessed, to a certain degree, the correct answer. He considered the scattering of one electron on another in the Born approximation and, using a procedure similar to the semi-classical Klein–Gordon electrodynamics, established a correspondence between the electron's transition from one state to another and certain classical expressions for the densities of electric charge and current. The electromagnetic field thus produced, which he did not quantize but treated classically, acted on the second electron inducing its quantum transition into a new state. Despite the non-rigorous and non-symmetrical derivation, Møller's final formula for the scattering looked very reliable: it was symmetrical with regard to both electrons, relativistically invariant, and, indeed, was later fully confirmed within the fundamental theory. For the matrix element of the transition, in which the initial states of the electrons are set by the variables $\mathbf{p}_0^{I}$, u_0^{I}, $\mathbf{p}_0^{II}$, u_0^{II} (u is the spin part of the electron wave function, $\mathbf{p}$ is its momentum), and the final states by the variables $\mathbf{p}^{I}$, u^{I}, $\mathbf{p}^{II}$, u^{II}, he obtained the expression

$$\Phi = \frac{e^{I}e^{II}\hbar^{5}c^{2}}{\pi} \frac{<u^{II}u^{I}|1-\boldsymbol{\alpha}^{I}\boldsymbol{\alpha}^{II}u_0^{II}u_0^{I}>}{c^{2}(\mathbf{p}_0^{I}-\mathbf{p}^{I})^{2}-(E^{I}-E_0^{I})^{2}}\,\delta(\mathbf{p}^{I}+\mathbf{p}^{II}-\mathbf{p}_0^{I}-\mathbf{p}_0^{II}), \qquad (4)$$

where $\boldsymbol{\alpha}$ is a vector composed of Dirac's matrices, and Roman numerals denote variables corresponding respectively to the first and the second electrons (Kragh 1992, Roqué 1992).

While he was pondering Møller's formula and how to incorporate it into quantum electrodynamics, Dirac found a new approach to the whole theory, which he formulated in the paper "Relativistic Quantum Mechanics" dated 24 March 1932 (Dirac 1932). He transformed Møller's method quite considerably; perhaps only the initial formulation of the problem remained somewhat similar. Dirac wrote a system of two equations for two electrons:

$$i\hbar\frac{\partial}{\partial t_1}\Psi = (H_1 + \varepsilon_1 V(x_1,t_1))\,\Psi,$$

$$(5)$$

$$i\hbar\frac{\partial}{\partial t_2}\Psi = (H_2 + \varepsilon_2 V(x_2,t_2))\,\Psi,$$

where H_1 and H_2 are Hamiltonian functions of free particles, ε_1 and ε_2 their charges, and V the field potential. Each of the two particles was characterized by its own time variable, that is why the theory later came to be called "the many-times theory." The two equations were connected through the field potential V and the total wave function Ψ, which were common to both equations. To solve them, Dirac put $t_1 = t_2 = t$ and added the equations to obtain

$$(i\hbar\frac{\partial}{\partial t} - H_1 - H_2 - \varepsilon_1 V(x_1,t) - \varepsilon_2 V(x_2,t))\,\Psi = 0. \qquad (6)$$

He solved (6) by the method of successive approximations, treating the last two terms in parentheses as perturbation.

Two peculiar features of Dirac's theory need to be mentioned. First, differing from his 1927 theory of radiation and from the Heisenberg–Pauli theory, only the Hamiltonian functions for two particles were present, while the third term, the Hamiltonian function of the field itself, was lacking. Secondly, no direct Coulomb interaction between particles was introduced in the theory; electrons interacted only through the field potential V. For the latter, Dirac added the third equation, which corresponded to the free field without charges:

$$\Delta V - \frac{1}{c^2}\frac{\partial^2 V}{\partial t^2} = 0, \qquad (7)$$

where V was quantized in the usual manner. Therefore, in Dirac's theory, no static potential was postulated, the field consisted only of radiation.

To justify this unusual formulation of the problem, Dirac devoted more than a half of his paper to philosophical speculations, which was in general very uncharacteristic of him. He tried to imitate the discourse of the Copenhagen school, referring to the principles of correspondence and observability, but, in my view, did not do it very convincingly. This mode of thinking was alien to him; his philosophical argumentation looks self-contradictory and produces an impression of being developed *post factum*. Without presenting it at length, I will mention, as an example, that in order to justify his special treatment of the field, Dirac argued that the field plays a special role in the very process of observation and "we cannot therefore suppose the field to be a dynamical system on the same footing as the particles and thus something to be observed in the same way as the particles. The field should appear in the theory as something more elementary and fundamental" (Dirac 1932, p. 454).

Dirac presented his new theory as resulting out of a stricter compliance with the correspondence principle. For the point of departure, he chose the classical picture of electrons interacting with each other by means of absorbing and emitting waves of radiation. The elementary quantum process of this theory is thought to be a quantum jump "from the field of ingoing waves to the field of outgoing waves." He drew a parallel between the matrix element of this process with matrix elements introduced by Heisenberg in 1925 and, more generally, between the situation in quantum electrodynamics in 1932 and the situation in the old quantum theory just prior to the creation of quantum mechanics, making explicit an analogy between his new proposal and Heisenberg's revolutionary paper of 1925.

The reason for such inflated claims was, most probably, the astonishing result of his calculations. Even though they could be viewed as preliminary, because Dirac considered only the simple case of non-relativistic particles moving in a one-dimensional space and interacting through scalar waves, the second-order approximation resulted in an equation for the Ψ that corresponded to the static force of attraction between the two electrons. This inspired high hopes in him that it would be possible, in the three-dimensional case, to derive the Coulomb field out of the picture of particles exchanging radiation waves. Dirac's astonishment and enthusiasm would not have been that great, had he known that it was possible to represent the Coulomb potential with the help of waves even in classical electrodynamics, and that this method had already been applied in quantum electrodynamics by Fermi in 1930. Fermi's paper, however, had been published in Italian and was still little known when Dirac was developing his theory. Fermi obtained the Coulomb potential from waves corresponding to the scalar potential φ and the longitudinal component of the vector potential $\mathbf{A}$ of the electromagnetic field (Fermi 1930). Dirac was apparently quite surprised when he discovered this possibility by himself and hoped to solve on this basis the problems plaguing quantum electrodynamics.

His paper, however, was only a sketch of a possible theory. A realistic quantum electrodynamics would have to be constructed in a three-dimensional space, with electromagnetic waves instead of the simple scalar potential, and with particles described by Dirac's relativistic equation for the electron. In two months after the publication of Dirac's proposal, Fock and Boris Podolsky tried to meet these requirements. Podolsky, an American, was then working in Kharkov at the Ukrainian Institute of Physics and Technology. Fock came to visit there from Leningrad, and in June 1932 they co-authored two papers on the further development of Dirac's new theory (Fock and Podolsky 1932a, 1932b). The first one extended the treatment to the three-dimensional case and managed to derive the Coulomb potential with the correct sign from scalar waves: two particles with the same electric charge repelled one another, while in Dirac's one-dimensional theory they attracted one another. In the second paper, Fock and Podolsky added relativistic wave equations for electrons and the electromagnetic, rather than scalar field.

In the quantization of the electromagnetic field, they encountered the same mathematical difficulty as Heisenberg and Pauli, and as Fermi before them: the field described by the Maxwell equations could not be quantized by the usual canonical method (Heisenberg and Pauli 1929, Fermi 1929, 1930). Their version of the solution suggested quantizing a more general field and then imposing an additional constraint in order to satisfy Maxwell's equations. This additional condition was understood as a restriction on the wave function Ψ rather than on the operators of the electromagnetic field itself. This offered the third method of the quantization of electromagnetic field, which was different from, but also had something in common with, the two earlier ones (Fock and Podolsky were aware of the Heisenberg–Pauli method but had only a vague notion of Fermi's ideas). The specific forms of the generalized field and of the additional condition could vary. In their June 1932 paper, Fock and Podolsky were still unable to solve all the remaining mathematical problems, and the issue of how to bring the method of field quantization to complete consistency would still be discussed in their subsequent correspondence with each other and with Dirac.[21]

Another big problem, the relativistic description of particles, was not brought to completion either. Fock hoped to derive the retarding interaction of two electrons in the approximation to the order of $(v/c)^2$. The desired result had already existed in a less rigorous treatment by Gregory Breit (Breit 1929). Fock's calculation did not agree with the Breit formula and looked unsatisfactory: apart from the plausible terms of the order $(v/c)^2$ it also included incomprehensible imaginary terms of the order v/c. In September, Podolsky corrected the derivation, and the result then agreed with the Møller formula, rather than with the approximate Breit formula. (Podolsky and Fock 1932).[22] The circle was thus completed: the formula which had given the initial impulse to the new theoretical proposal received through it a solid justification.

While the new theory was being developed mathematically, its revolutionary value was called in question. In April 1932, Dirac presented his first proposal, which was still in press, at a conference in Copenhagen. Rosenfeld also attended the conference; he waited until Dirac's paper came out and published a critique, proving that the new theory and the old one, by Heisenberg and Pauli, were mathematically equivalent, the implication being that both were equally inadequate. Despite conspicuously different basic assumptions and equations, their formalisms could be translated into one another by a canonical transformation (Rosenfeld 1932).[23]

Despite Rosenfeld's finding, Dirac, Fock, and Podolsky proceeded, by correspondence, to develop the theory further, and in September the three met together in Leningrad at a conference on the theory of metals. Dirac and Fock presented there reports on their latest results, which agreed and overlapped to a great extent. After a vacation in the Crimea, which he spent together with Piotr Kapitza, Dirac stopped in Kharkov in early October on his way back and promised Ivanenko to contribute a paper to the newly launched Soviet journal, *Physikalische Zeitschrift der Sowjetunion*. As it turned out, Podolsky was already writing a paper on the further development of the new theory, and they finally agreed upon the publication of a joint paper by three authors. Its text was written mainly by Podolsky in Kharkov, altered and approved through extensive correspondence with Fock in Leningrad and Dirac in Cambridge, and finally resulting in the famous Dirac–Fock–Podolsky theory, dated 25 October 1932 (Dirac, Fock, and Podolsky 1932).[24]

Dirac's proposal was carried there to a completion. Main improvements over the previous Fock–Podolsky papers belonged to Dirac. He gave a simplified (compared to Rosenfeld's) proof that the new and the old quantum electrodynamics were mathematically equivalent and corrected the additional condition in the method of the quantization of electromagnetic field, which resolved remaining contradictions.[25]

Eventually, all formulations of quantum electrodynamics (Heisenberg–Pauli, Fermi, Heisenberg (1931), and Dirac–Fock–Podolsky (1932)) proved to be equivalent representations of the same theory, despite motivations to find something radically new. Although the new quantum revolution did not happen, the Dirac–Fock–Podolsky version was in several respects better than the older one by Heisenberg and Pauli. Its relativistic invariance was explicit, due to the introduction of separate time variables for each of the particles, and did not have to be proved in a complicated way. It included the so-called interaction representation, which was more convenient for calculations in most cases. In the second edition of *The Principles of Quantum Mechanics* (1935), Dirac presented quantum electrodynamics on the basis of the Dirac–Fock–Podolsky paper, and later, in the 1940s, it would play an important role in the covariant formulation of quantum electrodynamics (Tomonaga 1973, Schweber 1994, p. 277).

9. 1933: New Times

The review written by Pauli in 1932 reflects the situation of crisis in quantum electrodynamics (Pauli 1933). In his judgement, only isolated fragments of relativistic quantum theory were reliable, while attempts at unifying them into a consistent general scheme failed. Pauli recognized Dirac's theory of radiation and the Dirac equation for the electron but rejected or viewed very skeptically second quantization, his own (with Heisenberg) version of quantum electrodynamics, and the hole theory. In 1925, Pauli wrote a similar critical review of the old quantum theory (Pauli 1926), but it became outdated already in press due to the appearance of the first papers on quantum mechanics. The situation practically repeated itself with the review of 1932, this time because of the discovery of the positron.

The first announcement of a new positively charged light particle by Carl Anderson in September 1932 (Anderson 1932) was not noticed by many. The recognition came in early 1933 after an almost simultaneous publication of Anderson's second paper and the paper by Patrick Blackett and Giuseppe Occhialini (Anderson 1933, Blackett and Occhialini 1933). The view that the new particle is nothing else but the antielectron predicted by Dirac became accepted very quickly and signified the beginning of a new stage in the development of quantum electrodynamics (de Maria and Russo 1985, Roqué 1997).

A number of difficulties were solved. Combined efforts by Fermi, Heisenberg, Pauli, Dirac, Fock, and Podolsky delivered satisfactory methods of the quantization of the electromagnetic field. The discovery of the positron strengthened the credibility of the Dirac electron and the hole theory. Divergent calculations in the second order of the perturbation theory were the most serious among the remaining problems. By applying more or less artificial rules, physicists were learning how to subtract infinite values from the results and to leave only sensible finite terms that could be compared with experiment. Entirely consistent rules of dealing with this problem would not be developed, however, until the late 1940s (Schweber 1994). In the meantime, the crisis in quantum electrodynamics did not disappear entirely but became more of a chronic disease. There was still a lot of dissatisfaction with the state of the theory and a number of further attempts were made to suggest some fundamental changes in its very foundations, among the most prominent ones, by Jordan in 1933–1934, by Born and Leopold Infeld in 1934–35, by Dirac in 1936 and 1938, and a few others. The ultimate remedy, however, was as evasive as before, and a more pragmatic approach towards using the theory despite its apparent shortcomings was gradually gaining in popularity.

The discovery of the positron changed the relationship between the theory of quantum electrodynamics and experiment. Up to that point, quantum electrodynamics had hardly produced anything new for experimental physicists. Its accomplishments consisted mainly of developing a new, more rigorous and consistent foundation and justification for the existing stock of experiments and empirical formulas, which were already explained or derived with the help of earlier, less fundamental approaches. Antielectron became its first major independent prediction and it opened up a whole vast area of new phenomena, in which quantum electrodynamics had obvious advantages over earlier theories, not only logical advantages but also heuristic ones.

Despite internal imperfections it became a working theory and was providing calculations of new effects that were, within a certain area, in reasonable agreement with the growing amount of experimental data (Heitler 1936). The main line of development, apart from calculating particular effects, was represented by developing subtraction methods to handle the divergences, infinite values which continued to appear in various calculations (Rueger 1992, Brown 1993). Meanwhile, a generational change occurred, and earlier leaders, including Dirac, did not play such a crucial role any longer. Dirac's two last papers of fundamental importance for quantum electrodynamics were dated 1933.

The first one was actually written in late 1932 (Dirac 1933). Dirac returned there to his favorite idea that the Hamiltonian formalism, in which time plays a special role, is not suitable for a fundamental relativistic theory. He tried the Lagrangian dynamics instead, showing how it could be taken over into the quantum theory. Although the Lagrangian formalism has not prevailed over the Hamiltonian one in quantum field theory, Dirac's paper had a long-term consequence: in 1948 Richard Feynman transformed its approach into his new formulation of quantum mechanics on the basis of path integrals (Schweber 1994, p. 390).

In the report presented at the Leningrad conference on nuclear physics in September and at the 7th Solvay Congress in Brussels in October 1933, Dirac put forward the idea of vacuum polarization. Formulated in the terms of his hole theory, a charged particle causes displacements in the sea of negative energy electrons, which produce an observable effect similar to screening: the electric charge of the particle appears smaller than it really is (Dirac 1934a,b). While developing this idea, Dirac suggested one of the first mathematical methods of subtraction, in which a finite result of calculations is obtained as a difference between two infinite values (Dirac 1934c). Although he, thus, can be considered as an early forerunner of the renormalization method, he never regarded it as a final solution to the difficulties of quantum electrodynamics. He remained skeptical of renormalization even after it had developed into a consistent formalism, valid in all approximations of the perturbation theory. His disappointment with quantum electrodynamics, which started in the early thirties, never totally disappeared (see Kragh 1990, ch.8).

10. Conclusions

Dirac, as we saw, made great contributions to the development of practically all basic concepts and methods of quantum electrodynamics.[26] The nature of his influence, however, varied from case to case. In some cases, he created foundations of broadly accepted methods (quantization of the electromagnetic radiation, the Dirac equation, the subtraction technique), in others, he had put forward an important initial idea but gave it up later (second quantization). He developed some other approaches virtually alone or with very few supporters (the corpuscular theory of radiation, the hole theory, the prediction of the positron, the many-times theory), ignoring strong criticisms. Some of his ideas were picked up and developed further by others only years later (the Lagrangian method in quantum theory, the monopole).

Although his personal contribution to the early quantum electrodynamics is larger than anybody else's, he did not dominate the theory as much as, say, Einstein did in the case of general relativity. And the development of the theory often did not live up to his expectations either. Metaphorically speaking, his works formed the tangent vector to the trajectory, but there also always existed normal acceleration, which deflected the trajectory from the direction in which Dirac was trying to move it.

Remarks scattered among Dirac's various papers allow a reconstruction of his ideal of a theory that corresponded to his physical worldview and aesthetic tastes. Of the two related but different tasks, the relativistic generalization of quantum mechanics and the quantum treatment of electrodynamical processes, he clearly considered the first one as more fundamental. One can more properly call his preferred approach "relativistic quantum mechanics" rather than "quantum electrodynamics," and this is reflected in the titles of his papers. Relativistic invariance was, for him, the basic requirement which had to be satisfied explicitly and directly, rather than through corrections or approximations. The way to achieve this, as he repeatedly stated, was to treat time and space variables alike in the fundamental equations. This condition, however, was very hard to meet, and he himself often had to put up with relativistic modifications of the Hamiltonian formalism.

In questions of physical worldview, in his attitude towards the wave-particle dilemma, which occupied the minds of many of his contemporaries,

Dirac was more inclined to view both matter and radiation in corpuscular terms. This is particularly apparent in his earlier papers and is represented by his tendency to understand radiation as an ensemble of photons, by his efforts to develop a relativistic mechanics of a single particle, and by his model of the sea of negative electrons. Even though he used mathematical formalisms of the wave theory and of quantized waves, in his physical interpretation and in discussions of obtained results, Dirac predominantly relied on corpuscular concepts.

His style was to attack fundamental issues while leaving aside calculations of concrete effects and applied problems. Dirac disliked phenomenological and hybrid theories and preferred to ground his work in some fundamental principle, such as relativistic invariance. Since fundamental requirements were difficult to satisfy all at once in one comprehensive theory, he dealt with them separately, one by one, in his various papers. Physical principles were especially important for him during the formulation of the problem and in the interpretation of final result, while in intermediate calculations he had a complete faith in mathematical formalism, even when formulas were counterintuitive.

Dirac's disagreements with his German colleagues were increasing throughout the period discussed in this paper. They often used and developed further his mathematical approaches, but did not share his physical interpretations and preferences. In particular, Dirac's corpuscular quantum mechanical theory of radiation and his hole theory had very few supporters. His somewhat isolated status within the field certainly contributed to his critical outlook.[27]

NOTES

[1] See the bibliography compiled by R. H. Dalitz and R. Peierls (1986) and Dirac (1995) for the papers of his last years with their titles speaking for themselves: "Does renormalization make sense?", "The requirements of fundamental physical theory," "The inadequacies of quantum field theory."

[2] The map of alternative approaches within the emerging quantum mechanics and of their relative popularity is given in Kojevnikov and Novik (1989). Dirac's fundamental contributions to quantum mechanics of that year are discussed in detail in Mehra and Rechenberg (1982), Kragh (1990), Darrigol (1992).

[3] Up to 1926, the history is covered in detail in Stuewer (1975).

[4] On the algebra of q-numbers, see Mehra and Rechenberg (1982) and Darrigol (1992).

[5] This was the so-called transformation theory, which later became the basis of Dirac's book *The Principles of Quantum Mechanics* (1930b). In Copenhagen, Bohr and Heisenberg still felt the need to understand and to interpret the theory and

accomplished this in 1927 with the help of "uncertainty" and "complementarity." For Dirac, however, quantum mechanics was completed even before these developments.

[6] He calculated coefficients for spontaneous and stimulated radiation in (Dirac 1927c) and the Kramers–Heisenberg dispersion formula in his next paper (Dirac 1927d). For a detailed analysis of Dirac's radiation theory see Jost (1972), Darrigol (1984), Kojevnikov (1988a).

[7] On the history of early quantum electrodynamics see Cini (1982), Miller (1994), Schweber (1994, Ch.1), and Schweber (1995).

[8] The reconstruction in Kojevnikov (1988a) actually shows that second quantization preceded the transformation theory and was initially connected with Dirac's earlier paper (Dirac 1926b). The motivation behind the quantization of the wave function was to make the value of the expression for the number of atoms in an excited state $N_r = a^*_r a_r$ an integer number. The transformation theory made it possible to derive the consequences of this proposal and to show that it leads to the description of a Bose ensemble, but it also destroyed the initial reason for it, because $a^*_r a_r$ began to be understood as probabilities and did not have to be integer. Therefore, Dirac did not mention this initial motivation when he published the idea of the second quantization in his 1927 theory of radiation.

[9] For the detailed history of the Dirac equation, see Kragh (1981b), Moyer (1981a), Kragh (1990, Ch. 3).

[10] In fact, the Klein–Gordon equation (2) can also be formally rewritten as a differential equation of the first order in time (see, e.g., Akhiezer and Berestetskiy 1981, pp. 9–10). But the main objection to Dirac's argument came with the understanding that relativistic quantum theory is an essentially many-body theory and that its strict formulation as a one particle theory is impossible (see Pauli and Weisskopf 1934).

[11] On the program of quantized waves and wave-particle duality, see Darrigol (1986), Kojevnikov (1990a) and in this volume, Schweber (1994, pp. 33–44).

[12] Pauli to Kramers, 7 February 1928, Pauli to Dirac, 17 February 1928 (Pauli 1979, pp. 432–5).

[13] Pauli to Klein, 18 February 1929 (Pauli 1979, p. 488).

[14] More on the 1930 crisis in quantum electrodynamics in Kojevnikov (1988b, pp. 113–116), Rueger (1992).

[15] For the text of Dirac's letter to Bohr see Moyer (1981b), for Dirac's Paris lectures see Dirac (1931b). More on the hole theory in: Moyer (1981b), Kobzarev (1990), Kragh (1990), Dirac and Tamm (1993).

[16] Weyl (1932, Ch. 4, §6). Pauli also came to the conclusion that the masses of the electron and the hole must be equal (Tamm to Dirac, 13 September 1930, in Dirac and Tamm 1993).

[17] Pauli to Meitner and others, 4 December 1930; Pauli to Klein, 12 December 1930 (Pauli 1985, pp. 39–47).

[18] For the history of the monopole, see Kragh (1981a), Krivonos (1986).

[19] Pauli to Peierls, 29 September 1931 (Pauli 1985, pp. 93–94).

[20] "Your theoretical prediction about the existence of the anti-electron . . . seemed so extravagant and totally new that you yourself dared not to cling to it and preferred rather to abandon the theory" (Tamm to Dirac, 5 June 1933, in Dirac and Tamm 1993). Tamm must have known this firsthand, because in the spring of 1931 he was in Cambridge and worked with Dirac on the theory of the monopole.

[21] Fock to Dirac, 7 July; Dirac to Fock, 19 July; Fock to Dirac, 16 October; Dirac to Podolsky, 2 November 1932 (Dirac and Fock 1990; Fock's Personal Papers in the Archive of the Russian Academy of Sciences, St. Petersburg Branch, # 1034)

[22] See also Podolsky to Fock, 26 September and 1 October 1932 (Dirac and Fock 1990). The same year, the Breit and Møller formulas were derived in Bethe and Fermi (1932) on the basis of Fermi's formulation of quantum electrodynamics.

[23] The two formulations correspond to two different representations of quantum theory, respectively the Schrödinger and the interaction representations. The characteristic feature of the interaction representation is that the equation of motion for the field operators has the form of the equation of motion for the free field, as if there are no charges (see Akhiezer and Berestetskiy 1981, p. 132). This is just the case in Dirac's equation (Eq. 7). Dirac's 1932 theory, however, uses the interaction representation only for the electromagnetic field, but not for operators describing particles.

[24] Podolsky to Fock, 9 October; Fock to Dirac, 16 October; Dirac to Podolsky, 2 November; Podolsky to Fock, 10 November; Dirac to Fock, 11 November; Podolsky to Dirac, 16 November; Podolsky to Fock, 24 November 1932 (Dirac and Fock 1990; Fock's Personal Papers in the Archive of the Russian Academy of Sciences, St. Petersburg Branch, # 1034).

[25] There was a debate between Podolsky and Dirac about the correct formulation of the equivalence proof (Dirac to Podolsky, 2 November; Podolsky to Dirac, 16 November; Podolsky to Fock, 24 November 1932). In addition to §1 of the joint paper, which belongs mostly to Dirac, Podolsky wrote §§5 and 7. Fock proved the equivalence of the additional condition in the field quantization with that given by Heisenberg and Pauli (Fock to Dirac, 16 October 1932, §6 of the joint paper).

[26] See also an appraisal in Schweber (1994, pp. 2, 70–72).

[27] The paper was originally published in Russian (Kojevnikov 1990b) in a volume on Paul Dirac which I edited together with Boris Valentinovich Medvedev and which developed from a 1986 conference held at the Institute for History of Science and Technology in Moscow. Since then, a number of important publications have appeared, including Helge Kragh's biography of Dirac, Dirac's *Collected Papers*, Sam Schweber's and others' works on the history of quantum electrodynamics, which I added to the list of references.

REFERENCES

Anderson, C. A. (1932). "The Apparent Existence of Easily Deflectable Positives." *Science* 76: 238–239.
—— (1933). "The Positive Electron." *Physical Review* 43: 491–494.

Akhiezer, A. I., and Berestetskiy, V. B. (1981). *Kvantovaya electrodinamika.* 3rd ed. [*Quantum Electrodynamics*] Moscow: Nauka.

Blackett, P. M. S., and Occhialini, G. P. S. (1933). "Some Photographs of the Tracks of Penetrating Radiation." *Royal Society of London. Proceedings* 139: 699–726.

Bethe, H., and Fermi, E. (1932). "Über Wechselwirkung von zwei Electronen." *Zeitschrift für Physik* 77: 296–306.

Born, Max (1926). "Das Adiabatenprinzip in der Quantenmechanik." *Zeitschrift für Physik* 40: 167–192.

Born, M., Heisenberg, W., and Jordan, P. (1926). "Zur Quantenmechanik, II." *Zeitschrift für Physik* 35: 557–615.

Breit, G. (1929). "The Effect of Retardation on the Interaction of Two Electrons." *Physical Review* 34: 553–573.

Bromberg, Joan (1976). "The Concept of Particle Creation before and after Quantum Mechanics." *Historical Studies in Physical Sciences* 7: 161–191.

Brown, Laurie, ed. (1993). *Renormalization: From Lorentz to Landau (and Beyond).* New York: Springer.

Cini, M. (1982) "Cultural Traditions and Environmental Factors in the Development of Quantum Electrodynamics (1925–1933)." *Fundamenta Scientiae* 3: 229–253.

Dalitz, R. H., and Peierls, R. (1986). "P. A. M. Dirac. 1902–1984." *Royal Society of London. Biographical Memoirs* 32: 139–185.

Darrigol, Olivier (1984). "La genèse du concept de champ quantique." *Annales de Physique Française* 9: 433–501.

—— (1986). "The Origin of Quantized Matter Waves." *Historical Studies in Physical and Biological Sciences* 16: 197–253.

—— (1992). *From C-numbers to Q-numbers: The Classical Analogy in the History of Quantum Theory.* Berkeley: University of California Press.

Darwin, C. G. (1927). "The Electron as a Vector Wave." *Royal Society of London. Proceedings* 116: 227–233.

de Maria, M., and Russo, A. (1985). "The Discovery of the Positron." *Rivista di Storia della Scienza* 2: 237–286.

Dirac, P. A. M. (1926a). "Relativity Quantum Mechanics with an Application to Compton Scattering." *Royal Society of London. Proceedings* 111: 405–423.

—— (1926b). "On the Theory of Quantum Mechanics." *Royal Society of London. Proceedings* 112: 661–677.

—— (1927a). "The Compton Effect in Wave Mechanics." *Cambridge Philosophical Society. Proceedings* 23: 500–507.

—— (1927b). "The Physical Interpretation of the Quantum Dynamics." *Royal Society of London. Proceedings* 113: 621–641.

—— (1927c). "The Quantum Theory of the Emission and Absorption of Radiation." *Royal Society of London. Proceedings* 114: 243–256.

—— (1927d). "The Quantum Theory of Dispersion." *Royal Society of London. Proceedings* 114: 710–728.

— (1928a). "The Quantum Theory of the Electron." *Royal Society of London. Proceedings* 117: 610–624.

— (1928b). "The Quantum Theory of the Electron, Part II." *Royal Society of London. Proceedings* 118: 351–361.

— (1928c). "Zur Quantentheorie des Elektrons." In *Quantentheorie und Chemie.* Hans Falkenhagen and S. Hirzel, eds. Leipzig: S. Hirzel, pp. 85–94.

— (1929). "Quantum Mechanics of Many-Electron Systems." *Royal Society of London. Proceedings* 123: 714–733.

— (1930a). "A Theory of Electrons and Protons." *Royal Society of London. Proceedings* 126: 360–365.

— (1930b). *The Principles of Quantum Mechanics.* Oxford: Clarendon Press.

— (1930c). "On the Annihilation of Electrons and Protons." *Cambridge Philosophical Society. Proceedings* 26: 376–385.

— (1931a). "Quantized Singularities in the Electromagnetic Field." *Royal Society of London. Proceedings* 133: 60–72.

— (1931b). "Quelquez problèmes de mécanique quantique." *Institut Henri Poincaré. Annales* 1: 357–400.

— (1932). "Relativistic Quantum Mechanics." *Royal Society of London. Proceedings* 136: 453–464.

— (1933). "The Lagrangian in Quantum Mechanics." *Physikalische Zeitschrift der Sowjetunion* 3: 64–72.

— (1934a). "Teoriya pozitrona" ["A Theory of the Positron"]. In *Atomnoye yadro* [*The Atomic Nucleus*]. M. P. Bronstein, V. M. Dukel'skiy, D. D. Ivanenko, and Yu. B. Khariton, eds. Leningrad and Moscow: Gostekhteorizdat, pp. 129–143.

— (1934b). "Theorie du positron." In *Structure et propriétés des noyaux atomiques.* Rapports et discussions du septième conseil de physique tenu à Bruxelles du 22 au 29 Octobre 1933. Paris: Gauthier-Villars, pp. 203–230.

— (1934c). "Discussion of the Infinite Distribution of Electrons in the Theory of the Positron." *Cambridge Philosophical Society. Proceedings* 30: 150–163.

— (1935). *The Principles of Quantum Mechanics.* 2nd ed. Oxford: Clarendon Press.

— (1959). "Sovremennoe sostoyanie reliativistskoy teorii elektrona." ["Contemporary State of the Relativistic Theory of the Electron."] *Institut istorii estestvoznaniya i tekhniki. Trudy* [*Institute for the History of Science and Technology. Transactions*] 22: 32–33.

— (1983). "The Origin of Quantum Field Theory." In *The Birth of Particle Physics.* L. M. Brown and L. Hoddeson, eds. Cambridge and New York: Cambridge University Press, pp. 39–55.

— (1995). *The Collected Works of P. A. M. Dirac, 1924–1948.* R. H. Dalitz, ed. Cambridge and New York: Cambridge University Press.

Dirac, P. A. M., and Fock, V. A. (1990). "Perepiska Diraka s V. A. Fokom. Materialy k istorii stat'i Diraka–Foka–Podol'skogo." ["Dirac's Correspondence with V. A. Fock. Documents on the History of the Dirac–Fock–

Podolsky Theory"] A. B. Kozhevnikov, V. Ya. Frenkel, N. I. Nevskaya, and
T. I. Efremidze, eds. In: Medvedev and Kozhevnikov 1990, pp. 177–197.

Dirac, P. A. M., Fock, V. A., and Podolsky, B. (1932). "On Quantum Electrodynamics." *Physikalische Zeitschrift der Sowjetunion* 2: 468–479.

Dirac, P. A. M., and Tamm, I. E. (1993–1996). *Paul Dirac and Igor Tamm.
Correspondence.* Part 1: 1928–1933; part 2: 1933–1936. A. B. Kojevnikov,
ed. Munich: Max-Planck-Institut für Physik. Preprints 93–80 and 96–40.

Fermi, E. (1929). "Sopra l'elettrodinamica quantistica." *Accademia dei Lincei.
Rendiconti* 9: 881–887.

— (1930). "Sopra l'elettrodinamica quantistica." *Accademia dei Lincei. Rendiconti* 10: 431–435.

Fock, V. A. (1932). "Konfigurationsraum und zweite Quantelung." *Zeitschrift für
Physik* 75: 622–647; 76: 852.

Fock, V. A., and Podolsky, B. (1932a). "Zur Diracschen Quantenelektrodynamik."
Physikalische Zeitschrift der Sowjetunion 1: 798–800.

— (1932b). "On the Quantization of Electromagnetic Waves and the Interaction
of Charges in Dirac's Theory." *Physikalische Zeitschrift der Sowjetunion* 1:
801–817.

Gordon, W. (1926). "Der Comptoneffekt nach der Schrödingerschen Theorie."
Zeitschrift für Physik 40: 117–133.

Heisenberg, W. (1931). "Bemerkungen zur Strahlungstheorie." *Annalen der Physik*
9: 338–346.

Heisenberg, W., and Pauli, W. (1929). "Zur Quantendynamik der Wellenfelder."
Zeitschrift für Physik 56: 1–61.

— (1930). "Zur Quantentheorie der Wellenfelder. II." *Zeitschrift für Physik* 59:
168–190.

Heitler, W. (1936). *The Quantum Theory of Radiation.* Oxford: Clarendon Press.

Jordan, P. (1927a). "Über eine neue Begründung der Quantenmechanik." *Zeitschrift für Physik* 40: 809–838.

— (1927b). "Über eine neue Begründung der Quantenmechanik." *Gesellschaft der
Wissenschaften zu Göttingen. Nachrichten*: 161–169.

— (1927c). "Zur Quantenmechanik der Gasentartung." *Zeitschrift für Physik* 44:
473–480.

— (1928). "Die Lichtquantenhypothese. Entwicklung und gegenwärtige Stand."
Ergebnisse der Exakten Wissenschaften 7: 158–208.

Jordan, P., and Klein, O. (1927). "Zum Mehrkörperproblem der Quantentheorie."
Zeitschrift für Physik 45: 751–765.

Jordan, P., and Pauli, W. (1928). "Zur Quantenelektrodynamik ladungsfreier
Felder." *Zeitschrift für Physik* 47: 151–173.

Jordan, P., and Wigner, E. (1928). "Über das Paulische Äquivalenzverbot." *Zeitschrift für Physik* 47: 631–651.

Jost, Res (1972). "Foundations of Quantum Field Theory." In *Aspects of Quantum
Theory.* A. Salam and E. P. Wigner, eds. Cambridge: Cambridge University
Press, pp. 61–77.

Klein, O. (1927). "Elektrodynamik und Wellenmechanik vom Standpunkt des Korrespondenzprinzips." *Zeitschrift für Physik* 41: 407–442.

— (1929). "Die Reflexion von Elektronen an einem Potentialsprung nach der relativistischen Dynamik von Dirac." *Zeitschrift für Physik* 53: 157–165.

Klein, O., and Nishina, Y. (1929). "Über die Streuung von Strahlung durch freie Elektronen nach der neuen relativistischen Quantendynamik von Dirac." *Zeitschrift für Physik* 52: 853–868.

Kobzarev, I. Yu. (1990). "K istorii positrona." ["On the History of the Positron"] In Medvedev and Kojevnikov 1990, pp. 21–34.

Kojevnikov, A. B. (1988a). "Dirak i kvantovaya teoriya izlucheniya." ["Dirac and the Quantum Theory of Radiation"] In *Einshteinovskiy sbornik 1984–85.* I. Yu. Kobzarev and G. E. Gorelik , eds. Moscow: Nauka, pp. 246–270.

— (1988b). "V. A. Fock i metod vtorichnogo kvantovaniya." ["V. A. Fock and the Method of Second Quantization"] In: *Issledovaniya po istorii fiziki i mekhaniki, 1988. (Studies in the History of Physics and Mechanics, 1988)* A. T. Grigorian, ed. Moscow: Nauka, pp. 115–138.

— (1990a). "Einshteinovskaya formula dlya fluktuatsii i korpuskulyarno-volnovoy dualizm." ["Einstein's Fluctuation Formula and the Wave-Particle Duality"] In *Einshteinovskiy sbornik 1986–87.* I. Yu. Kobzarev and G. E. Gorelik, eds. Moscow: Nauka, pp. 102–124.

— (1990b). "Kvantovaya elektrodinamika Diraka." ["Dirac's Quantum Electrodynamics"] In Medvedev and Kojevnikov 1990, pp. 44–66.

Kojevnikov, A. B., and Novik, O. I. (1989). "Analysis of Informational Ties in Early Quantum Mechanics (1925–1927)." In: *Acta Historiae Rerum Naturalium nec non Technicarum.* Iss. 20. Prague, pp. 115–159.

Kragh, Helge (1981a). "The Concept of the Monopole. A Historical and Analytical Case Study." *Studies in History and Philosophy of Science* 12: 141–172.

— (1981b). "The Genesis of Dirac's Relativistic Theory of Electrons." *Archive for History of the Exact Sciences* 24: 31–67.

— (1984). "Equation with Many Fathers: The Klein–Gordon Equation in 1926." *American Journal of Physics* 52: 1024–33.

— (1990). *Dirac. A Scientific Biography.* Cambridge: Cambridge University Press

— (1992). "Relativistic Collisions: The Work of Christian Møller in the Early 1930's." *Archive for History of the Exact Sciences* 43: 299–328.

Krivonos, L. N. (1986). "Istoriya gipotezy monopolya Diraka." ["The History of Dirac's Monopole Hypothesis"] *NTM. Schriftenreihe für Geschichte der Naturwissenschaften, Technik und Medizin* 23: 35–45.

Medvedev, B. V., and Kojevnikov, A. B. eds. (1990). *Pol' Dirak i fizika XX veka.* [*Paul Dirac and Twentieth-Century Physics*] Moscow: Nauka.

Mehra, Jagdish, and Rechenberg, Helmut (1982). *The Historical Development of Quantum Theory.* Vol. 4. New York: Springer-Verlag.

Miller, Arthur I. (1994). *Early Quantum Electrodynamics. A Source Book.* Cambridge: Cambridge University Press.

Møller, Christian (1931), "Über den Stoß zweier Teilchen unter Berücksichtigung der Retardation der Kräfte." *Zeitschrift für Physik* 70: 786–795.

Moyer, Donald Franklin (1981a). "Origins of Dirac's Electron, 1925–1928." *American Journal of Physics* 49: 944–948.

—— (1981b). "Evaluations of Dirac's Electron, 1928–1932." *American Journal of Physics* 49: 1055–1062.

Oppenheimer, J.R. (1930). "On the theory of Electrons and Protons." *Physical Review* 35: 562–563.

Pauli, W. (1926). "Quantentheorie." *Handbuch der Physik.* Bd 23. Berlin: Springer pp. 1–278.

—— (1927). "Zur Quantenmechanik des magnetischen Elektrons." *Zeitschrift für Physik* 43: 601–623.

—— (1933). "Die allgemeinen Prinzipien der Wellenmechanik." *Handbuch der Physik.* Bd 24, T. 1. Berlin: Springer, pp. 82–272.

—— (1979). *Wissenschaftlicher Briefwechsel mit Bohr, Einstein, Heisenberg u.a.* 2 vols. Karl von Meyenn, ed. New York: Springer-Verlag.

Pauli, W., and Weisskopf, V. (1934). "Uber die Quantizierung der skalaren relativistischen Wellengleichung." *Helvetica Physica Acta* 7: 709–731.

Podolsky, B., and Fock, V. A. (1932). "Derivation of Möller's Formula from Dirac's Theory." *Physikalische Zeitschrift der Sowjetunion* 2: 275–277.

Roqué, Xavier (1992). "Møller Scattering: A Neglected Application of Early Quantum Electrodynamics." *Archive for History of the Exact Sciences* 44: 197–264.

—— (1997) "The Manufacture of the Positron." *Studies in History and Philosophy of Modern Physics* 28: 73–129.

Rosenfeld, Leon (1931a). "Zur Kritik der Diracschen Strahlungstheorie." *Zeitschrift für Physik* 70: 454–462.

—— (1931b). "Zur korrespondenzmäßigen Behandlung der Linienbreite." *Zeitschrift für Physik* 71: 273–278.

—— (1931c). "Bemerkung zur korrespondenzmäßigen Behandlung des relativistischen Mehrkörperproblems." *Zeitschrift für Physik* 73: 253–259.

—— (1932). "Über eine mögliche Fassung des Diracschen Programms zur Quantenelektrodynamik und deren formalen Zusammenhang mit der Heisenberg–Paulischen Theorie." *Zeitschrift für Physik* 76: 729–734.

Rueger, Alexander. (1992). "Attitudes towards Infinities: Responses to Anomalies in Quantum Electrodynamics, 1927–1947." *Historical Studies in the Physical and Biological Sciences* 22: 309–337.

Schweber, Silvan S. (1994). *QED and the Men who Made it: Dyson, Feynman, Schwinger, and Tomonaga.* Princeton: Princeton University Press.

—— (1995). "Early Quantum Electrodynamics." *Studies in History and Philosophy of Modern Physics* 26: 201–211.

Slater, J. C. (1927). "Radiation and Absorption on Schrödinger's Theory." *National Academy of Sciences of the USA. Proceedings* 13: 7–12.

Small, Henry (1983). *Physics Citation Index, 1920–1929*. Philadelphia: Institute for Scientific Information.

Stuewer, Roger (1975). *The Compton Effect: Turning Point in Physics*. New York: Science History Publications.

Tamm, I. (1930). "Über die Wechselwirkung der freier Elektronen mit der Strahlung nach der Diracschen Theorie des Elektrons und nach Quantenelektrodynamik." *Zeitschrift für Physik* 62: 545–568.

Tomonaga, S. I. (1973). "Development of Quantum Electrodynamics." In *The Physicist's Conception of Nature*. J. Mehra, ed. Dordrecht: Reidel, pp. 404–412.

Weyl, H. (1972). *The Theory of Groups and Quantum Mechanics*. H. P. Robertson, trans. New York: Dutton.

Einstein's Zurich Colloquium

Boris Yavelov

1. Introduction

Late in July 1912 Albert Einstein, already a famous scientist, returned to Zurich from Prague (Illy 1979). He was called to his alma mater, the Federal Polytechnical Institute of Switzerland (Eidgenössische Technische Hochschule, hereafter ETH), to fill the post of an ordinary professor in the newly-established department of mathematical physics, part of Department VIII, which was responsible for training mathematics and physics teachers (Waser 1979).[1] Back in 1900, Einstein had graduated from the same unit, then known as Department VI-A. At that time his efforts to obtain any position at the "Poly" failed due to strained relations with H. F. Weber, the head of the physics department (see, e.g., Seelig 1960), but Weber's death on 24 May 1912 (Weiss 1912) removed that obstacle. Weber's former student, Pierre Weiss, who had been teaching at the ETH since 1902, filled Weber's post and obviously was instrumental in calling Einstein back from Prague. Seelig notes that it was Weiss's "bright idea" to get recommendations for Einstein from Henri Poincaré and Marie Curie (1960, p. 226)[2].

On 1 October 1912 Einstein was officially elected to the post that was to remain his for three semesters,[3] until he left for Berlin in spring 1914. Seelig (1960, p. 233ff) cites a list of the new professor's duties. From 1912 to 1913 his mandatory weekly load was seven hours. In addition, Einstein and Weiss led regular workshops with students (three times a month) on writing physics papers.[4]

Einstein freely took upon himself a weekly evening colloquium in physics. The meetings were open to all interested people, including students, and were organized around reports of new work and discoveries in physics. In point of fact, Einstein simply revived the custom he had himself established in 1909–1910, his first Zurich period. Here I shall focus on the 1912–1914 colloquia, which were important for the international community of physicists, because of the reputation of their leader.

Yuri Balashov and Vladimir Vizgin, eds., *Einstein Studies in Russia.*
Einstein Studies, Vol. 10. Boston, Basel, Birkhäuser, pp. 261–295.

The information about the colloquium meetings is, unfortunately, scanty and allows but a patchy reconstruction based on scattered facts and a few eyewitness accounts. Documents are of primary importance in such circumstances. Here is one of them.

2. Einstein's Postcard to Leonid Mandelstam

A centenary Festschrift of Leonid Mandelstam (1879–1944) (Rytov 1979, pp. 58–59) includes the text of Einstein's postcard mailed to Mandelstam, who was then in Strasbourg (Einstein to Mandelstam, 23 July 1913, Einstein 1993, Doc. 457, p. 540):

> Dear Mr. Mandelstam!
> I have just spoken at the colloquium about your beautiful work on surface fluctuations, of which Ehrenfest had told me earlier. I deeply regret your absence.
> With best wishes, your
>
> A. Einstein.

Einstein is referring to Mandelstam's "Über die Rauhigkeit freier Flüssigkeitsoberflächen" (1913). Since Einstein says that he "has just spoken at the colloquium," it would seem that the event took place on July 23.

And it was a Wednesday. It is natural to assume that a particular day of the week was reserved for these regular meetings. The fact that Wednesday was selected is remarkable; it seems that physicists at least could call that day "the all-European colloquium day"! Here are the facts.

1. MUNICH. A. SOMMERFELD'S COLLOQUIUM. In his letter to Paul Ehrenfest from Munich on 11 July 1909 Abram Ioffe wrote: "Now instead of talking in a café we have a colloquium every Wednesday" (Moskovchenko and Frenkel 1973, p. 269).

2. ST. PETERSBURG. EHRENFEST'S COLLOQUIUM. Ehrenfest wrote to Ioffe (probably in 1909): "We need to decide whether we want to meet Wednesdays or on some other day. I'm all for Wednesdays" (Moskovchenko and Frenkel 1973, p. 34). It was eventually decided in favor of Wednesday.

3. LEIDEN. EHRENFEST'S COLLOQUIUM. "Ehrenfest can be regarded the founding father of physical colloquia in the Netherlands. He himself presided over evening Wednesday colloquia in physics for twenty years" (de Haas-Lorentz 1957, p. 105; see also pp. 109, 112, 114, 138, 177).

4. THE BERLIN PHYSICAL COLLOQUIUM. This "Wednesday Colloquium" (Hund 1967), which, by the 1920s, had become a legend (Herneck 1970), was held at the Institute of Physics on the Reichstag embankment.

Einstein's postcard to Mandelstam allows one to determine the names of some of the participants in the colloquium, as there are over a dozen more signatures on it. Who were these people?

The Mandelstam Festschrift does not throw any light on it. Meanwhile, I have identified twelve of the signatures. They belong to Ivan Andreyev, David Reichinstein, Jun Ishiwara, Gunnar Nordström, Simon Ratnowsky, Karl Herzfeld, Johann Kern, Eva Bruins, Emil Baur, Catherine Frankamp, and Max Laue.[5] Einstein signed the postcard twice.

Of those on the list, only Max von Laue had, by that time, earned world fame for his discovery, in 1912, of the diffraction of X-rays by the crystal lattice. In the fall of that same year, he became an extraordinary professor at the University of Zurich to replace Peter Debye, who had moved to the University of Utrecht. Debye had come to Zurich in 1911 to take Einstein's position after the latter had been invited to the German University in Prague. ·

Since the rest of the identified names are less illustrious (or even totally unfamiliar to most readers), it is worthwhile citing here the information I have managed to collect about them. I shall also provide information on those of Einstein's colleagues in Zurich who did not sign the Mandelstam postcard but took part in the colloquia.

3. Participants in the Zurich Colloquia

Three names on the postcard belong to people who came from Russia: Andreyev, Reichinstein, and Ratnowsky.

IVAN ANDREYEV (1880–1919), a talented chemist who invented an industrial contact cycle for producing nitric acid that is still used, entered St. Petersburg University in 1899. Not long before graduation he was expelled for political reasons and went abroad to continue his studies. In 1906 he obtained a degree in chemical engineering from the Technische Hochschule in Karlsruhe. He was awarded his second degree in chemistry after he passed the exams for the complete course of study at St. Petersburg University two years later. Wishing to extend his expertise, he secured a business assignment abroad. Between mid-February and early August 1913 he was in Zurich, where he attended lectures at the ETH. Upon his return to Russia, he made significant contributions to chemical technology (see Gamburg 1945, Volkov 1980).

DAVID REICHINSTEIN (1882–1955) was born in Mogilev (where Leonid Mandelstam was born before him); he studied at Kiev University for some time before going to Germany, where, in 1906, he obtained a degree in chemical engineering (from Karlsruhe, just like Andreyev). In 1909 he studied at Strasbourg University, where Mandelstam was a *Privatdozent*. In 1911–1918 he taught electrochemistry at the University of Zurich. He came back home for a short stay and then found himself abroad again, first as a professor of chemistry in Czechoslovakia and then as a *Privatgelehrter* in Germany. In his last years, he was a science journalist (see Reichinstein 1953). He died in Zurich.

SIMON RATNOWSKY (1864–1945) was born in Stavropol and was educated at Swiss and French Universities. In 1910 he was made professor of physics at the University of Zurich; in 1913 he became a lecturer, and in 1921 a professor without chair at the same university.[6] Five years later he went to teach physics and mathematics in a cantonal school in Winterthur (Lexikon 1947, p. 255). In his famous history of physics, von Laue mentioned Ratnowsky among those who, between 1906 and 1910, experimentally proved the correctness of the relativistic momentum formula (Laue 1950, p. 22).

GUNNAR NORDSTRÖM (1881–1923) is a physicist well-known to historians of general relativity. In 1905 he graduated from the University of Helsinki, majoring in physics. The next year he went to Göttingen to study physical chemistry with Walther Nernst, but Minkowski drew him to the problems of relativity, which remained his main interest till the end of his short life. In 1910–1918 he was associate professor of theoretical physics at the University of Helsinki, and in 1919–1923 professor of physics at the Helsinki Polytechnical Institute. Nordström made every effort to spend his summer vacations at European centers of theoretical physics, such as Zurich, Berlin, and Leiden (see Isaksson 1985).

Three signatures on Einstein's postcard belong to Dutchmen.

JOHANN J. KERN (1880–?) was born in Leiden. Early in 1913 he defended his thesis on electrodynamics at the University of Zurich, under the supervision of his countryman Debye.

EVA DINA BRUINS (1885–?) and CATHERINE A. FRANKAMP were, in 1913, Weiss's post-graduate students at the ETH, working on magnetism. More likely than not they owed their studies in Zurich to close professional ties between Weiss and Heike Kamerlingh Onnes (see, e.g., Weiss and Kamerlingh Onnes 1910).

KARL HERZFELD (1892–1978), who was born in Vienna, made a name for himself in theoretical physics in the decade that pre-dated quantum mechanics. In 1910–1912 he was a student at the University of Vienna; in

1912 and 1913 he studied in Zurich, and in 1913 and 1914 in Göttingen. In 1919 he secured a position under Arnold Sommerfeld and Kasimir Fajans as a lecturer on theoretical physics and physical chemistry at the University of Munich. In 1926 he moved to the United States, where he remained for the rest of his life (see Wheeler 1979).

JUN ISHIWARA (1881–1947) can well be regarded as the first Japanese theoretical physicist. He graduated from the University of Tokyo in 1906 and since 1911 was teaching physics in Japan. In the 1910s he used every opportunity to come to Europe to improve his knowledge, visiting Munich, Zurich, and other centers of theoretical physics. His main works are in special and general relativity and in quantum theory (see Speziali 1972, pp. 542–543).

EMIL BAUR (1873–1944) was born in Ulm, Einstein's home town. From 1911 to the end of his life, he was professor of physical chemistry at the ETH (Treadwell 1944).

Who were other people that frequented Einstein's colloquium? Undoubtedly, there were many of them. No complete list of the colloquium participants is available, but we can easily identify five more members on the basis of direct evidence.

One of them, OTTO STERN (1888–1969), was probably closest to Einstein during those years. In 1912 he came to Prague to continue his study with Einstein. Together they moved to ETH, where Stern became a *Privatdozent* in physical chemistry (see Segré 1973, p. 216). Their closeness prompted talk about Stern as an "assistant" to the great man (see, for example, Clark 1971, p. 163). It seems strange that he was not among those who signed the postcard. Yet it is possible that the upper left signature belongs to him. Indeed, it looks like "Zöcky," and Pierre Speziali notes that Stern most probably had a nickname that sounded like "Sucki" (Speziali 1972, pp. 38–39). The fact that the signature comes immediately after that of Einstein would nicely correspond with Stern's status as an "assistant."[7]

The prominent mathematician ERNST ZERMELO (1871–1953) was a professor of mathematics at the University of Zurich from 1910 to 1916. He was well known for his fundamental contributions to set theory. The physicists knew him because of his famous debate with Ludwig Boltzmann in the late nineteenth century over the foundations of the kinetic theory of gases. In 1894–97 Zermelo worked in Berlin with Planck, and in 1899–1910 in Göttingen with Hilbert.

PAUL EHRENFEST, who was a friend of Einstein and a well-known physicist, took part in several colloquia.

It seems that F. TANK (b. 1890) was the most frequent participant in the colloquium. In those years he was a university student and Laue's assistant. Later he became professor of electrical engineering at the ETH.

ADRIAAN DANIEL FOKKER (1887–1972), a Dutch theoretical physicist, who studied under H. A. Lorentz and Ehrenfest, and who had just been appointed professor at Leiden University, frequented Einstein's Wednesday colloquia during the winter semester of 1913–14 (Casimir and de Groot 1972).

Finally, five more people probably belonged to the same circle. There is no direct evidence of their involvement, though there is little doubt that they attended colloquia.

MICHELE BESSO, Einstein's old friend, settled in Zurich at that time, having left the Bern Patent Office. Reichinstein later recalled:

> In colloquium meetings I made the acquaintance of a friend of Einstein's Mr. B., who had come to Einstein from Bern. Decidedly, B. is a fine, deep, very upright, and also very talented personality.
>
> After a time B. began to work on the theory of relativity with Einstein. But here Einstein was very exacting; he was obviously dissatisfied with B., and left him to his fate. (Reichinstein 1934, p. 46)

There is hardly any doubt that 'Mr. B.' refers to Besso.

In 1913, Einstein helped MIECZYSLAW WOLFKE (1883–1947), a Polish scientist, to get a position as a lecturer in theoretical and experimental physics at the ETH (Wolfke 1980). At that time, Wolfke was working on problems in the quantum theory of electromagnetic radiation.

HERMANN WEYL, a future celebrity in mathematics, came to work at the ETH in the winter semester of 1913–14. In 1908 he earned his doctorate at Göttingen, and in 1916 he started a lecture course on general relativity. Einstein and Weyl worked in close contact when they were colleagues at the ETH (see Reid 1976, p. 89).

MAX BEHACKER, an Austrian physicist who graduated from the University of Vienna in 1905, took up science in earnest only in 1913. He spent the summer semester in Berlin, where he studied under Planck. At the end of the semester he published an important paper on Nordström's theory of gravitation (Behacker 1913). In 1914 he obtained a position as assistant professor at the German University in Prague, but in July 1915 he was killed in action near Lvov. Philipp Frank, Behacker's university colleague (who replaced Einstein in the Department of Mathematics at the same university), wrote in an obituary that "Behacker spent the winter semester of 1913–14 in Zurich where he studied under Einstein" (Frank 1916).

AUGUST PICCARD (1884–1962), a Swiss physicist and designer, who earned fame with his flights into the upper atmosphere in a stratospheric balloon and with record immersions in a bathyscaphe, was an engineering student at the ETH. A man with an original mind, he found it hard to deal with professor Weber (just like Einstein before him); Weber failed him at the graduation examination in physics. Weiss, whose relations with Weber were also strained, came to his aid and the young man got his diploma. In this way Piccard became Weiss's assistant in the laboratory of magnetism. In order to improve the laboratory magnets, Weiss needed a creative engineer (see Tilgenkamp 1956, p. 18). In 1913 Piccard defended his doctoral dissertation *Magnetization of Water and Oxygen*. Weiss and Einstein submitted favorable reports on the dissertation (Eidgenössische 1972, p. 65). In that same year, 1913, Piccard began his teaching career by lecturing on methods of physical measurement at the ETH; in 1915 he became a *Privatdozent*, and in 1922 a professor. In the 1920s he conducted balloon experiments on topics having to do with Einstein's work.

Having identified the people who definitely or very probably were members of Einstein's "circle" in Zurich, let us note some characteristic features of this community that kept it together. One noticeable feature is the patchwork composition of the group: men and women; representatives of many countries: Switzerland, Austria, Germany, Russia, Finland, Poland, Japan; physicists, chemists, and mathematicians; distinguished professors as well as fresh doctors, beginning lecturers, and assistants, along with students still working on their first degrees.

Their ages ranged from twenty to sixty, though most of the group fell between twenty-five and thirty-five. The absence of those whom young Russian physicists used to called "generals" was conspicuous (Moskovchenko and Frenkel 1973, p. 269). There were no scientists who felt at home on the academic-administrative Olympus, who took their real service to science to be a state service and who bowed to the ossified academic hierarchy. All this was absent from the Zurich colloquium. It seems that the contrary was true: the circle professed non-conformism, democratic scientific and social views (see Feuer 1974, p. 4ff for more details), open skepticism towards the "generals," and indifference to ranks and honors. As regards Einstein himself and his closest friends Besso and Ehrenfest, this hardly needs to be proved. So let us look at the other participants. Detailed discussion or even mentioning of their works would lead us too far afield. Accordingly, we shall limit ourselves to those facts and eyewitness accounts that highlight the academic profile of the group and the human characteristics of its members.

STERN. His student, Isidore Rabi, wrote about Stern in *Physics Today*:

> To the best of my knowledge he never devoted himself to a minor question. From his earlier work on entropy and Nernst's heat theorem, to his demonstration of space quantization with the glorious Stern–Gerlach experiment, the reality of matter waves and the anomalous magnetic moment of the proton, Stern was always close to the basic problems of physics as they evolved. (Rabi 1969, p. 103)

He had the following to say about Stern's Hamburg period, when Pauli was his closest colleague at the university:

> Some of Pauli's great theoretical contributions came from Stern's suggestions, or rather questions. . . . Stern was one of the antistuffy generation of German professors who observed with a mixture of amusement and contempt the pomposity of their predecessors. . . . He was open to new ideas both in theory and in experimental techniques, as one would expect from one who had a long association with Einstein. . . . His manipulative skills were only fair. . . . On the other hand he was very inventive, ingenious and bold in overcoming experimental difficulties. (Rabi 1969, pp. 103, 105)

Stern's other student Emilio Segré adds some valuable details about the human side of his teacher:

> He was always invited to conferences because he always had something important to say.
>   He was easily accessible, at least by the German standards of the time, to students and postdoctoral fellows, with whom he had lunch regularly. He went frequently to the movies, but had to be told by his companion Pauli whether he had already seen the film or not.
>   . . .
>   Stern's extremely successful work at Hamburg came to an end in August 1933 with his resignation, which was caused by the Nazis. The pretext was the dismissal of his colleague and old pupil Estermann and an order to remove Einstein's portrait from his office. (Segré 1973, pp. 230, 229)

ZERMELO. Planck used to refer to Zermelo as "my talented pupil." In the academic community, he was widely known as a bold attacker on recognized authorities. Back in 1896 he launched an offensive against Boltzmann's statistical ideas in the form of an ingenious argument based on Poincaré's recurrence theorem. In 1904 he caused quite a stir with his claim that any set could be ordered. His works on set theory triggered a stream of objections from Poincaré, an irreconcilable enemy of the emerging set theory, whose prestige served as a firm foundation for anti-Cantorians.

In one of his lectures in Göttingen, where Zermelo was working at that time, Poincaré unleashed a storm of criticism against Cantor's theory and Zermelo's achievements, going so far as questioning Zermelo's famous theorem. The great French mathematician told his audience that "even if we admit the ingenuity of Zermelo's proof, still it should be trashed or better thrown out the window." Richard Courant later recalled that Zermelo, "an emotional and very original person, was both infuriated and despaired. If he had been a bit more dexterous, he would have probably killed Poincaré on the same day at the dinner table. But he was extremely clumsy" (1981, p. 162).

His former colleague in Zurich, George Polya,[8] offers some observations too:

> Ernest Zermelo . . . liked to spend some of his time in coffee-houses. His conversation there was interspersed with sarcastic remarks about his colleagues. Commenting on an address which had great success at a recent mathematical meeting, he criticized the speaker's style and eventually condensed his disapproval into two rules according to which, he mockingly asserted, that address must have been constructed:
> I. *You cannot overestimate the stupidity of your audience.*
> II. *Insist on the obvious and glide nimbly over the essential.*
> Zermelo's remarks were often witty; very unjust on the whole, but striking and revealing about some particular point. (Polya 1965, p. 141)

It is interesting to note that Zermelo's fascination with highly abstract mathematical ideas went hand in hand with his acute interest in physics and purely practical problems.[9] In 1929 he offered a method of comparative evaluation of the strength of participants in chess tournaments that was quite popular at one time. Among the papers left after his death one could see German versified translations from Homer along with a draft of a patent application for a gyroscopic stabilizer for bicycles and motorcycles (Rootsellar 1975).

REICHINSTEIN (1953). It seems that Reichinstein's scientific interests were rather broad. In his biography of Einstein (1932; English trans. 1934) he outlined the foundations of the theory of relativity, putting much emphasis on the philosophical and methodological issues. In 1940 he published a two-volume monograph on the problem of aging (Reichinstein 1940), a book on the philosophy of religion, and took an aggressive stance vis-à-vis quantum mechanics. His direct contacts with Einstein prompted a bold yet short-lived conception of "non-preferred systems" in electrochemistry, which was an attempt to use the methodology of Einstein's special relativity in chemistry by proclaiming the equivalence of all chemical systems.

Reichinstein's book, *Properties of the Volumes of Adsorption* (Reichinstein 1916), earned commendations from Einstein,[10] who wrote in a letter to Besso: "It is brilliantly written. You all could fill a few charming hours with it" (Einstein to Besso, 21 April 1916, Einstein 1998, Doc. 215, p. 283, trans. A. M. Hentschel). Later Einstein also wrote an introduction to another book by Reichinstein, *Surface Processes in the Living and Non-living Nature* (Reichinstein 1930), and derived one of its equations (for electrolysis with an alternating current).

BAUR (Treadwell 1944). It would be no exaggeration to say that Emil Baur was a romantic. He tried to extract gold from seawater and worked on the production of artificial diamonds. But the focus of his interests was "chemical cosmography": a set of projects aimed at determining the role of chemical processes in the origin and evolution of the universe. Chemical cosmography dealt with fundamental problems in stellar, solar, earth, and ocean chemistry. Baur's theoretical interests centered on issues that did not fit in the classical framework. He wrote extensively on one-way chemical equilibria, cyclical reactions, and periodical chemical reactions in homogeneous media. One even suspects that Belousov's idea of periodical chemical reactions was prompted by Baur's works. (Belousov graduated from the ETH in 1915. See Polishchuk 1984, p. 192.)

ISHIWARA was known at home as a poet who wrote in the lofty, classical Japanese tradition rooted in antiquity (see Speziali 1972, pp. 542–543).

FOKKER was famous as an active fighter for the purity of his native Dutch tongue, an enthusiast of the thirty-one-tone musical system, and an expert in mechanical toys. One of these toys, which he made himself for demonstrating the equivalence of gravity and inertia, was shown at a Zurich colloquium, but probably later, in the 1920s (see Casimir and de Groot 1972).

LAUE, born into a Prussian officer's family and a product of the classical traditions of academic life in Germany, was probably less inclined to romanticism and radicalism in scientific endeavors and to informality in communication than other members of the Zurich group. Being a consistent advocate of classical physics, Laue never showed much enthusiasm for general relativity and quantum theory. In 1959 he wrote:

> I must acknowledge that I often spoke to Einstein during the formative years
> of general relativity [apparently, he is referring to the Zurich period], but I

never quite understood his monologues. It was only later, when it all became perfectly clear, that I began to grasp, with veneration, the truth that had opened itself to him. (Quoted in Herneck 1974, p. 231)

He also recalled that, like many of his colleagues, he regarded general relativity as a mystery (Herneck 1979, p. 41). Towards the end of his life, von Laue, who was a straightforward and honest man, assessed his achievements rather modestly: "I am perfectly aware that my works were nothing more than a consistent application of established methods that simply had to be slightly changed to fit particular tasks" (Ewald 1979, p. 92). It seems quite typical of Laue's psychological make-up that, having met Einstein for the first time when he was still working in the Bern Patent Office, Laue "did not believe that this man could create the theory of relativity" (Herneck 1974, p. 231). On the other hand, Einstein, who was never a diligent student, would have been taken aback by Laue's remark that "he was never able to understand students who were late for lectures" (Laue 1956, p. 174).

4. The Colloquium Participants Speak

Let us now hear what these people (and we have mentioned twenty-one names) have had to say about the colloquia.

In his biography of Einstein, Carl Seelig quotes LAUE's recollections:

[W]hen I came to Zürich University in October 1912 as associate professor, I found Einstein as full professor of theoretical physics at the [ETH] and a vivid personal and scientific relationship ensued. One afternoon every week Einstein held a physics discussion on new developments in physics. Although it took place in the [ETH] physics building, the teachers and students of the university[11] naturally had access and appeared in great numbers, while the admirable experimental [ETH] physicist Pierre Weiss, stood apart with his students, most of whom spoke French.[12] After the discussion Einstein went, with any who cared to accompany him, to eat in the Kronenhalle restaurant.

At that time the general relativity theory was in its embryo stage and I still remember many disputes with Einstein on this subject. . . . Furthermore, at that time he was particularly interested in theoretical quanta questions and excited by Niels Bohr's 1913 atomic theory.[13] He collected around him a great number of pupils of whom Otto Stern . . . and K. F. Herzfeld . . . were the most important. The liveliest discussions took place on those summer days in 1913 when the temperamental Paul Ehrenfest visited Zürich. I can still see Einstein and Ehrenfest, having outstripped a horde of physicists, climbing the

Zürichberg[14] and Ehrenfest shouting out in jubilation: "I've understood it!"
(Seelig 1960, pp. 131–132)

EHRENFEST and his Russian wife, Tatyana (1876–1964) spent three or
four weeks in Zurich (Moskovchenko and Frenkel 1973, p. 128), from 18
June 1913 to early July (see Klein 1970, pp. 293–294). On 28 August 1913
Ehrenfest wrote to Ioffe about his stay in Zurich:

> [We had] very interesting discussions about gravitation and energy at absolute
> zero. Stern (a great talent!), Nordström (from Helsingfors), Ishiwara, and
> Herzfeld (he is 21) from Vienna were there too. Einstein is indeed without
> equal, and a very nice man too! (Moskovchenko and Frenkel 1973, p. 128)

The same letter reveals how the Mandelstam postcard was written: "On
the way to Zurich and back we stopped in Strassbourg to see Mandelstam"
(ibid.). Apparently Ehrenfest learned about Mandelstam's "beautiful work
on surface fluctuations" (Mandelstam 1913) in the middle of July 1913
from the author himself. (Mandelstam's article was received by *Annalen
der Physik* on 5 April 1913 and appeared in the June issue.) It seems that
Einstein and Mandelstam knew each other, at least by correspondence, even
before the postcard was written. The informal opening, "Dear Mr. Mandelstam," and Einstein's remark "I deeply regret your absence" also suggest
it. The academic community of the time tended to use more official forms
of addressing a colleague whom one did not know; and Mandelstam was a
professor. It should also be noted that during his Strasbourg period
(1899–1914), Mandelstam made frequent trips to Switzerland (see
Papaleksi 1979, p. 20).

The entries about the Zurich visit in Ehrenfest's diary (see Klein 1970,
pp. 294–295) tell about a colloquium discussion of a new interpretation of
the X-ray diffraction by crystals suggested by Bragg. After the colloquium
its participants (including, besides Einstein and Ehrenfest, Laue and
Zermelo) continued their debates in the neighborhood café (probably
"Kronenhalle"). This particular colloquium took place on Tuesday, June
24, not the usual Wednesday. Most probably, the date had to be changed
because Ehrenfest had to spend the next day, Wednesday, June 25, outside
Zurich. In his diary, he also tells about his meetings with Besso and
Grossmann.

ANDREYEV's report of his trip abroad gives information about one more
paper discussed at the Zurich colloquium in summer 1913:

My studies of theoretical physics in Zurich boiled down to reading and attending lectures of Prof. Einstein and his colloquium. It was also frequented by local professors and assistants and by Prof. Einstein's closest students. They would give papers on diverse subjects in physics and physical chemistry. One session was devoted to my paper, "Chemical Effect of Ultra-Violet Rays on Gases." (Quoted in Volkov 1980)

It is interesting to note that Andreyev motivated the need for his Zurich trip as follows:

Setting off abroad for studies, I believed it urgent to improve my knowledge in theoretical physics, in particular, to familiarize myself with the ideas that would open up a new era in chemistry. I am here referring to the so-called *Quantentheorie*, an electronic theory of electricity, radiation, etc. [sic] Prof. Einstein, who is now heading the department of theoretical physics at the Zurich Polytechnical, is presently considered an outstanding theoretical physicist. This is why I decided to go to Zurich. (Quoted in Volkov 1980)

REICHINSTEIN told more than the others about the Zurich colloquium. He was one of Einstein's closest friends during the first and second Zurich periods. He wrote a very interesting scientific biography of the great physicist (Reichinstein 1932; English trans. 1934). His hero never objected to it.

Reichinstein began attending Einstein's colloquia in 1910. He later recalled that he had been struck by Einstein's low calm voice and deeply-set thoughtful eyes, " which did not seem at all teacherish," and by his cordiality.

At these meetings for which an evening per week was set aside, new papers were read aloud, this being followed by a discussion. After the colloquium, Einstein went to a café together with those with whom he was most intimate.

Here the discussion was continued and the conversation became more informal. . . . Imperceptibly Einstein became my teacher in the direct sense of the word.

. . .

Once, I recollect, Einstein could not come to the café with us after the weekly colloquium. He said that his wife had wash-day, and that he had to take care of the baby.

. . .

Altogether I gave three lectures in Einstein's colloquium. I remember my second lecture especially, at which not only my colleagues, but also the students who were working on their dissertations with me, were present.

. . .

I shall never forget how once during colloquium, when the discussion was as good as finished, Einstein suddenly exclaimed: "What is light?" Deep silence followed. We did not want to disturb the thoughts of the master. (Reichinstein 1932, pp. 24, 26, 49, 39)

Here is what Reichinstein recalled about a dispute between Einstein and Laue on general relativity: "When Einstein lectured on its first elements during colloquium, Max von Laue, who was present, would not agree with the theory. Einstein replied playfully: 'It is not at all necessary that every physicist should agree with it!'" (1934, p. 47). Another time Ehrenfest was present and Reichinstein recalled him saying, "It is strange Einstein always gets the results he wants from his work" (1934, p. 47).

5. Bohr's Atom

Let us look more closely at one memorable event in the history of the colloquium.

In his 28 August 1913 letter to Ioffe sent from Leiden to St. Petersburg Ehrenfest told about his visit to Einstein in Zurich. He wrote:

I am in despair over Bohr's "Quantum Mechanical Inferences from the Balmer Law" (in *Philosophical Magazine*): if the Balmer series can be derived in *this way* [Ehrenfest's emphasis], then I should throw away all of physics and follow suit myself" (Moskovchenko and Frenkel 1973, p. 129).

Ehrenfest is referring here to the first part of Bohr's seminal trilogy, "On the Constitution of Atoms and Molecules," which appeared in early July 1913 (Bohr 1913a), where, for the first time, Bohr presented his quantum theory of atomic spectra. Three days earlier, Ehrenfest expressed the same attitude in a letter to Lorentz, who was then in Haarlem: "Bohr's work on the quantum theory of the Balmer formula (in the *Phil. Mag.*) has driven me to despair. If this is the way to reach the goal, I must give up doing physics" (Ehrenfest to Lorentz, 25 August 1913, as quoted in Klein 1970, p. 278).

How and when did Ehrenfest become familiar with Bohr's article? We can provide a rather precise time frame: between early July (when he returned from Zurich to Leiden) and August 25 (the date of his letter to Lorentz). Sure enough, by that time many other physicists had also "noticed" Bohr's contribution to the *Philosophical Magazine* (though their reactions were often quite different from Ehrenfest's). Sommerfeld referred to it as early as July. In a postcard sent from Munich to Bohr on September

4, Sommerfeld wrote: "I thank you very much for sending me your extremely interesting work. . . . Although I am for the present still rather skeptical about atom models in general, nevertheless the calculation of this constant [Rydberg's constant] is indisputably a great achievement" (quoted in Rosenfeld and Rüdinger 1967, pp. 55–56).[15] According to Walther Gerlach, Friedrich Paschen, who worked in Tübingen, was ecstatic about Bohr's work already in summer 1913: "This sets the agenda for physics for the coming three decades" (quoted in Gerlach 1979, p. 94).

Bohr's friend, George de Hevesy, who worked in the Manchester Laboratory, congratulated him on his impressive achievement. In Göttingen, on the other hand, the new theory was given a hostile reception (see Rosenfeld and Rüdinger 1967, especially p. 56ff). Half a century later, Bohr recalled that "outside the Manchester group [my ideas] were received with much skepticism" (quoted in Hoffmann and Dukas 1972, p. 175).

There are reasons to believe that Ehrenfest first became familiar with Bohr's work when it was discussed at his Leiden colloquium. Most probably, when browsing through the journals that had arrived during his visit to Zurich,[16] he made a note of Bohr's article and asked one of his students to make a report on it at one of the coming meetings. It seems that he himself had not read the article at that time; otherwise he would have quoted its title correctly in his letter to Ioffe. Klein also notes that there are no traces of Ehrenfest's study of Bohr's theory in his notebooks (see Klein 1970, pp. 278–279). Apparently, he listened to the report and formed his impression on the basis of it.

Bohr's article was discussed at the Zurich colloquium as well. In a biographical essay written in 1969 one of Stern's students described the following episode: "After having discussed the paper on a long walk [when the colloquium ended], they [Stern and von Laue] decided to take an oath, 'If this nonsense of Bohr should in the end prove to be right we will quit physics'" (Segré 1973, p. 218). Later, in 1973, the story was confirmed by Heisenberg, who referred to his conversation with Stern (Heisenberg 1975, p. 390). Stern and Laue's reaction is remarkably close to that of Ehrenfest, even in its verbal expression! Were all of them influenced from Leiden?

This was probably the case. The Dutchman Fokker was the one who presented Bohr's article at the Einstein colloquium. We know it from his student Abraham Pais: "In 1913–1914 [Fokker] studied with Einstein in Zurich and there he gave his first colloquium on Bohr's theory of the hydrogen atom. Einstein, Laue and Stern were among the audience. Einstein did not react immediately, but rather kept a meditative silence"[17] (Pais 1967, p. 222).

When did this colloquium take place? Fokker later wrote: "I defended my thesis in 1913 and remained in Zurich for the winter semester to work with Einstein" (Fokker 1955). The title page of his thesis (Fokker 1913) bears the date of its defense: 24 October 1913.

Obviously, Bohr's work was discussed in Zurich not earlier than October 24. During his stay in Zurich Fokker was very close to Ehrenfest. This is evidenced by the warm acknowledgment in his thesis: "It was you, Ehrenfest, who gave me strength in the last year through the generous manifestations of your cordial interest and the lively scientific exchanges in your home" (Fokker 1913, p. 1).[18] Undoubtedly Fokker was aware of Ehrenfest's famous opinion of Bohr's ideas, which Ehrenfest shared not only with Lorentz and Ioffe. The similarity between Stern and Laue's and Ehrenfest's reactions could then be explained by Fokker's spicing his report with his friend's *bon mot*.

Tank, another participant in the memorable colloquium, offered a different account of the meeting (see Jammer 1966, p. 86). According to him, when the discussion was coming to an end, Laue protested: "This is all nonsense! Maxwell's equations are valid under all circumstances" (quoted in Jammer 1966, p. 86). Then Einstein stood up and said: "Very remarkable! There must be something behind it. I do not believe that the derivation of the absolute value[19] of the Rydberg constant is purely fortuitous" (ibid.). Max Jammer, who had heard about this episode from Tank in 1964, concludes: "In fact, Einstein recognized the importance of Bohr's theory as soon as he became acquainted with it" (Jammer 1966, p. 86). It seems that this apologetic conclusion (which is still widely accepted by historians of science) is incorrect. In fact, it is known that Einstein first learned about Bohr's theory at least a month earlier.

Einstein took part in the *Naturforscherversammlung* in Vienna (21–27 September 1913) where he met Hevesy. By that time, Hevesy had already worked two years with Rutherford in Manchester. Einstein and Hevesy had known each other from Hevesy's days at the ETH. Hevesy gave identical and fairly detailed accounts of his conversation with Einstein at the meeting in two letters he sent from Vienna to Bohr (on 23 September; see Bohr 1981, p. 532) and to Rutherford (on 14 October; see Eve 1939, p. 225). He reports that, being asked about Bohr's theory, Einstein replied that "it is a very interesting one, important one if it is right and so on and he had very similar ideas many years ago but had no pluck to develop it [and publish it, as the letter to Rutherford said]" (Hevesy to Bohr, 23 September 1913, Bohr 1981, p. 532).[20] But when Hevesy told Einstein about a new success scored by Bohr's theory, the interpretation of the Pickering–Fowler spectral series as belonging to the ionized helium, Einstein was truly astonished.[21]

He said that this theory is then one of the greatest discoveries (see Eve 1939, p. 225). "'Than [sic] the frequency of the light does not depand [sic] at all on the frequency of the electron'. . . . And this is an *enormous achiewement* [sic; emphasis by Hevesy]. The theory of Bohr must then be wright [sic]" (Hevesy to Bohr, 23 September 1913, Bohr 1981, p. 532).

Thus Einstein had been acquainted with Bohr's theory even before the Zurich colloquium. Why, then, did he remain silent (according to Fokker) or limit himself to a short remark that really boiled down to "there should be something behind all this"? This remark is probably the key. Willingly or not, Einstein betrayed his resistance to the kind of argumentation that "stood behind" Bohr's results. "The frequency of light does not depend on the electron frequency" was his summary of the most radical claim of the new theory, which he offered in his conversation with Hevesy. It seems that this claim must have been unacceptable to Einstein because it was in obvious contradiction with the demand that all natural phenomena must have a causal explanation.[22] And if so, then there should be something else "behind" Bohr's interpretation of the Rydberg constant, something very different from his own explanation, which, therefore, is hardly worth a serious examination.

It seems, at first glance, that Fokker and Tank contradict one another. Yet there is a natural explanation of this seeming contradiction: Einstein did not take part in the lively discussion but believed it his duty, as the head of the colloquium, to sum up the proceedings. Fokker, no mean theoretician himself, remembered the fact that Einstein was silent throughout the discussion, while Tank, who was less versed in theory, was duly impressed with the great man's concluding remark.

6. Was There an Einstein School?

Reichinstein recalls that immediately after Laue's talk at the Zurich colloquium focused on his famous work on X-ray interference, Einstein launched a discussion by offering to the audience a clear and concise summary ("substance and soul") of what they had just heard.

> Everybody left the lecture filled not so much with what Laue had said, as with the remarks Einstein had passed on it. I walked beside Professor Baur and he repeated what he had said at the time of making Einstein's acquaintance at the café: "Einstein extemporizes on the most intricate problems of physics with as much ease as if he were talking about the weather. Others need a lot of time

and have to work hard to merely understand and digest every one of these problems he was talking about." (Reichinstein 1934, p. 45)

Reichinstein mentions immediately below that when Baur met Einstein for the first time, he could not help exclaiming: "How can he possess so much knowledge of scientific literature at this early age?" (Reichinstein 1934, p. 45). Reichinstein also quotes a physicist from Prague whose field was far removed from Einstein's interests. He was amazed to get valuable suggestions from a colleague who, he knew, had never worked in his area but was able to grasp important things in the course of their conversation (see ibid., p. 46).

Reichinstein himself was astonished to see, in his very first encounters with Einstein, how he "was immediately able to express any scientific problem in mathematical language, for instance in the form of a differential equation" (Reichinstein 1934, p. 38). Reichinstein also notes that Einstein's critical comments and his objections to the commonly-accepted logic invariably impressed his audience. His remarks of this sort were frequent and could be heard in the strangest of circumstances: not only at colloquia, but also during walks with his colleagues or students, in a café, and in light conversations. He often clothed his objections in a jocular tone. "His joking, his good humor, his wit, were beyond description" (1934, p. 48). Fokker recalled that "on his way from the colloquium Einstein used to have beer with the rest of us or buy a bratwurst and eat it up right there in the street" (Fokker 1955).

As a rule, Einstein's colloquia were crowded, the atmosphere less official than was usual at the time, and the discussions quite friendly and lively. Einstein was invariably well-disposed toward anyone wishing to discuss the latest news in science, be it a well-known scientist or just a student (see Clark 1971, p. 152). According to Frank, who knew Einstein well, the great scientist was never irritated (unlike many of his colleagues) by naive questions put to him by dilettantes and showed a great deal of patience trying to explain things to them. Indeed, he was not inclined to exaggerate the difference between a professional scientist and a lay person interested in science. He seems to have enjoyed sorting out a most amateurish objection (see Frank 1947, p.117 ff). "His attitude toward students was characterized chiefly by his friendliness and readiness to help them. When a student really had a problem in which he was profoundly interested, even if it was a very simple one, Einstein was ready to devote any amount of time and effort to help him solve it" (Frank 1947, p. 118). Reichinstein later wrote that Einstein's attitude to young scientists and students resembled very little the usual treatment that a teacher was expected to accord his pupils.

> [The relations] were much more intimate, more friendly. Einstein often spoke about quite intimate subjects and this not only when one of us was alone with him, but on walks where four or five people were present. . . . It is interesting that as soon as a colleague of his own rank—a Zürich university professor, for instance Professor Zermelo—was present, Einstein became more reserved. Of course we also poked fun and laughed, but the conversation did not overstep a certain limit. With us, who were really only a few years younger than he himself, he was much more free and unconstrained. The feeling of a social distance between us who were assistants (only a few of us were unsalaried lecturers at that time) and himself, was at that time unknown to him. It was only later on, in Berlin, that this feeling of distance became noticeable. (Reichinstein 1934, p. 48)

These accounts (and they are numerous) might produce an impression that the Zurich Wednesdays proceeded in an idyllic atmosphere of permissiveness and lack of criticism. This was hardly the case: Einstein never agreed with what he considered wrong; his disputes with Bohr and Nernst prove the point beyond any doubt. Even Reichinstein's generally apologetic book has many examples showing that Einstein was far from following the principle of "non-resistance." One of the students who was writing his thesis under Einstein and attended his colloquium complained that "Einstein [was] out to vex us all" (Reichinstein 1934, p. 41).

We have already discussed the subjects that cropped up at Einstein's colloquia. Not surprisingly, the participants concentrated on the most pressing physical issues. It is somewhat surprising that much attention was paid to chemistry. There were at least three professional chemists in the group and at least four sessions were devoted to chemical problems.[23] There could be several reasons for this interest in chemistry. First, early in this century, physics was much closer to chemistry than now. Second, chemistry was prominent in Zurich, both at the University and the ETH.[24] Third and most important, the colloquium leader had wide scientific interests himself (see, for example, Frenkel and Yavelov 1981). Robert Rompe suggests the same in his recollections of Einstein's contribution to Laue's Berlin physical colloquium in Berlin in the 1920s:

> Some [colloquium] topics were selected according to Einstein's wishes. . . . Strangely, he was very interested in the electrical piano designed by Nernst (an ancestor of the electronic instruments favored by young people today) and asked von Laue to let Nernst speak at the seminar. (Rompe 1979)

It is probable that, from time to time, the Zurich colloquium plunged into subjects far removed from "normal" science. Here is what Frank wrote about Einstein's attitude to Felix Ehrenhaft:

> Ehrenhaft . . . was a man of the *direct experience*. He believed only what he saw, and constantly found isolated phenomena that did not fit into the grand scheme. For this reason he was frequently regarded with disdain, especially by persons who accepted the general scheme as an article of faith. A man like Einstein, who had himself brought these general principles to life, always felt mysteriously attracted whenever he heard of irregularities. Even though he did not believe that they existed, yet he suspected that there might be the germs of new knowledge in these observations. (Frank 1947, p. 175)

What fruits did Einstein's colloquium bear? A complete answer is hardly possible. But even a cursory glance at what the colloquium participants published between 1912 and 1915 shows that many of them (e.g., Herzfeld, Kern, Nordström, Ratnowsky, Fokker, and Ehrenfest) were quite active. Quite a few of their contributions were close to Einstein's main interests: quantum theory (including the quantum theory of specific heats and the problem of "zero-point energy"), special relativity, gravitation theory. By itself, this is hardly conclusive evidence of the colloquium's influence: many physicists not connected with Zurich in any way were actively working in the same areas.

But on closer inspection, the publications of the authors mentioned above do bear the marks of the Zurich discussions. It is likely that, while in Zurich, Ehrenfest spoke at the colloquium about his work on adiabatic invariants that he had completed in May 1913 (see Moskovchenko and Frenkel 1973, p. 125). His next two articles on the topic, which were finished later in 1913, contain these acknowledgments: "I do not know how to resolve the difficulties described in §§ 2, 3, and 4, the most serious one being indicated by Einstein" and "This example was suggested by Herzfeld in discussion" (Ehrenfest 1913a and 1913b). In his article on the theory of dissociation (dated "Zurich, February 1914"), Stern emphasized that his research was prompted by "Mr. Ehrenfest's oral question" (Stern 1914). Herzfeld mentioned Einstein's and Stern's suggestions in his work on the electrochemistry of diluted solutions (dated November 1913; see Herzfeld 1913a) and referred to a very interesting comment by Einstein in the article (dated 28 March 1913) "On the Electron Theory of Metals" (Herzfeld 1913b),[25] which was widely quoted at the time. In his publication on the theory of radiation, which appeared in early 1914, von Laue mentioned "an important comment by Einstein made in a private conversation" (Laue 1914).

It is quite probable that Einstein himself benefited from his communications with colleagues. Thus, for example, he included Stern's theory of the vapor pressure of mono-atomic solids (dated May 1913; see Stern 1913) in the lecture course on statistical mechanics he gave at the ETH in the summer semester of 1913 (see Einstein 1985).

Further evidence of the colloquium's influence is provided by Einstein's contacts with Weiss, contacts whose very existence might be doubted given Laue's above-quoted remark about how Weiss remained "aloof" in the colloquium. In December 1912 Einstein and Stern sent their joint article on the hotly-debated problem of zero-point energy for publication (Einstein and Stern 1913). In proof-reading they added a note saying that Weiss had drawn their attention to a circumstance that seemed to corroborate their arguments. On 20 June 1913, upon his arrival in Zurich, Ehrenfest wrote to V. K. Arkadiev in Moscow: "Yesterday I told Weiss and Einstein about your research. They were both very interested in your experiments and your idea" (quoted in Polivanov 1954).[26] Weiss referred to his conversations with Einstein about magnetism more than once. Thus in his article on the nature of the molecular field[27] (February-March 1914) he acknowledged: "I am extremely grateful to Mr. Einstein, who sided with the ideas presented in this article after he had kindly discussed them with me" (Weiss 1914, p. 212).

Fokker's relations with Einstein are another case in point. Having defended his thesis (on the Brownian motion of the electron in the electromagnetic radiation field) in Leiden on October 24, Fokker then traveled to Zurich.[28] On November 20 he sent a short article (Fokker 1914a) summarizing his research in Leiden for publication. He continued his studies in Zurich and obtained more general results. Later he wrote:

> The first thing that Einstein said after he looked at my generalization was: "What will happen if I use other coordinates?" Invariance was his touchstone and primary interest. At that time he was totally immersed in his work on the theory of invariants. (Fokker 1955, p. 127)

Fokker concluded his next article (dated "Zurich, 11 December 1913") with thanks to Professor Einstein "who inspired and stimulated me" (1914b). In 1915, Planck described this work as very important:

> Does quantum theory have a role to play in pure electrodynamics, when the electromagnetic radiation interacts with the electric charge or dipole? Or is its

significance reduced to the laws of collision of material particles? A year ago this question had no answer. Today we have an answer that seems to be final. It supports the first option mentioned above and is a new argument in favor of quantum theory. A. D. Fokker has calculated the stationary state of the system of rigid rotating dipoles in a given radiation field. This calculation, performed in strict accordance with the principles of classical electrodynamics, has unequivocally pointed to results that are in direct contradiction with experiment. (Planck 1915)

The gratitude Fokker expressed to his mentor was in line with tradition. But it is hard to imagine that his fundamental work failed to draw Einstein's attention, especially given that it developed Einstein's favorite fluctuation approach (see, in this connection, Yavelov 1985).

As is well known, in 1911–1915 Einstein was preoccupied with general relativity (see, e.g., Vizgin 1981). Naturally enough, this subject figured prominently in the Zurich discussions. More on this below.

Can one speak of the existence of "Einstein's school" in Zurich in 1912–1914? This may sound like a heretical question to ask, as the view is still widespread that Einstein was a lonely scientist who had minimal, if any, interest in collaboration with others and in educating students (see, e.g., Mikulinsky 1977, p. 229). Everybody agrees that he never strove to set up a school of his own (see ibid., pp. 158, 263). I am not going to debate the notion of "scientific school" but will approach the question empirically.

There is no doubt that, from the late 1920s onwards, Einstein could be described as a lonely scientist. However, he cannot be considered an "anti-collectivist" in the first half of his life, when all his major discoveries were made. Let us focus on his work in 1912–1914.

First, out of twenty-four articles he published during that period four were co-authored: two with Grossmann, one with Stern, and one with Fokker. They comprise seventeen percent of the total. Taking it into account that some other papers were shorter contributions to ongoing discussions, this figure rises to twenty-two percent.[29] Even a superficial study of Poggendorf's handbook shows that in the "theoretical physics" section, Einstein had the second largest number of jointly authored works between 1912 and 1914. He comes after Born who was the record-holder with his forty-four percent.[30] (For comparison, only about ten percent of Sommerfeld's papers were co-authored, while famous Woldemar Voigt, with his large retinue of students, had just five per cent of his works jointly authored.)

Second, there are no grounds to believe that Einstein preferred solitude when he was younger or that he wanted to distance himself from the international physics community. His schedule for the fall of 1913, for example, includes the annual meeting (September 7–10) of the Swiss Society of Natural Scientists in Frauenfeld (50 km away from Zurich), the *Naturforscherversammlung* in Vienna (21–28 September), and the second Solvay Congress in Brussels (27–31 October). Einstein gave talks at each of these meetings and actively participated in discussions. During the period under discussion (1912–14) he wrote a huge number of letters on scientific subjects.[31] This was the time when, to use Reichinstein's words, "a pilgrimage of physicists and chemists-theoreticians to Einstein" began. The number of visitors was even greater when Einstein moved to Berlin (Reichinstein 1934, p. 46). In summer and fall 1913 Einstein was host to Planck, Nernst, Ehrenfest, Marie Curie, the chemist Fritz Haber, the astronomer Erwin Freundlich, Andreyev, Nordström, Ishiwara, Fokker, and Herzfeld. Undoubtedly, there were other visitors as well.[32] It seems that no other European center of theoretical physics could boast a similar "pilgrimage" at that time.

Third, one should bear in mind that Einstein's "minimal interest in educating students," the image that later gained currency, was hardly evident during his Zurich period. Indeed, his letters to friends bear indications of the contrary:

> I take my lectures very seriously, so that I must spend much time on their preparation.
>
> . . .
>
> I like my new profession very much. I have quite an intimate relationship with my students and hope that I'll be able to stimulate a good many of them. I have already been able to give suggestions about doctoral theses. (Einstein to Laub, 31 December 1909, Einstein 1993, Doc. 196, pp. 226–227, trans. by A. Beck)

And there is more evidence of Einstein's outstanding gift as a teacher (see, for example, Seelig 1960, pp. 234–240).

In fall 1912 Einstein wrote to Sommerfeld: "Now I am totally engrossed in the problem of gravitation. . . . Never before has the process been so painful" (Einstein to Sommerfeld, 29 October 1912, Einstein 1993, Doc. 421, p. 505). This work, however, never interfered with his attention to other physical problems, nor did it prevent him from teaching. Frank, who came to Prague to succeed Einstein, later wrote that Einstein's students enthusiastically told him that their professor had invited them to come to him with their problems whenever they felt they needed his advice. "I shall always be able to receive you. . . . You will never disturb me, since

I can interrupt my work at any moment and resume it immediately as soon as the interruption is past" (Frank 1947, p. 118).

Stern had the following to say about Einstein as a teacher: "I have a great deal to thank Einstein for. Not only did he teach me to see the real problems in the quantum theory, but his whole attitude towards physics and his encouraging grasp of my problems had a decisive influence on my development as a physicist" (in a letter to Seelig of April 1952, quoted in Seelig 1960, p. 216). We have already quoted Reichinstein on this subject.

Thus there appears to be reason to think that there was an Einstein school in 1912–1914 associated with his Zurich colloquium. Indeed, there was the great man eager to share his knowledge and ideas, and lavishly endowed with teaching abilities and a desire to communicate with his colleagues. There were enthusiastic students who were fully aware of their teacher's intellectual potential. And finally, there were "lessons" in the form of fascinating and fruitful scientific discussions organized and stimulated by Einstein himself.

And yet there was no scientific school in Zurich in the sense in which one speaks of the schools of Sommerfeld, Bohr, Born, and Landau. Why?[33] One plausible (and perhaps unexpected) explanation is that the Zurich school had no time to take shape. Its outlines began to emerge only in late 1913, after Einstein's intention to move to Berlin had become known.

7. A Planned Onslaught on General Relativity

Let us return to the Zurich discussions of 1912–1914, now focusing on their "gravitational component."

In summer 1913, Ehrenfest wrote to Lorentz:

> My visit to Zurich proved very interesting. Since, besides Einstein, Nordström (from Helsingfors) is also staying here, you can learn a lot about their gravitational theories in a few hours. These two theories are competing with each other. Einstein boldly asserts that there can be no third alternative:[34] If the light rays curve during the solar eclipse next year, Einstein is right; if they remain straight, Nordström's theory is correct!!![35] (Quoted in Isaksson 1985, p. 42)

This letter was written on July 2, when Ehrenfest was still in Zurich.

The fall 1913 issue of *Annalen der Physik* includes papers by two members of the Zurich colloquium who signed the postcard to Mandelstam: Nordström's "On the Theory of Gravitation from the Standpoint of the

Relativity Principle" (1913) and Ishiwara's "On the Principle of Least Action in the Electrodynamics of Moving Ponderable Bodies" (1913). The first was dated "Zurich, July 1913" and the second "Zurich, 15 July 1913." Since the editors received both articles on July 24, there is every reason to believe that Nordström had finished his by July 15 as well. The Japanese theoretician did not mention his contacts in Zurich directly, yet the very content of the article makes it clear that he had started revising his former position as an ally of Abraham, whose theory he had supported, against Einstein, in their dispute on gravitation (see Vizgin 1981, pp. 176–178).

In Nordström's paper the hallmark of the Zurich colloquium can be detected on the very first page: "All the above-mentioned theoretical ambiguities can be removed with the help of a plausible suggestion for which I am indebted to Mr. von Laue and Mr. Einstein" (Nordström 1913). Vladimir Vizgin has pointed out another bit of evidence of Nordström's conversations with Einstein in Zurich that can be found in the same paper (see Vizgin 1981, p. 243). The paper developed what is known as "Nordström's second theory," which was instrumental in formulating general relativity. It took shape in summer 1913 under the obvious influence of the Zurich contacts and was an improved version of "Nordström's first theory," a scalar theory of gravitation he had introduced back in fall 1912. Nordström advocated the scalar theory until his arrival in Zurich. At that time, he did not see any way to meet the demand for a strict equality between the inertial and gravitational masses in his first theory (see Vizgin 1981, p. 189).

Fokker, whom Lorentz dispatched to Zurich to do "postdoctoral" work on fluctuation theory and quantum theory, became interested in the theory of gravitation and remained true to this interest throughout his life. Towards the end of his stay in Zurich, in February 1914, he co-authored, with his great teacher, the paper, "Nordström's Theory of Gravitation from the Points of View of the Absolute Differential Calculus" (Einstein and Fokker 1914). Together, they brought the second Nordström's theory to perfection. They derived its basic equations in a very simple way, directly from the condition of the constant velocity of light and Lorentz covariance (see Vizgin 1981, p. 250). This was an important step towards the final tensor equations of general relativity of November 1915 (see Isaksson 1985).

I have already mentioned Behacker as a possible participant in Einstein's colloquium. The paper in which he derived the laws of free fall and of planetary motion in the framework of Nordström's first theory was marked "Geneva, 1 September 1913" (Behacker 1913). It is doubtful that Behacker had contacts with the relativists in Zurich. First, Geneva and

Zurich were seven hours apart by train. Second and more importantly, Nordström's first theory had been abandoned in Zurich as of July. Behacker's main conclusion was that, when the terms of the order of $(v/c)^2$ and higher were neglected, the laws of motion in Nordström's theory coincided with the classical ones. But primary importance seemed to lie precisely in relativistic corrections to the classical laws.

Be that as it may, Behacker became interested in the gravitation theory, soon realized that Zurich was the leader in the field, and came to work with Einstein for the 1913–14 winter semester (see Frank 1916). Nordström derived the laws of motion from his second theory in January 1914, upon his return to Helsinki (Nordström 1914). It should be noted that neither Behacker nor Nordström calculated the precession of Mercury's perihelion. On the other hand, between 1912 and 1914, Einstein did not make any such calculations either (see, e.g., Vizgin 1981, p. 252).

Thus in summer 1913, Einstein led a planned onslaught on general relativity in Zurich. In the winter semester of 1913–14, two talented theoreticians, Fokker and Behacker, who had come as "postdocs," were working side by side with Einstein. It appears that this micro-community had the potential to become a relatively numerous and productive "gravitational school," a leader in its field. The name of the great man attracted many young physicists; it seems that Lorentz and Ehrenfest were prepared to send more post-graduate students to Zurich. It should be noted that Lorentz, who enjoyed high prestige and popularity in the Netherlands, was in close contact with Einstein. Being actively engaged in the same field, he followed Einstein's progress in general relativity with great interest.[36]

8. Conclusion

But the process of formation of Einstein's school in Zurich was cut short at an early stage.

There is an old Swiss proverb, "If God wants to do good to a Swiss, He gives him a home in Zurich." Undoubtedly, Einstein felt the truth of it. On 18 November 1911 he wrote to Grossmann from Prague: "I am overjoyed with a prospect of coming back to Zurich. I have even sacrificed an invitation to Utrecht University to it" (Einstein to Grossmann, 18 November 1911, Einstein 1993, Doc. 307, p. 351). In early 1912 a group of professors from the ETH submitted to the administration a request to offer a position to Einstein. They wrote: "It is known that Einstein is not happy in Prague and that the Professor would very gladly change it for another university. . . . [H]e has always had a great predilection for Zürich" (quoted in Seelig 1960, p. 232). On February 2 Einstein triumphantly informed the Sterns: "Two days ago (halleluiah!) I got an offer from the Polytechnical in Zurich. . . . All of us, the old folks and the cub bears [the children] are delighted" (Einstein to Alfred and Clara Stern, 2 February 1912, Einstein 1993, Doc. 352, pp. 402–403). Earlier in 1912 Einstein declined an offer from the University of Vienna, which attempted to seduce him with a "princely salary" (see Hoffmann and Dukas 1972, p. 99). But in fall 1913, Planck and Nernst arrived in Zurich with the intention of persuading Einstein to move to Berlin on fabulous conditions (see Hoffmann and Dukas 1972, pp. 100–101). After much painful deliberation, Einstein agreed.[37]

It was very hard for him to leave Zurich; he went to great effort to become its citizen, he left behind many friends, and he liked the environment and lifestyle. He approved of the atmosphere of liberalism and internationalism that reigned in Zurich. His letters of that period are bitterly ironic: "[O]n Easter I leave for Berlin as an Academy-man without any obligations, like a living mummy in a way" (Einstein to Jakob Laub, 22 July 1913, Einstein 1993, Doc. 455, p. 538; trans. by A. Beck). Einstein said to Stern: "These two [Planck and Nernst] look at me like people who are shopping for a rare postage stamp" (Papaleksi 1979, p. 19). Returning from the farewell party in his favorite "Kronenhalle," Einstein said to his colleague of ETH days: "The gentlemen in Berlin are gambling on me as if I were a prize hen. As for myself I don't even know whether I'm going to lay another egg" (Clark 1971, p. 173).

Einstein must have been reluctant to abandon his child, the Zurich colloquium, with its atmosphere of easy friendliness. He never attempted to set up a similar colloquium in the years to come.[38]

Acknowledgments. The author wishes to thank V. Vizgin, G. Gorelik, I. Kobzarev, and V. Frenkel for their useful comments and suggestions.

NOTES

[1] The mathematician Marcel Grossmann (1878–1938), an old friend of Einstein's, headed the department. It was on his initiative that Einstein was invited to ETH (see Seelig 1960, p. 225).

[2] Hereafter quotations from (Seelig 1960) are given in M. Savill's translation (Seelig 1956).

[3] The winter semester of 1912–13; the summer semester of 1913; and the winter semester of 1913–14.

[4] The author is grateful to Dr. J.-P. Siedler, Director of the Zurich Polytechnical Library, Dr. B. Glaus, the head of the department of historical collections there, and H. T. Lütstorf, an employee in the Library, who kindly provided him with the course schedules at the "Poly" for 1912–1914.

[5] Other signatures belong to R. A. Biegel, Ernst Keller, Heinrich Löwy, Sophie Rotszajn, and Otto Stern. See Einstein (1993, Doc. 457, note 5, p. 540). —*Editors.*

[6] The chair in physics was offered to Erwin Schrödinger. It is interesting to note that, besides Ratnowsky, another scientist from Russia was contending for the same position: Paul Epstein, a student of P. N. Lebedev and Arnold Sommerfeld (Rasche and Staub 1979).

[7] According to the editors of Einstein 1993, Stern did in fact sign the postcard to Mandelstam. See Einstein (1993, Doc. 457, p. 540, note 5).—*Editors.*

[8] Since 1914, a *Privatdozent* at the "Poly."

[9] In 1902 Zermelo translated into German Gibbs's *Elementary Principles in Statistical Mechanics.*

[10] In 1918, this book was reviewed in *Uspekhi fizicheskikh nauk* by P. Lazarev (1878–1942) who noted: "The results obtained by the author have not been generally recognized, yet they are of some interest" (Lazarev 1918, p. 249).

[11] The ETH and the University of Zurich stand close to one another. This seems to facilitate contacts between the two educational establishments and played an important role in the history of physics. It was at one of the joint colloquia early in 1925 that Professor Debye of the ETH prompted Professor Schrödinger of the University to derive an equation that was to form the basis of quantum mechanics (see Bloch 1976).

[12] Einstein and Weiss maintained friendly relations. Von Laue is quite correct here in emphasizing the issue of language. As one would expect, German was the colloquium's working tongue. At the same time, as already noted, Weiss's students from the Netherlands attended the discussion of Mandelstam's paper. Therefore, there is no reason to think that Weiss's students ignored the colloquia. Being born in Alsace, Weiss naturally had no problems with German. Stern was probably right when he wrote: "Pierre Weiss never took part in discussions. He forbade everyone to smoke, and that was awful. . . . However, no one could forbid anything to Einstein" (quoted from Jost 1979). Let us note in passing that von Laue was a smoker himself (see Seelig 1960, p. 130).

[13] On Laue's position in these discussions, see below.

[14] A green mountain at what was then the outskirts of Zurich. Einstein lived in a sunny quarter near it (see Seelig 1960, p. 233).

[15] Later, Bohr often ridiculed Sommerfeld for his skepticism "about atom models in general." These models (of the Bohr type) quickly became the focus of Sommerfeld's own research.

[16] Ehrenfest gives us a hint in his letter to Ioffe in late 1912: "By the way, I have just subscribed to *Phil. Mag.*" (Moskovchenko and Frenkel 1973, p. 100).

[17] Pais adds: "In 1914 Fokker spent six weeks with Rutherford, where he met Bohr. Bohr asked everyone: 'Do you believe it?'" (meaning his theory) (Pais 1967, p. 222).

[18] It is also notable that Fokker's thesis contained numerous references to Ehrenfest (and Einstein).

[19] That is, the value expressed in terms of universal constants.

[20] To camouflage his skepticism, Einstein used the formula: "very interesting and important if true" (see, for example, Pais 1979, p. 906). Einstein's answer implies that he was acquainted with Bohr's theory even before he came to the Vienna congress.

[21] It was commonly believed that the series belonged to hydrogen. But then the positions of the spectral lines did not conform with Bohr's theory. By ascribing the Pickering–Fowler spectral lines to ionized helium, Bohr proved that his theory fitted the experimental data to the highest degree, thereby turning an alleged anomaly into a weighty confirmation instance for his theory. His article (Bohr 1913b), dated 8 October 1913, appeared in *Nature* on 23 October 1913.

[22] Einstein was, of course, by no means alone in having such an attitude at the time. See e.g., Frenkel and Yavelov (1985, p. 27).

[23] Reichinstein's three lectures (the first of them read in 1911), which dealt with oscilloscopic measurements of the thickness of absorbed strata (see Reichinstein 1953, p. 14), and Andreyev's report.

[24] Suffice it to say that four of the future Nobel Prize winners in chemistry were working in Zurich at the time: Alfred Werner (1866–1919), Richard Willstäter (1872–1942), Leopold Ruzicka (1887–1976), and Hermann Staudinger (1881–1965) who were awarded the prize in 1913, 1915, 1939, and 1953 respectively.

[25] Einstein was profoundly interested in electrical conductivity of metals. See (Yavelov 1980).

[26] Ehrenfest refers to Arkadiev's work in which he discovered an absorption of high-frequency electromagnetic waves in ferromagnetic substances that was sensitive to frequency.

[27] The concept of molecular field was introduced by Weiss in 1907. This is probably the greatest achievement associated with his name.

[28] His thesis introduced the equation fundamental for the theory of random processes (Fokker 1913, p. 53). Today it is known as the "Fokker–Plank equation" or the "Einstein–Fokker–Planck equation" (see Gel'fer 1981, p. 366).

[29] This count is based on the [incomplete—*Editors*] list of Einstein's works in (Einstein 1967, pp. 578–593).

[30] Count based on the list of Born's works in (Born 1977, pp. 267–274). His record is explained by the fact that, in 1912, he and Theodore von Karman introduced a fundamental theory of oscillations of the crystal lattice that was developed in a series of joint publications. See (Born and Karman 1913).

[31] On Einstein's numerous scientific contacts during his stay in Prague see (Illy 1979).

[32] In August 1913 Einstein visited Kamerlingh Onnes, who was spending his leave in Switzerland not far from Zurich (Prosdij 1959).

[33] Arguably, the process of the formation of scientific schools in theoretical physics did not begin to develop until the mid-1910s. It was in the early 1920s that the schools of Sommerfeld, Born, and Ehrenfest gained momentum (see, respectively, Pauling 1951, Kemmer and Schlapp 1971, Frenkel 1977). In this respect, as well as institutionally, theoretical physics was lagging behind experimental physics, chemistry, and biological sciences.

[34] In September 1913, at the Vienna *Naturforscherversammlung*, Einstein was far from being equally "bold." He said he was not prepared to regard these two theories as the only possible alternatives and added: "However, I dare say that in the present state of our knowledge, they are the most *natural* ones" (1913, p. 1251; Einstein's emphasis).

[35] In his paper submitted to the Amsterdam Academy on February 22 (just before leaving for Zurich) Ehrenfest set out to prove that Einstein's equivalence principle was wrong "in application to all stationary fields of gravitation except for a very special class" (1913c). One can surmise that, while planning his visit to Einstein, he was intending to discuss this work with him. But when he found himself in Zurich, at the front line of the gravitational problem, Ehrenfest probably realized that his objections to the equivalence principle were outdated. In any case, there seems to have been no responses to his work in the literature.

[36] It was probably thanks to Lorentz that the Netherlands practically led the world in accepting Einstein's ideas on general relativity at that time. In other countries they were received much less enthusiastically.

[37] This decision was probably prompted by the increasing discord in his family (see, e.g., Trbuhović-Djurić 1983).

[38] Einstein gave regular seminars on statistical mechanics in Berlin in the 1920s. Notably, Eugene Wigner and Leo Szilard used the word 'seminar' rather than 'colloquium' to describe these meetings (see Wigner 1980, p. 461 and Weart and Szilard 1978, p. 9). They were intended solely for students. As Wigner later recalled, Einstein "wanted to establish contact with his young colleagues, because he wanted to know about their ideas and attitudes" (Wigner 1980, p. 462).

REFERENCES

Behacker, Max (1913). "Der freie Fall und Planetenbewegung in Nordströms Gravitationstheorie." *Physikalische Zeitschrift* 14: 989–992.

Bloch, Felix (1976). "Heisenberg and the Early Days of Quantum Mechanics." *Physics Today* 29: 23–27.

Bohr, Niels (1913a). "On the Constitution of Atoms and Molecules." *Philosophical Magazine* 26: 1–25.

—— (1913b). "The Spectra of Helium and Hydrogen." *Nature* 92: 231–232.

—— (1981). *Collected Works.* Vol. 2. *Work on Atomic Physics (1912–1917).* Ulrich Hoyer, ed. Amsterdam: North-Holland.

Born, Max (1977). *Razmyshleniya i vospominaniya fizika.* [*Reflections and Recollections of a Physicist*] Moscow: Nauka.

Born, Max and Karman, Theodor (1913). "Zur Theorie der spezifischen Wärme." *Physikalische Zeitschrift* 14: 15–19.

Casimir, H. B. G., and de Groot, S. R. (1972). "Adriaan Daniel Fokker." *Koninklijke Nederlandse Akademie van Wetenschappen. Jaarboek*: 114–118.

Clark, Ronald W. (1971). *Einstein: The Life and Times.* New York and Cleveland: The World Publishing Company.

Courant, Richard (1981). "Reminiscences from Hilbert's Göttingen." *Mathematical Intelligencer* 3: 154–164.

de Haas-Lorentz, Geertruida-Luberta (1957). "Reminiscences." In *H. A. Lorentz: Impressions of His Life and Work.* Geertruida-Luberta de Haas-Lorentz, ed. John C. Fagginger, trans. Amsterdam: North-Holland, pp. 82–120.

Eidgenössische (1972). *Eidgenössische Technische Hochschule Zürich. Dissertationsverzeichnis 1909–1971.* Zurich.

Ehrenfest, Paul (1913a). "Bemerkung betreffs der spezifischen Wärme zweiatomiger Gase." *Deutsche Physikalische Gesellschaft. Verhandlungen* 15: 451–457.

—— (1913b). "Een mechanisch theorema van Boltzmann en zijne betrekking tot de quantentheorie." *Koninklijke Akademie van Wetenschappen te Amsterdam. Wis- en Natuurkundige Afdeeling. Veslagen van de Gewoone Vergaderingen* 22 (1913–1914): 586–593.

—— (1913c). "On Einstein's Theory of the Stationary Gravitational Field." *Koninklijke Akademie van Wetenschappen te Amsterdam. Section of Sciences. Proceedings* 15: 1187–1191.

Einstein, Albert (1913). "Zum gegenwärtigen Stande des Gravitationsproblems." *Physikalische Zeitschrift* 14: 1249–1262.

—— (1967). *Sobranie nauchnykh trudov.* [*Collected Works*] Vol. 4. Moscow: Nauka.

—— (1985). *Statistische Mechanik.* Zurich: ETH-Bibliothek.

—— (1993). *The Collected Papers of Albert Einstein.* Vol. 5, *The Swiss Years: Correspondence, 1902–1914.* Martin J. Klein, et al., eds. Princeton: Princeton University Press.

—— (1998). *The Collected Papers of Albert Einstein.* Vol. 8, *The Berlin Years: Correspondence, 1914–1918. Part A: 1914–1917.* Robert Schulman, et al., eds. Princeton: Princeton University Press.

Einstein, Albert, and Fokker, Adriaan Daniel (1914). "Die Nordströmsche Gravitationstheorie vom Standpunkt des absoluten Differentialkalküls." *Annalen der Physik* 44: 321–328.

Einstein, Albert, and Stern, Otto (1913). "Einige Argumente für die Annahme einer molekularen Agitation beim absoluten Nullpunkt." *Annalen der Physik* 40: 551–560.

Eve, Arthur Stewart (1939). *Rutherford: Being the Life and Letters of the Rt. Hon. Lord Rutherford, O. M.* New York: Macmillan.

Ewald, P. (1979). "M. von Laue: Mensch und Werk." In *Feier der 100. Geburts-tage von Albert Einstein, Otto Hahn, Lise Meitner, Max von Laue.* Stuttgart: Wissenschaftliche Verlagsgesellschaft, pp. 85–89.

Feuer, Lewis Samuel (1974). *Einstein and the Generations of Science.* New York: Basic Books.

Fokker, Adriaan Daniel (1913). *Over Brown'sche bewegingen in het stralingsveld en waarschijnlijkheids-beschouwingen in de stralingstheorie.* Dissertation. Haarlem.

— (1914a). "Über Brownsche Bewegungen im Strahlungsfeld." *Physikalische Zeitschrift* 15: 96–98.

— (1914b). "Die mittlere Energie rotierenden elektrischen Dipole im Strahlungs-feld." *Annalen der Physik* 43: 810–820.

— (1955). "Albert Einstein. 14 Maart 1879–18 April 1955." *Nederlandsch Tijd-schrift voor Natuurkunde* 21: 125–129.

Frank, Philipp (1916). "Max Behacker." *Physikalische Zeitschrift* 17: 41–43.

— (1947). *Einstein: His Life and Times.* George Rosen, trans. Shuichi Kusaka, ed. New York: Alfred A. Knopf.

Frenkel, Viktor Ya. (1977). *P. Ehrenfest.* Moscow: Atomizdat.

Frenkel, Viktor Ya., and Yavelov, Boris E. (1981). *Einshtein-izobretatel'.* [*Einstein the Inventor*] Moscow: Nauka.

— (1985). "O nekotorykh istoriko-fizicheskikh aspektakh opytov Einsteina–de Gaasa." ["On the History of the Einstein–de Haas Experiment"] In *Einsteinov-skiy sbornik, 1980–1981.* I. Yu. Kobzarev and G. E. Gorelik, eds. Moscow: Nauka, pp. 10–36.

Gamburg, D. Yu. (1945). "Osnovopolozhnik azotnoy promyshlennosti." ["The Founder of the Nitric Acid Industry"] *Uspekhi khimii* 14: 239–252.

Gel'fer, Ya. M. (1981). *Istoriya i metodologiya termodinamiki i statisticheskoy fiziki.* [*History and Methodology of Thermodynamics and Statistical Physics*] Moscow: Vysshaya shkola.

Gerlach, Walter (1979). "Erinnerungen an A. Einstein. 1908–1930." *Physikalische Blätter* 35: 93–102.

Heisenberg, Werner (1975). "Development of Concepts in the History of Quantum Theory." *American Journal of Physics* 43: 389–394.

Herneck, Friedrich (1970). *Bahnbrecher des Atomzeitalters. Grosse Naturforscher v. Maxwell bis Heisenberg.* 5th ed. Berlin: Der Morgen.

— (1974). *Pionery atomnogo veka.* Moscow: Progress. Translation of Herneck 1970.

— (1979). *Max von Laue.* Leipzig: Teubner.

Herzfeld, Karl (1913a). "Zur Elektrochemie äusserst verdunnter Lösungen, im besonders radioaktiver Stoffe." *Physikalische Zeitschrift* 14: 29–32.

— (1913b). "Zur Elektronentheorie der Metalle." *Annalen der Physik* 41: 27–52.

Hoffmann, Banesh, and Dukas, Helen (1972). *Albert Einstein: Creator and Rebel.* New York: Viking.

Hund, Friedrich (1967). *Geschichte der Quantentheorie.* Mannheim: Bibliographisches Institut.

Illy, József (1979). "Albert Einstein in Prague." *Isis* 70: 76–84.

Isaksson, E. (1985). "Der finnische Physiker Gunnar Nordström und sein Beitrag zur Entstehung der allgemeinen Relativitätstheorie Albert Einstein." *NTM-Schriftenreihe zur Geschichte der Naturwissenschaft, Technik und Medizin* 22: 29–52.

Ishiwara, Jun (1913). "Über das Prinzip der kleinsten Wirkung in der Elektrodynamik bewegter ponderable Körper." *Annalen der Physik* 42: 986–1000.

Jammer, Max (1966). *The Conceptual Development of Quantum Mechanics.* New York: McGraw-Hill.

Jost, Res (1979). "Einstein und Zürich, Zürich und Einstein." *Naturforschende Gesellschaft Zürich. Vierteljahrsschrift* 124: 7–23.

Kemmer, N. and Schlapp, R. (1971). "Max Born (1882–1970)." *Biographical Memoirs of Fellows of the Royal Society of London* 17: 17–52.

Klein, Martin J. (1970). *Paul Ehrenfest. Vol. 1. The Making of a Theoretical Physicist.* Amsterdam: North-Holland; New York: American Elsevier.

Laue, Max von (1914). "Die Freiheitsgrade von Strahlenbündeln." *Annalen der Physik* 42: 497–524.

— (1950). *History of Physics.* Trans. Ralph Oesper. New York: Academic Press.

— (1956). "My Physical Career." In *History of Physics.* P. S. Kudryavtsev, ed. Moscow: Gostekhteorizdat, pp. 165–200.

Lazarev, P. P. (1918). "Review of *Die Eigenschaften des Adsorptionsvolumens,* by David Reichinstein." *Uspekhi fizicheskikh nauk* 1: 249.

Lexikon (1947). *Biographisches Lexikon verstorbener Schweizer.* Zurich: Schweizerische Industrie-Bibliothek, Departement Lexikon.

Mandelstam, Leonid I. (1913). "Über die Rauhigkeit freier Flüssigkeitsoberflächen." *Annalen der Physik* 41: 609–624.

Mikulinsky, S. R., ed. (1977). *Shkoly v nauke. [Schools in Science]* Moscow: Nauka.

Moskovchenko, N. Ya., and Frenkel, Viktor Ya., eds. (1973). *Ehrenfest-Ioffe. Nauchnaya perepiska 1907–1933 gg. [Ehrenfest-Ioffe. Scientific Correspondence 1907–1933]* Leningrad: Nauka.

Nordström, Gunnar (1913). "Zur Theorie der Gravitation vom Standpunkt des Relativitätsprinzips." *Annalen der Physik* 42: 533–554.

— (1914). "Die Fallgesetze und Planetenbewegungen in der Relativitätstheorie." *Annalen der Physik* 43: 1101–1110.

Pais, Abraham (1967). "Reminiscences from the Post-war Years." In *Niels Bohr: His Life and Work as Seen by His Friends and Colleagues.* Stefan Rozental, ed. Amsterdam: North-Holland, pp. 215–226.

— (1979). "Einstein and the Quantum Theory." *Reviews of Modern Physics* 51: 863–914.

Papaleksi, N. D. (1979). "Kratkiy ocherk zhizni L. I. Mandel'shtama." ["A Brief Biography of L. I. Mandelstam"] In (Rytov 1979, pp. 5–53).

Pauling, Linus (1951). "Arnold Sommerfeld." *Science* 114: 383–384.

Planck, Max (1915). "Über Quantenwirkungen in der Elektrodynamik." *Königlich Preussische Akademie* (Berlin). *Physikalisch-Mathematische Klasse. Sitzungsberichte*: 512–519.

Polya, George (1965). *Mathematical Discovery: On Understanding, Learning, and Teaching Problem Solving*, vol. 2. New York: Wiley.

Polishchuk, V. (1984). "Na obshchikh osnovaniyakh." ["On Common Grounds"] *Novy Mir*, no. 4: 183–207.

Polivanov, K. M. (1954). "V. K. Arkadiev." *Izvestiya AN SSSR. Seriya fiziki* 18: 307–311.

Prosdij, B. A. van (1959). "Some Letters from A. Einstein to H. Kamerlingh Onnes." *Janus* 48: 41–43.

Rabi, Isidore I. (1969). "Otto Stern. An Obituary." *Physics Today* 22 (October): 103–105.

Rasche, G., and Staub, H. H. (1979). "Physik und Physiker an der Universität Zürich." *Naturforschende Gesellschaft Zürich. Vierteljahsschrift* 124: 205–219.

Reichinstein, David (1916). *Die Eigenschaften des Adsorptionsvolumens.* Zurich: Leemann.

—— (1930). *Grenzflächenvorgänge in der unbelebten und belebten Natur.* Leipzig: J. A. Barth.

—— (1932). *A. Einstein. Sein Lebensbild und seine Weltanschauung.* Berlin: Selbstverlag.

—— (1934). *A. Einstein: A Picture of His Life and His Conception of the World.* M. Juers and D. Sigmund, trans. Prague: Stella Publishing House. Translation of Reichinstein 1932.

—— (1940). *Das Problem des Alterns und die Chemie der Lebensvorgänge.* 2 vols. Dielsdorf; Zurich: H. Akerets.

—— (1953). *Wissenschaftliche Bibliographie.* Zurich: Verlag Aristoteles.

Reid, Constance (1976). *Courant in Göttingen and New York: The Story of an Improbable Mathematician.* New York and London: Springer-Verlag.

Rompe, Robert (1979). "Einshtein i berlinskie fiziki v kontse 20-h godov." ["Einstein and the Berlin Physicists in the late 1920's"] *Priroda*, no. 3: 54–57.

Rootsellar, B. (1975). "E. Zermelo." In *Dictionary of Scientific Biography*, vol. 16. C. C. Gillispie, ed. New York: Scribner, pp. 613–616.

Rosenfeld, Léon, and Rüdinger, Erik (1967). "The Decisive Years, 1911–1918." In *Niels Bohr: His Life and Work as Seen by His Friends and Colleagues.* Stefan Rozental, ed. Amsterdam: North-Holland, pp. 38–73.

Rytov, S. M., ed. (1979). *Akademik L. I. Mandel'shtam: K 100-letiyu so dnya rozhdeniya.* Moscow: Nauka.

Seelig, Carl (1956). *Albert Einstein: Leben und Werk eines Genius unserer Zeit.* Zurich: Europa Verlag.

—— (1960). *Albert Einstein: A Documentary Biography.* Mervyn Savill, trans. London: Staples Press Limited.

Segré, Emilio (1973). "Otto Stern." *National Academy of Sciences of the USA. Biographical Memoirs* 43: 215–236.

Speziali, Pierre, ed. (1972). *Albert Einstein–Michele Besso. Correspondance, 1903–1955.* Paris: Hermann.

Stern, Otto (1913). "Zur kinetische Theorie des Dampfdrucks einatomiger fester Stoffe und über die Entropiekonstante einatomiger Gase." *Physikalische Zeitschrift* 14: 629–633.

—— (1914). "Zur Theorie der Gasdissoziation." *Annalen der Physik* 42: 497–524.

Tilgenkamp, Erich (1956). *Reisen in ungewöhnlichen Räume. Eine autorisierte Biographie*, vol. 1. Berlin: Verlag Neues Leben.

Treadwell, W. D. (1944). "Emil Baur." *Naturforschende Gesellschaft Zürich. Vierteljahrsschrift* 89: 222–224.

Trbuhović-Djurić, Desanka (1983). *Im Schatten Albert Einstein. Das tragische Leben der Mileva Einstein-Marić.* 2nd ed. Bern: P. Haupt.

Vizgin, Vladimir P. (1981). *Relyativistskaya teoriya gravitatsii: istochniki i formirovaniye, 1900–1915.* [*Relativistic Theory of Gravitation: Sources and Formation, 1900–1915*] Moscow: Nauka.

Volkov, V. A. (1980). "I. I. Andreev—sozdatel' kontaktnogo sposoba polucheniya azotnoy kisloty." ["I.I. Andreyev, the Inventor of the Contact Method of Nitric Acid Production"] *Voprosy istorii estestvoznaniya i tekhniki*, no. 3: 120–122.

Waser, P. G. (1979). "Eröffnungsansprache des Rektors der Universität Zürich." *Naturforschende Gesellschaft Zürich. Vierteljahrsschrift* 124: 3–6.

Weart, Spencer R., and Szilard, Gertrud Weiss, eds. (1978). *Leo Szilard: His Version of the Facts. Selected Recollections and Correspondence.* Cambridge, Massachusetts: MIT Press.

Weiss, Pierre (1912). "Heinrich Friedrich Weber. Nekrologe." *Naturforschende Gesellschaft Zürich. Vierteljahrsschrift* 57: 596–604.

—— (1914). "Sur la nature du champ moleculaire." *Archives des sciences physiques et naturelles* 37: 105–106, 201–213.

Weiss, Pierre, and Kamerlingh Onnes, Heike (1910). "Researches on Magnetism at Very Low Temperatures." *Leiden Communications*, no. 114.

Wheeler, John Archibald (1979). "Karl Herzfeld." *Physics Today* 32: 99.

Wigner, Eugene P. (1980). "Thirty Years of Knowing Einstein." In *Some Strangeness in the Proportion: A Centennial Symposium to Celebrate the Achievements of Albert Einstein.* Harry Woolf, ed. Reading, Massachusetts: Addison-Wesley, pp. 461–468.

Wolfke, K. (1980). "Wspomnienie o Ojcu, Mieczyslawie Wolfke." *Postepy Fizyki* 31: 551–557.

Yavelov, Boris (1980). "Einstein i problema sverkhprovodimosti." ["Einstein and Superconductivity"] In *Einshteinovskiy sbornik, 1977.* V. L. Ginzburg, B. G. Kuznetsov, and U. I. Frankfurt, eds. Moscow: Nauka, pp. 158–186.

—— (1985). "K istori problemy elektricheskikh fluktuatsiy." ["On the History of the Problem of Electrical Fluctuations"] In *Issledovaniya po istorii fiziki i mekhaniki.* A. Grigoryan, ed. Moscow: Nauka, pp. 143–160.

"What May Happen to a Man Who Thinks a Great Deal but Reads Very Little"

Viktor Frenkel and Boris Yavelov

In his biography of Einstein, first published in 1952, Carl Seelig says:

> It is not well known that, at that time, Einstein took a run at aircraft design. The German airforce was then still armed with slow and clumsy Tauben and biplanes, which, by comparison with the faster aircraft of Entente, were rather archaic.[1] In an attempt to close the gap, the Luftverkehrsgesellschaft in Berlin-Johannisthal called on academics to contribute to the technical improvement of the airforce. Einstein was one of the few who agreed. He undertook a new airfoil design intended for serial production. Eberhard [see below], the chief test pilot, was to make a test flight; he was no longer young and treated the fruit of the famous theoretician's efforts with suspicion. "We'll see how the hare to going to run," he grumbled to the inventor before the flight. After a long start, the plane took to the air, but the flight was far from being spectacular. The plane was waddling like a "pregnant duck." The pilot was overjoyed to find himself on firm ground in one piece. Being a man of principle, he suggested that another pilot should test the aircraft. The second pilot was Hanuschke, who later recounted the episode to us [i.e., to Seelig]: "The result was no better: As an aircraft designer, the great theoretician Einstein failed conspicuously." And his design was buried forever. (Seelig 1960, pp. 251–252)

Friedrich Herneck, however, calls the story told by Seelig into question: "It is hardly probable that in 1916 [?] Einstein made an unsuccessful attempt to improve the design of military [?] airplanes, as one of his biographers asserts" (Herneck 1966, p. 129). Einstein scholars generally agree that Seelig's book is a reliable source: No one has so far been able to find any deliberate distortions of facts or signs of unscrupulousness in his work. On the other hand, Herneck is also respected throughout the world as a

Yuri Balashov and Vladimir Vizgin, eds., *Einstein Studies in Russia.*
Einstein Studies, Vol. 10. Boston, Basel, Birkhäuser, pp. 297–306.
© 2002 The Center for Einstein Studies.

conscientious historian and as the author of widely-read biographies of prominent physicists. Who, then, is right? Did this episode really happen or was it just one of the numerous anecdotes connected with the famous name?

In support of his doubts, Herneck notes that, first, there is no evidence that Einstein participated in aircraft design; and second, this course of action would run counter to his consistently pacifist stance. We shall discuss both points. First, is there really "no evidence," as Herneck claims?

In 1955 the magazine *Interavia* published an article entitled "Professor Einstein's 'Folly'," which reproduced the letters that Paul Ehrhardt and Einstein exchanged in 1954.

Paul Georg Ehrhardt (1889–1961) was a famous pilot of German aviation's "heroic age." He made his first flight back in 1909 as a passenger of the famous Orville Wright. He paid 500 gold franks to be in the air for two-and-a-half minutes. In 1912 he became a professional pilot and in 1917 was appointed head of the experimental department of the Berlin aircraft design firm Luftverkehrsgesellschaft (LVG). Starting in the mid-1920s, he tried his hand at writing. His first book, a utopian novel entitled *The Last Power*, was translated into Russian in 1926. He combined writing with flying until 1954, when he was involved in a serious accident. The well-known catalogue, *Deutsches Literatur-Lexicon*, mentions a dozen of his works, mostly connected with aviation.

Here are some fragments from his letter written to Einstein on 26 August 1954 (EA 59-556, as quoted in Folly 1955):[2]

> Dear Professor,
>   My only object in writing this letter is to recall what was for me an unforgettable experience, though you perhaps may not recollect it. . . . When I took over technical management of the LVG's experimental section in the spring of 1917, my duties included the thankless task of dealing with offers from inventors. . . . I was therefore not exactly enthusiastic when, one day, I found on my desk a several-page document of this kind, written by hand, to make matters worse. . . . However, my first glance through the weighty manuscript showed me that the writer had far greater knowledge of theoretical physics than I had. . . .

Ehrhardt was referring to the letter in which Einstein suggested a new airfoil design that later came to be known as the "Katzenbuckelfläche" ("cat's back wing"). Being in doubt about his own ability to evaluate the idea, Ehrhardt sent the letter to the LVG experts in mechanical stresses. A special conference was then arranged, at which Einstein gave his arguments why this wing profile allowed the maximum lift with the minimal thrust and the zero angle of attack.

Ehrhardt's letter continues (EA 59-556, as quoted in Folly 1955):

A few weeks later, the "cat's back airfoil" had been fitted to the normal fuselage of a LVG biplane, and I was confronted with the task of testing it in flight. In those days every first take-off in a new type was a wager. I had supervised construction with mounting skepticism and expressed the fear that the machine would react to the lack of angle of incidence in the wing by dropping its tail and would thus presumably be obliged to take off in an extremely unstable attitude. Unfortunately the skeptic in me proved to be right, for I hung in the air like a "pregnant duck" after take-off and could only rejoice when, after flying painfully down the airfield, I felt solid ground under my wheels again just short of the airfield fence at Adlershof. The second pilot had no greater success: not until the cat's back airfoil was modified to give it an angle of incidence could we venture to fly a turn, but even now the pregnant duck had merely become a lame duck. . . .

Ehrhardt also mentioned that the unsuccessful experiment was followed by a rewarding informal meeting at which Einstein described his relativity theory to the LVG administration.

Einstein promptly replied to Ehrhardt's letter (EA 76-235, as quoted in Folly 1955):[3]

Princeton, 7.IX.54

Dear Mr. Ehrhardt,

I recall quite clearly the events you describe so humorously in your letter. That is what can happen to a man who thinks a lot but reads little. It is a strange fact that physicists did not understand the essence of flight until they themselves learned to fly, although they had had the flight of birds, in particular the soaring flight of birds of prey, before their eyes since the time of Olim and although the theoretical principles had been known since Euler's time. Bernoulli's equation, [which] when gravity is ignored, can be written:

$$(1/2)\rho v^2 + p = \text{const.}$$

was simply applied. Inside a flow, pressure p varies inversely with the square of the velocity (v^2).

It struck me that pressure could be obtained on a guiding surface by shaping the latter so that the speed of flow is different on the two sides of the surface. I thus arrived at this shape:

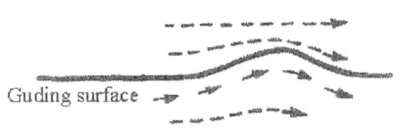

Top: local high speed
(negative pressure)

Bottom: local low speed
(positive pressure)

Although it is probably true that the principle of flight can be most simply explained in this way it by no means follows that it is wise to construct a wing in such a manner! You mentioned one of the reasons in your letter: the great extent to which torque depends on an oblique attitude. Nature knew well enough why she made birds' wings rounded in front and sharp-edged behind! Moreover it is quite unnecessary and even entails major losses to make the guiding surface cover the whole length of the bulge in the flow. A shape of the following kind is much better and also creates an effective bulge in the flow.

The bulge in the flow is caused by the sharp edge at the end of the wing.

I had not worked this out in full, but had stopped at my first idea. I also had not realized that, in order to avoid vortex and friction losses, the surfaces in contact with the moving air should not be bigger than is necessary to obtain the bulge in fluid motion.

I have to admit that I have often been ashamed of my folly of those days, but got a great deal of pleasure from your good-natured letter.

With best wishes.

Your
A. Einstein

This shows that Seelig came close, even textually, to Ehrhardt's account. The inaccurate dates, the errors in the spelling of the pilot's name and his age (Seelig said that he was not young, while, in fact, he was barely thirty) are understandable as Seelig relied on the word-of-mouth testimony retold some thirty years after the event. Of course, the fact that Ehrhardt was a science fiction writer and that the article was published in a magazine not at all subject to accepted canons of archival rigor might still give rise to some doubts. But such doubts would hardly be justified. First, the article includes Einstein's handwriting; and second, Seelig learned about this episode not from Ehrhardt, but from Hanuschke, the second pilot; and third, the Einstein scholarship and his own works clearly show that he was always interested in the problems of aero- and hydrodynamics. Sure enough, his interest in these matters remained amateurish and was never at the center of his attention.

According to a story told by Vero Besso, the son of Einstein's closest friend, as early as 1904 or 1905, the future grand man of physics made a kite for him, which they tossed outside Bern. Many years later, Vero was unable to recall who actually tossed it, yet he distinctly remembered that Einstein had explained to him in detail why the kite flew (see Speziali 1979, p. 13). In the 1920s, Einstein wrote papers on the meandering of rivers (Einstein 1926)[4] and on the wind ship design known as "Flettner's rotor ship" (Einstein 1925). And there is also evidence that Einstein gave some useful suggestions to a prominent American yacht designer, W. Starling Burgess (see Frenkel and Yavelov 1981, p. 127).

Based upon his extensive study of the archives of the Berlin Academy of Sciences, Hans-Jürgen Treder, a well-known physicist and historian of physics, concluded that, at the time of his move to Berlin in April 1914, Einstein was employed by the Prussian Academy of Sciences not as a theoretical physicist but rather as an expert in aviation technology. His task was to evaluate new aircraft design proposals submitted to the Academy. (See Treder 1979, p. 51)

But the strongest evidence of Einstein's interest in aero- and hydrodynamics undoubtedly comes from his note, "Elementary Theory of Water Waves and of Flight" (Einstein 1916), which appeared in *Die Naturwissenschaften* on 25 August 1916. As already noted, Einstein made his sudden debut as an aircraft designer about the same time.

This note considers the flow of an incompressible non-viscous fluid and explains the physical meaning of Bernoulli's equation

$$\tfrac{1}{2}\,\rho v^2 + p = const.$$

Einstein used two simple examples to illustrate the equation's application and gave a very simple and elegant explanation of water waves. He employed similar considerations to explain the "cause of the wing's lift" and supplied a diagram showing a flow of a fluid around plate W with a bulge in the middle. The local cross section of the flow beneath the plate is greater than above it. Therefore, according to Bernoulli's equation, the pressure at point U is greater than at point O. The difference between the two pressures causes the lift. It is interesting to note that, in his September 1954 letter to Ehrhardt, Einstein explained this phenomenon in the same terms and included a similar diagram.

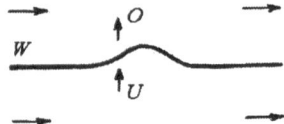

Einstein began his note with the question, "What accounts for the carrying capacity of the wings of our flying machines and of the birds soaring through the air in their flight?" and concluded it as follows: "We then have the supporting wing of a flying machine or of a bird soaring without moving its wings" (Einstein 1916, p. 509, 510; trans. by A. Engel). It seems plausible (and Einstein himself admitted this much in the letter to Ehrhardt) that the bold idea of increasing lift by giving the plane wing's the shape of the raised back of an irritated cat came to Einstein when he was thinking about the physics of flight.

This puts all of the evidence regarding Einstein's "aircraft design" efforts in place. There is no doubt that Seelig's account is essentially correct. As to the apparent inconsistency between Einstein's well-known pacifist sentiments and his preoccupation with the improvement of the design of the military aircraft—there is no reason to regard this as a serious problem.

Einstein's sharing his ideas with LVG can be attributed to the fact that the firm itself had solicited advice from the academic community. But Seelig, of course, grossly exaggerated the extent of Einstein's involvement when he wrote that "Einstein set out to design a new plane for serial production." He had nothing of the sort in mind; in fact, at that time, new aircraft was already the product of collective effort of many engineers. He just stumbled upon what he thought to be an interesting novel idea and he immediately decided to share it with pilots and aircraft makers. He did not even take the trouble to type his letter. In short, this doesn't look like "designing a new plane for serial production" at all!

One also should not forget that, at that time, aviation was not yet looked upon as a weapon of mass destruction; besides, there is no evidence that it was a *military* plane that Einstein wanted to improve. Finally, LVG was never an important player in Germany's production of military aircraft (Moedebeck 1923).

There is another indication that Einstein himself was not really serious about his new airfoil proposal. In his letter to Ehrhardt, he said, "That is what can happen to a man who thinks a lot but reads little" (EA 76-235, as quoted in Folly 1955). In the present context, "reading little" could only mean that Einstein never bothered to find out whether his suggestion was sound in the eyes of the experts.

One might think that this does not quite square with the opening of Einstein's 1916 article: "What accounts for the carrying capacity of the wings of our flying machines and of the birds soaring through the air in their flight?" But the next two sentences speak for themselves: "There is a widespread lack of clarity on this question. I must confess that I could not find anywhere in the specialized literature even the simplest answer" (Einstein 1916, p. 509; trans. by A. Engel).

It is common knowledge that Einstein was not a master of bibliographical search. More likely than not, he limited himself to a few leaflets and randomly selected books and articles. Otherwise he would have discovered that the problem of the lift wing had been solved. Back in 1906, Nikolai E. Zhukovsky published his ground-breaking paper "On Adjoint Vortices" (reprinted in Zhukovsky 1949) containing the famous theorem about the wing lift. Naturally enough, Einstein could not have read the article in Russian, which, furthermore, was published in a rather "exotic" serial, the *Proceedings of the Department of Physics of the Moscow Society of the Friends of Natural Sciences, Anthropology, and Ethnography*. But Zhukovsky's most important results were made public in the paper "On the Profiles of Carrying Surfaces of Airplanes," which appeared in 1910 in *Zeitschrift für Flugtechnik und Motorluftschiffahrt* (Zhukovsky 1910). In 1916 Zhukovsky's *Aerodynamics* was published in French (Zhukovsky 1916), language with which Einstein did not have any problems.

It should be noted that the German mathematician Wilhelm Martin Kutta arrived at a similar theory of the lift at about the same time as Zhukovsky and independently of him. He published his results in full in the respectable *Proceedings of the Bavarian Academy of Sciences* (Kutta 1910).

Obviously, Einstein never got as far as this kind of "special literature"; it is equally obvious that he failed even to glance through *Zeitschrift für Flugtechnik und Motorluftschiffahrt*, which was filled with references to Zhukovsky and Kutta and which frequently published articles by Ludwig Prandtl, a major authority in hydro- and aerodynamics (including the theory of flight), who was a professor at Göttingen University and the head of the Göttingen aerodynamic laboratory.

In all justice, there were very few serious works on the theory of flight at the time, and they were hard to find in the ocean of amateur proposals and pseudoscientific writings. Both popular and scientific journals were overwhelmed with them at that age of the aviation boom. Suffice it to say that quite a few authors attempted to derive a theory of flight from observations of birds. They were mostly biologists, physiologists, and ornithologists, mostly unfamiliar with the methods of the exact sciences, and their writings greatly contributed to the overall confusion about the matter. For example, in 1914 *Die Naturwissenschaften* (the same journal in which Einstein's 1916 note appeared) published four articles that compared the airplane and the bird, three of them by August Pütter, a professor of physiology from Bonn.[5] The fourth article was also "biological" in nature.

Understandably, Einstein concluded that "there [was] a widespread lack of clarity on this question." He was probably struck by the fact that nobody tried to explain the lift with the help of Bernoulli's equation! One could even speculate (and the letter to Ehrhardt supports this speculation) that it was precisely this "discovery" of a straightforward and elegant explanation that moved Einstein to suggest a new airfoil.

To be sure, this explanation is entirely correct and it is widely used today in popular literature and elementary textbooks. But it is too simplistic to base aircraft design on. In his letter to Ehrhardt, Einstein acknowledged: "Although it is probably true that the principle of flight can be most simply explained in this way it by no means follows that it is wise to construct a wing in such a manner!" He added that he had "often been ashamed of my folly of those days" (EA 76-235, as quoted in Folly 1955).

But in fact, the real shame should have fallen on the LVG designers: After all, Einstein made his suggestion as a private person and an amateur in airplane design. Being professionals, the LVG designers should have displayed better knowledge of the special literature; instead, they betrayed their total ignorance.

It seems that despite his important-sounding title as the "technical chief of the LVG experimental department," Ehrhardt saw Bernoulli's equation for the first time in Einstein's letter. Obviously, he got lost in the simple physics of the letter and, having little idea of what was going on there, forwarded the letter to the "experts in mechanical stresses"! The simple fact that the author was called upon to explain his elementary derivations suggests that the "experts in mechanical stresses" also had no idea of the theory of flight.

Acknowledgments. The authors thank professor L. G. Loitsyansky for his comments and interest in the article. The editors gratefully acknowledge permission to publish fragments of Ehrhardt's letter to Einstein and Einstein's response granted by The Albert Einstein Archives, The Jewish National and University Library, The Hebrew University of Jerusalem, Israel, and The Einstein Papers Project.

NOTES

[1] This is not true. When the First World War broke out, the German airforce was superior to that of the Entente. In fact, the Fokker monoplanes had no competitors in the enemy airforce. The situation changed towards the end of 1916, when the real rivalry began.

[2] Fragments from Ehrhardt's letter to Einstein and Einstein's response are quoted in *Interavia*'s translation, by kind permission of the editor of this magazine. The editors thank Dr. Gayane Tavrizyan for her assistance in checking the translation.

[3] The item number of the English translation in *Interavia* quoted here is EA 76-234.

[4] Interestingly, Einstein's son and a well-known specialist in hydrodynamics G. A. Einstein was an expert in this problem, which is not yet completely solved. See (Popov 1977, p. 30).

[5] He was one of the editors of *Die Naturwisssenschaften* (the second was Arnold Berliner, a physicist and Einstein's friend) and was probably responsible for the publication of Einstein's note.

REFERENCES

Einstein, Albert (1916). "Elementare Theorie der Wasserwellen und des Fluges." *Die Naturwissenschaften* 4: 509–510.

—— (1925). "El buque de Flettner." *La Prensa*, 13 April 1925.

—— (1926). "Die Ursache der Mäanderbildung der Flußläufe und des sogenannten Baerschen Gesetzes." *Die Naturwissenschaften* 14 (12 March, Heft 11): 223–224.

Erhard, H. (1914). "Der Flug der Tiere." *Die Naturwissenschaften* 2 (10 April, Heft 15): 357–363.

Folly (1955). "Professor Einstein's 'Folly.'" *Interavia* 10: 684–685.

Frenkel, Viktor Ya., and Yavelov, Boris E. (1981). *Einshtein-izobretatel'*. [*Einstein the Inventor*] Moscow: Nauka.

Herneck, Friedrich (1963). *Albert Einstein*. Leipzig: BSB Teubner.

—— (1966). *Albert Einstein*. Moscow: Progress. Translation of Herneck 1963.

Kutta, W. M. (1910). "Über eine mit den Grundlagen des Flugsproblems." *Bayerische Akademie der Wissenschaften. Mathematisch-naturwissenschaftliche Abteilung. Sitzungsberichte*: 92–105.

Lexikon (1971). *Deutsches Literatur-Lexikon*. Bern: Francke.

Moedebeck, Hermann (1923). *Moedebeck Taschenbuch für Flugtechniker und Luftschiffer*. 4th ed. Berlin: M. Krayn.

Popov, I. V. (1977). *The Puzzle of the River Bed*. Leningrad: Gidrometeoizdat.

Pütter, August (1914a). "Vogel und Flugzeug." *Die Naturwissenschaften* 2 (11 September, Heft 37): 861–865.

— (1914b). "Die Leistungen der Vögel im Fluge. I, II." *Die Naturwissenschaften* 2 (17 July, Heft 29): 701–705; 2 (24 July, Heft 30): 725–779.

Seelig, Carl (1960). *Albert Einstein. Leben und Werk eines Genies unserer Zeit.* Zurich: Europa Verlag.

Speziali, Pierre (1979). "Einstein's Friendship With Michele Besso." In *Albert Einstein: A Centenary Volume.* A. P. French, ed. Cambridge, Massachusetts: Harvard University Press, p. 13.

Treder, Hans-Jürgen (1979). "Albert Einstein an der Berliner Akademie der Wissenschaften." In *Albert Einstein in Berlin, 1913-1933,* vol 1. H-J. Treder, ed. Berlin: Akademie-Verlag, pp. 7–78.

Zhukovsky, Nikolai E. (1910). [N. Joukowsky] "Über die Konturen der Tragflächen der Drachenflieger." *Zeitschrift für Flugtechnik und Motorluftschiffahrt*: 281.

— (1916). *Bases théorique de l'aéronatique. Aérodynamique; cours professé à l'École impériale technique de Moscou, par N. Joukowski.* Translated from Russian by S. Drzewiecki. Paris: Gauthier-Villars.

— (1949). "O prisoedinyonnykh vikhryakh." ["On Adjoint Vortices"] In *Sobranie sochineniy* [*The Collected Works of N. E. Zhukovsky*], vol. 4. Moscow: Leningrad: Gostekhteorizdat, pp. 69–91.

Index

Smolin, Lee, 144
Smuts, Jan, 110
Sommerfeld, Arnold, 65, 161, 204,
    220, 265, 274, 275, 282–284,
    288, 290
Sonin, A. S., 163
Speziali, Pierre, 84, 265, 270, 301
Stachel, John, 7, 13, 47
Stadler, Friedrich, 47
Staub, H. H., 288
Staudinger, Hermann, 289
Stein, Howard, 47, 84
Steklov, Vladimir, 8
Stepanovskiy, Y. P., 175
Stern, Otto, 265, 268, 271, 272, 275,
    276, 280–282, 284, 287, 288
Stewart, T., 174
Stoops, R., 127
Stuewer, Roger, 251
Synge, John, 30, 35
Szilard, Leo, 290

Tabiryan, N. V., 174
Talmud, D., 153
Tamarkin, Yakov, 9, 11, 12, 14
Tamm, Igor, 240, 252, 253
Tank, F., 266, 276, 277
Tavrizyan, Gayane, 13, 305
Teilhard de Chardin, Pierre, 110
Terzian, Y., 145
Thirring, Hans, 160
Thirring, Walter, 112
Thomson, William, 188
Thorne, Kip, 31
Tilgenkamp, Erich, 267
Timoreva, A., 154
Tomonaga, S. I., 247
Trautman, Andrzej, 26, 27, 29, 31
Trbuhović-Djurić, Desanka, 290
Treadwell, W. D., 265, 270
Treder, Hans-Jürgens, 47, 301
Trkal, V. A., 8, 9, 11
Tsvetkov, V., 166–169

Uhlenbeck, George, 235
Urani, John, 145

van der Waerden, Bartel Leendert,
    222
Vasiliev, A. V., 84
Vavilov, Sergei, 154
Vizgin, Vladimir, 12, 13, 17, 19, 22,
    26, 37, 47, 65, 66, 70, 83, 84, 96,
    149, 155, 161, 175, 282, 286,
    287
Voigt, Woldemar, 151, 152, 163,
    174, 282
Volkov, V. A., 263, 273

Waller, Ivar, 238
Waser, P. G., 261
Weart, Spencer, 290
Weber, Heinrich Friedrich, 261, 267
Weinberg, Steven, 111
Weiss, Pierre, 261, 264, 271, 281,
    288, 289
Weisskopf, V., 252
Wergeland, H., 213
Werner, Alfred, 289
Weyl, Hermann, 37, 91, 96–100,
    103–105, 127, 131, 155, 158,
    172, 175, 240, 252, 266
Wheaton, Bruce, 221
Wheeler, John Archibald, 7, 31, 47,
    76, 265
Whitehead, Alfred North, 108–110
Whittaker, Edmund, 138
Wien, Wilhelm, 189, 190, 221
Wigner, Eugene, 142, 215, 234, 235,
    237, 290
Willstäter, Richard, 289
Wolff, Christian, 206
Wolfke, Mieczyslaw, 198, 200–202,
    266
Wolters, Gereon, 47, 48, 70, 80, 82,
    84